D0990059

COLLEGE ALGEBRA

COLLEGE ALGEBRA

Ross R. Middlemiss
Professor of Mathematics
Washington University

New York Toronto London
McGRAW-HILL BOOK COMPANY, INC.
1952

COLLEGE ALGEBRA

Library of Congress Catalog Card Number: 51-12629

VIII

THE MAPLE PRESS COMPANY, YORK, PA.

PREFACE

The course in college algebra is usually regarded as the basic course in mathematics, not only for students of engineering and the other sciences, but for liberal arts students as well. If the course is to meet the needs of these groups, it should contribute to the development of the student in several ways.

In the first place, it must of course provide the needed training in the essential techniques—in such matters as handling fractions, solving equations, and using logarithms. But this is not enough. It should contribute in a vital way to his basic thinking habits. The presentation should be such that he will develop a proper respect for precise definitions. It should be such that he will develop the habit of stating his hypotheses carefully, and such that as he progresses he will come to see clearly the difference between a theorem and its converse. It is true that some of our students may never, after finishing the course, use a determinant or even solve a quadratic equation. But surely all of them—scientists, sociologists, and even politicians—can profit from the training in straight thinking that should be a part of the instruction in these topics.

The present book does not differ significantly from others in essential content. It covers the topics that are usually taught in a standard course in college algebra. The main effort of the author has been directed toward a presentation that will make the additional contribution indicated above.

Thus, for example, when the student solves the equation $3x + 5 = 26$ by subtracting 5 from both members and then dividing both members of the resulting equation by 3, he is asked to observe that, in view of the axioms in regard to subtracting equals from equals and dividing equals by equals, he can make the following assertions:

$$\begin{aligned} &\text{If } 3x + 5 = 26, &\text{then}\quad 3x &= 21; \\ &\text{if} \quad\ \ 3x = 21, &\text{then}\quad x &= 7. \end{aligned}$$

He is thus taught that his operations have proved the assertion, "If $3x + 5 = 26$, then $x = 7$" and that this means that *no number other than 7 could be a root.* It is emphasized that the assertion, "If $x = 7$, then $3x + 5 = 26$" is quite a different matter, and that the truth of this assertion may be tested by starting with $x = 7$ and reversing the above steps, or by computing the value of $3x + 5$ under the hypothesis that $x = 7$.

When these ideas are pointed out early, and repeated at appropriate places, the student gets a training that is of lasting value not only in connection with his future work in mathematics but in almost every subject that he may subsequently study. The writer would be among the last to advocate rigor, especially at this stage, for the sake of rigor itself. His contention is rather that we should utilize the excellent opportunity presented by the course in college algebra to instill into the student the valuable habit of stating his hypotheses clearly and analyzing his assertions carefully. The student who has been trained to work mechanically may be completely baffled when, in attempting to solve the equation

$$\sqrt{4x + 5} = 2\sqrt{x + 1}, \tag{1}$$

he arrives successively at the equations

$$4x + 5 = 4(x + 1); \tag{2}$$

$$5 = 4. \tag{3}$$

He may be able to conclude nothing except that the method which usually "works" has mysteriously failed in this case. The student with proper training should see immediately that, if there is a number x for which (1) is true, then, for this same number x, (2) and (3) must be true. He should thus see clearly the proof of the fact that (1) cannot be true for any number x.

It is of course unnecessary to insist that the student analyze formally every step in the solution of every problem. But if he has not been trained in such a way that he will quickly draw the proper conclusion in a case like that above, then much of what he is doing must be rather meaningless to him. A sound understanding of the idea of equivalent equations can be obtained by means of the type of reasoning advocated here.

One of the things that is essential to a better presentation of the topics in college algebra is a better understanding, on the part of the student, of the number system. The first chapter of this text consists of four lessons in which an attempt is made to teach the essential ideas regarding the system of real numbers while at the same time reviewing the basic operations. In the first lesson only the natural numbers are considered, and the fundamental laws governing the basic operations with these numbers are explained. In the second lesson fractions of the form p/q, where p and q are natural numbers, are introduced, and definitions of equality, sum, product, etc., are given. In the third lesson the signed numbers are introduced, and in the fourth lesson positive integral powers and roots are defined and irrational numbers are introduced. In each of these lessons the student solves problems in which he must keep in mind the fact that the letters represent numbers of the particular class under discussion—and he is frequently asked to specify the restrictions under which certain operations can be performed within the system. This serves, among other things, to emphasize the fact that the letters represent numbers—a fact that students too often forget.

The second chapter is devoted to operations with polynomials, including special products and simple factoring, and to operations with rational fractional expressions.

Imaginary numbers are introduced in the third chapter when needed for the solution of quadratic equations. The complex number system is obtained simply by adjoining to the system of real numbers the one additional symbol i endowed with the following properties: It may be used with real numbers and letters representing real numbers in the operations of addition, subtraction, multiplication, and division, according to precisely the rules that would apply if it represented another real number, and subject to the one additional rule that $i^2 = -1$.

At the end of Chap. VII on negative and fractional exponents there is a discussion of the exponential function and its graph. Some kind of foundation is thus laid for the subject of logarithms which constitutes the next chapter.

In the chapter on mathematical induction the usual order of the two parts of the proof has been reversed. In teaching his own classes, the writer followed for years the instructions given by his textbooks and demonstrated first the truth of the theorem for a few special cases—for $n = 1, 2$, and 3, perhaps. When he then took the truth of the theorem for $n = k$ as a hypothesis, he was nearly always confronted with such questions as, "Doesn't k have to be one of the numbers for which you have tried it out?" and "For how many cases must you try it out before you can assume that it is true for $n = k$?" A better understanding of the main idea appeared to result when the procedure was reversed. We were then able to start out by saying that we did not know, and for the first part of our task did not care, whether the given theorem was true for *any* value of n—that our immediate concern was merely to determine the truth or falsity of the assertion that *if* the relation *were* true for some value of n, which we agreed to denote by k, then it *would be* true for the case in which $n = k + 1$.

Most teachers seem to agree that statement problems are valuable primarily because of the training they afford in translating verbal statements into mathematical language. The experience of the author indicates that one can accomplish about as much in this respect by assigning a few relatively straightforward problems as he can by having the student spend a lot of time unraveling one complicated puzzle after another. The statement problems given here should perhaps accomplish most of the good that can be derived from such problems without creating a wrong impression as to the main purposes of a course in college algebra.

A feature that may be of interest to many teachers is the fact that the book has been written so as to fall naturally into 48 lessons of about the length that one usually associates with the standard one-hour class for which the student is expected to spend approximately two hours in home study. A suggested lesson schedule follows the table of contents. It gives the text pages and sections covered by each lesson, lists briefly the topics involved, and gives three different problem selections to accompany the lesson. This helps the instructor to lay out his course in accordance with the amount of time at his disposal, and relieves him of the job of selecting a representative list of problems for each lesson.

ROSS R. MIDDLEMISS

ST. LOUIS, MO.
January, 1952

CONTENTS

SUGGESTED LESSON SCHEDULE

This book falls naturally into 48 lessons of the length normally associated with the standard one-hour class for which the student is expected to spend approximately two hours in home study. The following schedule gives the text pages and sections covered by each lesson, lists briefly the topics involved, and gives three approximately equivalent problem selections, any one of which would make a reasonable assignment.

Even in the case of a good class that can maintain a pace of one lesson per class period, the instructor will probably want to plan his course so as to spend extra time on certain topics that are of especial importance to his particular group. For example, if he wishes to emphasize mathematical induction, he will want to devote an extra day to lesson 26, and perhaps assign another one of the problem groups listed with this lesson. If he wishes to stress inequalities, he will similarly wish to spend two days on lesson 16.

In the case of a class that is not well prepared, it may be desirable to devote an extra day to each of several of the early lessons in order to make certain of a good foundation for the later lessons. For some classes it may be desirable to schedule a review for one class period out of every five or six, and assign for this period a couple of problems from each of the lessons being reviewed.

The time allotted to the course in most colleges and universities is not sufficient to enable one to cover all 48 lessons. In the standard one-semester course of three units, one might hope to cover from 36 to 40 of the lessons with an average class.

LESSON SCHEDULE

CHAPTER I. REAL NUMBERS. FUNDAMENTAL OPERATIONS

LESSON 1: Secs. 1–3, pages 1–7

Topics. Definitions and laws concerning fundamental operations with natural numbers. Prime numbers. L.C.M. and H.C.F.

Problem Assignments. Pages 6–7

A: 1, 3, 8, 18, 21, 26, 27, 30.

B: 2, 4, 9, 13, 22, 23, 28, 29.

C: 1, 5, 11, 15, 20, 25, 28, 30.

LESSON 2: Sec. 4, pages 7–13

Topics. Numbers of the form p/q where p and q are natural numbers. Basic operations with fractions.

Problem Assignments. Pages 12–13

A: 2, 7, 10, 13, 18, 22, 27, 28.

B: 3, 6, 11, 15, 19, 23, 26, 30.

C: 5, 8, 12, 16, 20, 24, 25, 31.

LESSON 3: Secs. 5–7, pages 14–21

Topics. Signed numbers and rules for basic operations with signed numbers. The system of rational numbers. Directed lines. Additional problems involving operations with fractions.

Problem Assignments. Pages 19–21

A: 3, 12, 13, 16, 24, 27, 32.

B: 5, 9, 14, 18, 23, 29, 34.

C: 8, 10, 13, 20, 26, 30, 35.

LESSON 4: Secs. 8–12, pages 21–29

Topics. Positive integral powers and roots. Irrational numbers. Proof of irrationality of $\sqrt{2}$. Decimal representation. Summary.

Problem Assignments. Pages 27–29

A: 1, 4, 10, 20, 23, 27, 34, 36.

B: 2, 9, 13, 21, 24, 28, 33, 35.

C: 3, 5, 11, 14, 22, 29, 32, 37.

CHAPTER II. OPERATIONS WITH POLYNOMIALS AND RATIONAL FRACTIONAL EXPRESSIONS

LESSON 5: Secs. 13–17, pages 30–37

Topics. Constants and variables. Algebraic expressions. Polynomials and basic operations with polynomials. Special products. Factoring.

Problem Assignments. Pages 36–37

A: 1, 3, 7, 14, 19, 25, 30, 34, 43, 47.
B: 2, 4, 6, 15, 20, 24, 29, 36, 44, 51.
C: 1, 5, 9, 16, 21, 23, 31, 37, 45, 54.

LESSON 6: Sec. 18, pages 38–41

Topics. Operations with rational fractional expressions. Problems that involve factoring in connection with the simplification of fractions.

Problem Assignments. Pages 40–41

A: 2, 6, 13, 16, 26, 32, 34, 35.
B: 3, 7, 12, 18, 27, 32, 33, 37.
C: 4, 9, 11, 20, 29, 30, 36, 38.

CHAPTER III. EQUATIONS

LESSON 7: Secs. 19–22, pages 42–49

Topics. Introduction—identities and conditional equations. Solving an equation. Some essentials of logic—hypothesis, conclusion, necessary and sufficient conditions. Equivalent equations.

Problem Assignments. Pages 48–49

A: 1, 2, 7, 13, 16, 22, 25, 33, 42.
B: 1, 4, 6, 14, 18, 21, 26, 39, 43.
C: 1, 3, 9, 15, 17, 20, 29, 36, 45.

LESSON 8: Secs. 23–25, pages 50–57

Topics. Definition of rational integral equation. General case of a linear equation. Solving a quadratic (case of real roots only) by factoring and by completing the square.

Problem Assignments. Pages 55–57

A: 2, 4, 13, 18, 26, 41, 43, 47, 56.
B: 1, 2, 5, 11, 19, 27, 38, 45, 53.
C: 2, 6, 14, 21, 31, 42, 44, 52, 55.

LESSON 9: Sec. 26, pages 57–62

Topics. Complex numbers. Basic operations with imaginary numbers. Additional problems on solving quadratics. Imaginary roots.

Problem Assignments. Pages 61–62

A: 3, 8, 12, 17, 23, 31, 34, 39, 47.
B: 2, 7, 11, 18, 24, 31, 35, 38, 50.
C: 5, 10, 14, 16, 30, 31, 37, 40, 45.

LESSON 10: Secs. 27–29, pages 63–70

Topics. The quadratic formula. The character of the roots. Sum and product of the roots. Equations that lead to quadratics.

Problem Assignments. Pages 68–70

A: 1, 5, 12, 19, 29, 30, 35, 41, 58.
B: 2, 7, 15, 22, 25, 33, 36, 45, 61.
C: 3, 6, 14, 23, 26, 34, 37, 46, 63.

LESSON 11: Sec. 30, pages 70–75

Topics. Equations of higher degree. Statement of the fundamental theorem. Solution of certain special types of equations of higher degree.

Problem Assignments. Pages 74–75

A: 1, 7, 12, 17, 21, 27, 34, 40.
B: 2, 4, 11, 18, 24, 28, 32, 38.
C: 3, 6, 13, 19, 25, 31, 33, 37.

CHAPTER IV. FUNCTIONS AND GRAPHS

LESSON 12: Secs. 31–32, pages 76–82

Topics. Rectangular coordinates. The graph of an equation.

Problem Assignments. Pages 81–82

A: 1, 3, 5, 9, 13, 17, 23, 29.
B: 2, 4, 6, 10, 12, 16, 24, 30.
C: 4, 5, 7, 11, 14, 18, 20, 32.

LESSON 13: Secs. 33–36, pages 82–89

Topics. Definition of function. Graph of a function. Graphical solution of an equation in one unknown. Applications.

Problem Assignments. Pages 87–89

A: 1, 4, 7, 12, 23, 32.
B: 2, 5, 8, 17, 26, 33.
C: 3, 6, 7, 19, 28, 35.

CHAPTER V. SYSTEMS OF EQUATIONS

LESSON 14: Secs. 37–39, pages 90–97

Topics. Systems of two linear equations in two unknowns. Graphical interpretation. Discussion of the general case.

Problem Assignments. Pages 94–97

A: 1, 2, 4, 7, 15, 22, 27, 29*c*, 30.
B: 1, 3, 4, 8, 14, 19, 25, 29*d*, 31.
C: 1, 2, 4, 10, 16, 20, 26, 29*e*, 32.

LESSON 15: Secs. 40–41, pages 97–105

Topics. Systems of three linear equations in three unknowns. Systems involving nonlinear equations.

Problem Assignments. Pages 102–105

A: 1, 7, 10, 14, 24, 36, 43, 49.
B: 2, 8, 11, 15, 25, 37, 44, 47.
C: 3, 6, 13, 18, 31, 38, 45, 46.

CHAPTER VI. INEQUALITIES

LESSON 16: Secs. 42–45, pages 106–114

Topics. Definitions of absolute and conditional inequalities. Basic theorems on inequalities Solution of inequalities.

Problem Assignments. Pages 112–114

A: 2, 7, 12, 18, 26, 36, 48, 54.

B: 4, 9, 13, 19, 27, 39, 47, 51.

C: 3, 10, 15, 22, 24, 38, 50, 55.

CHAPTER VII. NEGATIVE AND FRACTIONAL EXPONENTS. THE EXPONENTIAL FUNCTION

Lesson 17: Secs. 46–48, pages 115–120

Topics. Negative integers and zero as exponents. Rational fractions as exponents. General rules concerning rational exponents.

Problem Assignments. Pages 119–120

A: 3, 7, 12, 24, 29, 37, 40, 41.

B: 4, 9, 11, 19, 27, 38, 40, 42.

C: 6, 10, 15, 22, 31, 35, 40, 44.

Lesson 18: Secs. 49–50, pages 121–125

Topics. Irrational exponents. The exponential function and its graph. Approximate solutions of exponential equations by means of a graph.

Problem Assignments. Pages 124–125

A: 1, 4, 9, 14, 17, 19.

B: 2, 6, 12, 15, 18, 21.

C: 3, 7, 13, 16, 17, 22.

CHAPTER VIII. LOGARITHMS

Lesson 19: Secs. 51–55, pages 126–131

Topics. Introduction. Significant digits and approximate numbers. Definition of logarithm. Basic theorems concerning logarithms.

Problem Assignments. Pages 130–131

A: 1, 3, 8, 9, 13, 17, 18, 24.

B: 1, 4, 5, 10, 13, 16, 19, 25.

C: 1, 6, 7, 11, 13, 15, 20, 28.

Lesson 20: Secs. 56–59, pages 131–137

Topics. Common logarithms; characteristic and mantissa. Interpolation. Finding a number whose logarithm is given.

Problem Assignments. Pages 136–137

A: 1, 4, 5, 7, 15, 21, 23, 27, 35, 41.

B: 2, 3, 5, 12, 14, 20, 24, 29, 37, 42.

C: 1, 3, 5, 10, 18, 19, 25, 34, 40, 41.

Lesson 21: Sec. 60, pages 137–141

Topics. Computation with logarithms.

Problem Assignments. Pages 140–141

A: 1, 7, 15, 21, 27, 31, 35.

B: 3, 8, 16, 22, 29, 32, 34.

C: 4, 10, 17, 23, 30, 33, 36.

LESSON 22: Secs. 61–64, pages 141–147

Topics. Change of base. Natural logarithms. The logarithmic function and its graph. Exponential and logarithmic equations.

Problem Assignments. Pages 146–147

A: 1, 9, 13, 16, 20, 32, 36, 41.
B: 2, 10, 14, 17, 23, 34, 37, 39.
C: 3, 12, 15, 18, 26, 35, 36, 38.

CHAPTER IX. VARIATION

LESSON 23: Sec. 65, pages 148–152

Topics. The terminology of variation.

Problem Assignments. Pages 151–152

A: 2, 4, 7, 12, 13, 16.
B: 1, 5, 8, 11, 14, 18.
C: 3, 6, 9, 10, 15, 17.

CHAPTER X. PROGRESSIONS

LESSON 24: Secs. 66–69, pages 153–157

Topics. Definition of a sequence. Arithmetic progressions. Formulas for the nth term and for the sum of n terms.

Problem Assignments. Pages 156–157

A: 1, 5, 11, 13, 20, 28, 37, 38, 41.
B: 3, 6, 11, 12, 23, 29, 36, 39, 44.
C: 4, 10, 11, 17, 26, 30, 36, 40, 42.

LESSON 25: Secs. 70–74, pages 157–164

Topics. Geometric progressions. Formulas for the nth term and for the sum of n terms. Infinite geometric progressions. Periodic decimals.

Problem Assignments. Pages 162–164

A: 2, 7, 12, 15, 21, 24, 34, 38.
B: 4, 8, 11, 14, 18, 25, 35, 40.
C: 5, 9, 13, 17, 19, 27, 32, 41.

CHAPTER XI. MATHEMATICAL INDUCTION

LESSON 26: Secs. 75–76, pages 165–171

Topics. Proofs by means of mathematical induction.

Problem Assignments. Pages 170–171

A: 2, 5, 12, 17, 23, 26, 34.
B: 1, 7, 15, 18, 22, 25, 35.
C: 3, 9, 16, 19, 24, 27, 33.

CHAPTER XII. THE BINOMIAL THEOREM

LESSON 27: Secs. 77–79, pages 172–176

Topics. The binomial formula for positive integral exponents. Definition of the symbol $n!$ The general term in the expansion of $(a + b)^n$.

Problem Assignments. Pages 175–176
A: 2, 6, 16, 22, 31, 35, 39, 43.
B: 4, 13, 18, 24, 30, 36, 38, 41.
C: 5, 10, 20, 25, 33, 34, 37, 45.

Lesson 28: Secs. 80–81, pages 176–180
Topics. Proof of the binomial formula for positive integral exponents by means of mathematical induction. The binomial series.
Problem Assignments. Page 180
A: 1, 6, 11, 14, 21.
B: 2, 5, 10, 15, 23.
C: 3, 4, 9, 17, 26.

CHAPTER XIII. COMPOUND INTEREST AND ANNUITIES

Lesson 29: Secs. 82–83, pages 181–185
Topics. Definitions. Formulas for compound amount and present value.
Problem Assignments. Pages 184–185
A: 1, 4, 8, 10, 13, 16, 19.
B: 2, 5, 7, 11, 14, 17, 20.
C: 3, 4, 9, 12, 15, 18, 21.

Lesson 30: Secs. 84–86, pages 185–190
Topics. Annuities. Amount and present value of an annuity.
Problem Assignments. Pages 189–190
A: 1, 5, 9, 13, 17.
B: 2, 6, 10, 14, 18.
C: 4, 7, 12, 15, 16.

CHAPTER XIV. THEORY OF EQUATIONS

Lesson 31: Secs. 87–89, pages 191–198
Topics. Introduction. Synthetic division. The remainder theorem and the factor theorem.
Problem Assignments. Pages 197–198
A: 1, 5, 9, 10, 15, 22, 28
B: 2, 8, 9, 11, 14, 18, 29.
C: 4, 6, 9, 12, 16, 24, 30.

Lesson 32: Secs. 90–92, pages 198–204
Topics. Number of roots. Identical polynomials. Formation of an equation whose roots are given.
Problem Assignments. Pages 202–204
A: 2, 8, 11, 16, 18, 19, 27, 31, 36.
B: 4, 9, 13, 15, 18, 20, 28, 32, 38.
C: 5, 7, 12, 17, 18, 23, 29, 33, 40.

LESSON 33: Secs. 93–95, pages 204–210

Topics. Theorems regarding imaginary roots and a certain type of irrational root. Variations in sign. Descartes' rule of signs.

Problem Assignments. Pages 209–210

A: 2, 4, 12, 17, 21, 23, 28.

B: 1, 7, 13, 18, 22, 24, 27.

C: 3, 8, 14, 19, 21, 25, 29.

LESSON 34: Secs. 96–98, pages 210–215

Topics. Theorems regarding integral roots and rational roots. Bounds for the roots.

Problem Assignments. Pages 214–215

A: 2, 7, 11, 15, 19, 22.

B: 3, 6, 12, 18, 20, 23.

C: 4, 9, 13, 16, 21, 27.

LESSON 35: Secs. 99–100, pages 215–222

Topics. Fundamental theorems concerning irrational roots. Continuity of a polynomial. Computation of irrational roots by the method of successive linear approximations.

Problem Assignments. Pages 220–222

A: 1, 5, 10, 14.

B: 2, 6, 11, 15.

C: 1, 7, 12, 16.

LESSON 36: Sec. 101, pages 222–229

Topics. Transformation to diminish the roots of an equation by a given constant.

Problem Assignments. Pages 226–229

A: 2, 5, 8, 11, 14, 21.

B: 3, 6, 8, 11, 15, 22.

C: 4, 7, 8, 11, 17, 23.

LESSON 37: Sec. 102, pages 229–234

Topics. Horner's method.

NOTE: If lesson 35 has been omitted, Sec. 99 should be studied in connection with this lesson.

Problem Assignments. Pages 232–234

A: 1, 8, 11, 16.

B: 2, 10, 12, 17

C: 3, 9, 15, 18.

LESSON 38: Sec. 103, pages 234–238

Topics. Algebraic solution of the cubic.

Problem Assignments. Page 238

A: 1, 4.

B: 2, 7.

C: 3, 6.

CHAPTER XV. DETERMINANTS

LESSON 39: Secs. 104–105, pages 239–247

Topics. Determinants of order two and three. Systems of two linear equations in two unknowns and systems of three linear equations in three unknowns.

Problem Assignments. Pages 245–247

A: 1, 5, 11, 18, 24, 29*a*, 31.

B: 2, 7, 12, 19, 27, 29*b*, 30.

C: 3, 4, 15, 21, 25, 29*c*, 32.

LESSON 40: Sec. 106, pages 247–250

Topics. Determinants of order n. The double-subscript notation.

Problem Assignments. Pages 249–250

A: 1, 3, 7, 14, 16.

B: 2, 4, 8, 12, 17.

C: 1, 5, 10, 11, 18.

LESSON 41: Secs. 107–108, pages 251–256

Topics. Some properties of determinants.

NOTE: If the proofs of the theorems on determinants are to be stressed, an additional day should be devoted to Probs. 15 to 20.

Problem Assignments. Pages 255–256

A: 1, 6, 9, 11.

B: 3, 5, 10, 12.

C: 2, 4, 7, 14.

LESSON 42: Secs. 109–110, pages 256–263

Topics. Systems of n linear equations in n unknowns. Homogeneous equations. Systems in which the number of equations is not equal to the number of unknowns.

Problem Assignments. Pages 262–263

A: 2, 7, 11, 15.

B: 3, 8, 10, 16.

C: 6, 9, 12, 19.

CHAPTER XVI. COMPLEX NUMBERS

LESSON 43: Secs. 111–112, pages 264–269

Topics. Review of fundamental definitions and basic operations. Graphical representation of complex numbers. Graphical addition and subtraction.

Problem Assignments. Pages 268–269

A: 2, 8, 13, 27, 31, 44, 46, 53.

B: 5, 9, 17, 26, 35, 42, 45, 52.

C: 3, 10, 21, 28, 34, 41, 47, 50.

LESSON 44: Secs. 113–115, pages 269–274

Topics. Review of the principal trigonometric functions. Trigonometric form of a complex number. Multiplication and division of complex numbers in trigonometric form.

Problem Assignments. Pages 273–274

A: 2, 6, 15, 21, 29, 36, 44, 49, 55.

B: 3, 8, 16, 24, 31, 38, 42, 50, 56.

C: 5, 9, 18, 25, 34, 39, 45, 53, 55.

LESSON 45: Sec. 116, pages 275–279

Topics. Powers and roots of complex numbers. DeMoivre's theorem.

Problem Assignments. Page 279

A: 1, 7, 11, 15, 21.

B: 4, 8, 12, 17, 23.

C: 2, 6, 13, 18, 22.

CHAPTER XVII. PERMUTATIONS AND COMBINATIONS. PROBABILITY

LESSON 46: Secs. 117–120, pages 280–285

Topics. Introduction. Fundamental principle. Permutations of n distinct objects and of n objects some of which are alike. Circular permutations.

Problem Assignments. Pages 284–285

A: 1, 5, 7, 10, 11, 16.

B: 2, 4, 8, 13, 14, 17.

C: 3, 6, 9, 12, 15, 18.

LESSON 47: Secs. 121–124, pages 285–292

Topics. Combinations. Division of n objects into two or more groups. Combinations and the binomial coefficients.

Problem Assignments. Pages 291–292

A: 2, 5, 10, 12, 15, 22.

B: 1, 6, 11, 13, 16, 21.

C: 3, 7, 9, 14, 19, 20.

LESSON 48: Secs. 125–128, pages 292–300

Topics. Probability. Statistical probability. Mutually exclusive events. Independent and dependent events.

Problem Assignments. Pages 299–300

A: 1, 6, 7, 10, 13, 19.

B: 2, 5, 8, 11, 14, 16.

C: 3, 4, 9, 12, 15, 20.

CHAPTER I

REAL NUMBERS. FUNDAMENTAL OPERATIONS

1. ***Introduction.*** The student who is now beginning the study of college algebra has had from one to two years of algebra in high school. He has learned how to perform certain kinds of operations with what he rather vaguely calls "algebraic expressions" or "algebraic quantities," and he may have spent many hours in translating "word problems" into equations and solving these equations. Unfortunately, he may have done these things in a more or less mechanical fashion, and he may have learned very little about the underlying fundamental principles. For example, most students know that the product of two negative numbers is positive, but few could give any reason for this state of affairs. Many students who have used the relation

$$a^2 - b^2 = (a + b) \cdot (a - b)$$

as a formula in connection with exercises in factoring are surprised to discover much later that "it works even if a and b are numbers" and that one can actually find the value of $37^2 - 33^2$ by writing

$$37^2 - 33^2 = (70) \cdot (4) = 280.$$

It is perhaps largely because of this lack of any real understanding that many students have forgotten much of what they learned about algebra in high school by the time they start to college.

The purpose of this chapter is to provide the student with a review of such fundamental operations as adding and multiplying fractions or solving simple equations and, at the same time, give him a better understanding of the number system with which our subject is concerned. In this connection he must of course realize that man did not arrive upon the earth and promptly "discover" a number system—complete with rules for using same. The symbols that we call numbers, and the rules for reckoning with them, were devised by man. The beginnings were very crude and probably consisted only of symbols denoting "one," "two," etc., and not extending beyond ten. Advances came gradually, and in many cases these advances were slow in gaining acceptance. For example, there was a period of several hundred years during which negative numbers were used (rather hesitantly) by some mathematicians but avoided entirely by others who regarded them as fictitious. Even as late as the early part of the seventeenth century the whole idea of using negative numbers was viewed with suspicion by many people who were well versed in the mathematics of that time.

In making this review we shall follow to some extent the pattern in which our number system developed. Thus we shall first consider a number system consisting only of the *natural numbers* or *positive integers*, and note particularly the basic laws that apply to the fundamental operations of addition, subtraction, multiplication, and division within this system.

Next, we shall consider a system consisting of all numbers of the form a/b where a and b are natural numbers. We shall give definitions of sum, product, etc., for this system and try to see why these definitions are the natural ones. In this section the student will have an opportunity to review things that he may have forgotten in regard to operations with fractions.

Following this, we shall introduce what we call the *signed numbers* or *directed numbers* (the positive and negative numbers and the number *zero*). Again we shall find it necessary to decide upon definitions of sum, product, etc., for numbers of this system. We shall see, for example, that it is quite natural that our definitions should be such that $(-2)(-3) = +6$.

Finally, we shall discuss very briefly the *irrational numbers* and conclude with a summary of the basic rules that apply to the fundamental operations within the system of *real numbers*. The so-called *imaginary numbers*, which the student has encountered and about which he probably has only some vague ideas, will be discussed briefly when they are first needed in the text (page 57), and more fully in a separate chapter (Chap. XVI).

2. *The natural numbers.* The natural numbers are the symbols 1, 2, 3, 4, 5, 6, · · · ,* the three dots being used to mean "and so on." In regard to the fundamental operations with these numbers, we point out the following basic facts:

ADDITION. Let a and b be any two numbers or *elements* of the system. There is associated with them a definite third number of this system called their *sum* and denoted by $a + b$ or $b + a$. a and b are called *terms* of the sum. The operation of finding the sum is called *addition*, and the fact that it always exists within the system is expressed by saying that the system is *closed with respect to addition.*

EQUALITY. It may happen that two operations with numbers (even though they are different operations with different numbers) have as their result the *same* number. Thus the operations $2 + 5$ and $4 + 3$ yield the same number. We indicate this by saying that $2 + 5$ *equals* $4 + 3$, and writing

$$2 + 5 = 4 + 3.$$

This statement is called an *equation*, and the symbol $=$ is called the *sign of equality*. If a is any number of the system, we of course agree that $a = a$ and that a is not equal to any other number of the system. The symbol $\neq$ means *is not equal to.*

The statement

$$x + 3 = 7$$

is a true statement of equality if $x = 4$, but false if x is any other number of the system. We say that $x = 4$ *satisfies* the equation or is a *solution* of the equation. We also call the number 4 a *root* of the equation.

SUBTRACTION. This is defined in terms of addition: *The symbol* $\mathbf{a} - \mathbf{b}$ *denotes the number* $\mathbf{x}$ *such that* $\mathbf{b} + \mathbf{x} = \mathbf{a}$, *if such a number exists in the system.* Thus $26 - 14 = 12$ because $14 + 12 = 26$.

The number $a - b$ is called the *difference* of a and b, and the operation of finding it is called *subtraction*. The system is not closed with respect to subtraction. Thus the symbol $9 - 12$ is meaningless because there exists no natural number x such that $12 + x = 9$.

* Zero may be included but it will be more convenient for us to introduce it later.

If the number $a - b$ does exist in the system, we say that $\boldsymbol{a}$ *is greater than* $\boldsymbol{b}$ (written $\boldsymbol{a} > \boldsymbol{b}$) and that $\boldsymbol{b}$ *is less than* $\boldsymbol{a}$ (written $\boldsymbol{b} < \boldsymbol{a}$).

MULTIPLICATION. In the system of natural numbers this can also be defined in terms of addition. The symbol

$$a \times b \qquad \text{or} \qquad a \cdot b \qquad \text{or} \qquad ab$$

means $b + b + b + \cdots + b$ where there are a of the b's. Thus $5 \cdot 7$ means $7 + 7 + 7 + 7 + 7 = 35$. The result is called the *product* of a and b, and a and b are called *factors* of the product. The operation of finding the product is called *multiplication*, and from the property of closure with respect to addition it follows that the system is closed with respect to multiplication.

Figure 1

DIVISION. This is defined in terms of multiplication: *The symbol* $\boldsymbol{a} \div \boldsymbol{b}$ (*read* $\boldsymbol{a}$ divided by $\boldsymbol{b}$) *denotes the number* $\boldsymbol{x}$ *such that* $\boldsymbol{bx} = \boldsymbol{a}$, *if such a number exists in the system.* The number x is called the *quotient of* a and b, and the operation of finding it is called *division.* The system is of course not closed with respect to division. For example, the symbol $20 \div 7$ is meaningless because there exists no natural number x such that $7x = 20$.

The following basic laws apply to the above fundamental operations with natural numbers:

I. COMMUTATIVE LAW FOR ADDITION. *If* $\boldsymbol{a}$ *and* $\boldsymbol{b}$ *are any two numbers of the system, then* $\boldsymbol{a} + \boldsymbol{b} = \boldsymbol{b} + \boldsymbol{a}$. This is a result of our agreement to associate the same number with these symbols.

II. COMMUTATIVE LAW FOR MULTIPLICATION. *If* $\boldsymbol{a}$ *and* $\boldsymbol{b}$ *are any two numbers of the system, then* $\boldsymbol{a} \cdot \boldsymbol{b} = \boldsymbol{b} \cdot \boldsymbol{a}$. For example, by definition,

$$5 \cdot 3 = \text{the sum of 5 threes} = 3 + 3 + 3 + 3 + 3 = 15;$$
$$3 \cdot 5 = \text{the sum of 3 fives} = 5 + 5 + 5 = 15.$$

Thus the operations are different but the results are the same. We may interpret the first operation as finding the total number of dots in the rectangular array of Fig. 1 by regarding it as consisting of 5 vertical columns with 3 dots in each column, and the second as doing the same thing by regarding it as consisting of 3 horizontal rows with 5 dots in each row.

III. Associative law for addition. *If a, b, and c are any numbers of the system, then* $a + (b + c) = (a + b) + c$. It should be recalled here that we use parentheses, brackets, etc., for the purpose of grouping together quantities that are to be regarded as single units in the calculation. Thus $3 + (4 + 5)$ means $3 + 9$, while $(3 + 4) + 5$ means $7 + 5$. The above law states that the sum is independent of the way in which the numbers are grouped.

IV. Associative law for multiplication. *If a, b, and c are any numbers of the system, then* $a \cdot (b \cdot c) = (a \cdot b) \cdot c$. For example, $2(3 \cdot 4) = 2(12) = 24$, and $(2 \cdot 3) \cdot 4 = 6 \cdot 4 = 24$.

V. Distributive law. *If a, b, and c are any numbers of the system, then* $a \cdot (b + c) = a \cdot b + a \cdot c$. Thus

$$\begin{aligned} 3(4 + 5) &= (4 + 5) + (4 + 5) + (4 + 5) \\ &= (4 + 4 + 4) + (5 + 5 + 5) \\ &= 3 \cdot 4 + 3 \cdot 5.^* \end{aligned}$$

3. *Even and odd numbers. Prime numbers.* A natural number is said to be *even* if it has 2 as a factor. A natural number that is not even is called *odd*. Thus 2, 4, and 6 are even while 1, 3, and 5 are odd.

A natural number is called a *prime number* if it is greater than 1 and has no factor except itself and 1. Thus 2, 17, and 41 are prime numbers.

It can be proved that every nonprime natural number greater than 1 can be expressed as a product of prime numbers in one and only one way. For example,

$$78 = 2 \cdot 3 \cdot 13 \qquad \text{and} \qquad 476 = 2 \cdot 2 \cdot 7 \cdot 17.$$

The *highest common factor* (H.C.F.) of a given set of natural numbers is the largest number that is a factor of all members of the given set. Thus the H.C.F. of the numbers 12, 20, and 36

* We agree, in connection with this symbol, that the multiplications are to be performed first unless the contrary is indicated by parentheses. Thus $3 + 4 \cdot 5$ means $3 + 20$ while $(3 + 4) \cdot 5$ means $7 \cdot 5$.

is 4. The H.C.F. of any given set is equal to the product of the prime factors that are common to all the numbers of the set, each taken the smallest number of times that it occurs in any of the numbers:

Example

$$\left.\begin{array}{r}84 = 2 \cdot 2 \cdot 3 \cdot 7\\ 132 = 2 \cdot 2 \cdot 3 \cdot 11\\ 180 = 2 \cdot 2 \cdot 3 \cdot 3 \cdot 5\end{array}\right\} \text{H.C.F.} = 2 \cdot 2 \cdot 3 = 12.$$

The *lowest common multiple* (L.C.M.) of a given set of natural numbers is the smallest number that has each of the given numbers as a factor. Thus the L.C.M. of the numbers 3, 4, and 8 is 24. The L.C.M. of any given set is equal to the product of all the different prime factors in the numbers, each taken the greatest number of times that it occurs in one of the numbers.

Example

$$\left.\begin{array}{r}12 = 2 \cdot 2 \cdot 3\\ 20 = 2 \cdot 2 \cdot 5\\ 36 = 2 \cdot 2 \cdot 3 \cdot 3\end{array}\right\} \text{L.C.M.} = 2 \cdot 2 \cdot 3 \cdot 3 \cdot 5 = 180.$$

PROBLEMS

1. Is the set of all even numbers closed with respect to addition? The set of all odd numbers?

2. Is the set of all natural numbers less than 1,000 closed with respect to addition? With respect to multiplication?

3. If a and b are natural numbers, does there always exist a natural number x such that $ax = b$? If your answer is "no," specify some additional condition on a and b under which the equation $ax = b$ will always have a solution within the system of natural numbers.

4. If a and b are natural numbers, under what condition does the equation $a + x = b$ have a solution in the system of natural numbers?

5. One sometimes adds a column of numbers from top to bottom and then checks the result by adding from bottom to top. Which of the basic laws (I to V) is he using?

6. Find all prime numbers that are less than 50.

Express each of the following numbers as a product of prime numbers:

7. 74.	**8.** 212.	**9.** 114.	**10.** 184.
11. 253.	**12.** 192.	**13.** 1,309.	**14.** 1,221.
15. 1,722.	**16.** 1,619.	**17.** 493.	**18.** 6,241.

Find the L.C.M. and H.C.F. of each of the following sets of numbers:

19. 8, 12, 18.
20. 14, 21, 35.
21. 6, 12, 24, 48.
22. 7, 13, 19.
23. 150, 210, 240.
24. 30, 65, 105.
25. 16, 72, 128, 256.
26. 2,460, 3,260.

27. If x and y represent natural numbers, which of the following expressions invariably represent natural numbers: $x + y$; xy; $x - y$; $x \div y$?

28. If a, b, and c represent natural numbers, which of the following expressions invariably represent natural numbers: $2a + b - c$; $a + b + 2c$; $ab + ac - bc$; $a(b + c) + b(a + c)$?

29. Using the definition of product for the system of natural numbers, and the commutative and associative laws for addition, prove the distributive law $a(b + c) = ab + ac$.

30. Using the distributive law and the commutative law for multiplication, prove that

$$(x + y)(a + b) = ax + ay + bx + by,$$

where the letters represent any natural numbers.

4. *Fractions.* For the present we shall use the word *fraction* to mean a symbol of the form $\frac{a}{b}$ or a/b (read a over b) where a and b are natural numbers. a is called the *numerator* and b the *denominator* of the fraction.

The symbol is to be regarded as denoting a of b equal parts into which something is divided. Thus $3/4$ denotes three of four equal parts of something. The symbol $15/4$ denotes fifteen "quarters," and our rules for operating with fractions should of course be such that this is equivalent to $3 + \frac{3}{4}$. (In arithmetic this is written $3\frac{3}{4}$ and read three *and* three-fourths.)

We may agree that the fraction $a/1$ is to be regarded as identical with the natural number a. The system of all fractions then contains all the natural numbers.

It is of course up to us to decide upon the conditions under which we shall call two fractions equal, and upon what we shall mean by the sum, product, etc., of two fractions. Our definitions must not contradict those that have already been given for natural numbers and, if possible, they should be such that the basic laws I to V (page 4) are preserved. Early in his career, the reader became acquainted with these definitions but we review here the essential facts:

EQUALITY. We of course want to regard any fraction as equal to itself. We also want to regard $6/8$ as equal to $3/4$ because six of eight equal parts would be essentially the same as three of four equal parts. We set up the following definition: *If* a/b *and* c/d *are two fractions, then*

$$\frac{a}{b} = \frac{c}{d} \text{ if and only if } ad = bc.$$

For example,

$$\tfrac{14}{35} = \tfrac{6}{15} \qquad \text{because } 14 \cdot 15 = 35 \cdot 6.$$

In particular, if a/b is any fraction and x is any natural number, then

$$\frac{ax}{bx} = \frac{a}{b}.$$

Thus $\frac{14}{35} = \frac{2 \cdot 7}{5 \cdot 7} = \frac{2}{5}$. A fraction is said to be in *lowest terms* if its numerator and denominator have no common factor other than 1. We have here reduced the fraction $14/35$ to lowest terms by "canceling" the H.C.F. of numerator and denominator.

MULTIPLICATION. We set up the following definition: *The product of two fractions is the fraction whose numerator is the product of the two numerators and whose denominator is the product of the two denominators: i.e., if* a/b *and* c/d *are two fractions then*

$$\frac{a}{b} \cdot \frac{c}{d} = \frac{ac}{bd}.$$

Reasons for this definition can readily be given. For example if the dimensions of a rectangle are 4 ft. and 3 ft., the product of these numbers gives its area. If we want this still to be true when the dimensions are fractions, we must have the above definition of product. Thus, if the dimensions are $\frac{3}{4}$ ft. and $\frac{2}{3}$ ft., the area is

$$\tfrac{3}{4} \cdot \tfrac{2}{3} = \tfrac{6}{12} = \tfrac{1}{2} \text{ sq. ft.}$$

Again, if one buys 3 yd. of cloth at \$4 per yard, the product $3 \cdot 4$ gives the total cost. If he buys $\frac{2}{3}$ yd. at \$$\frac{3}{4}$ per yard, the cost is $\frac{2}{3} \cdot \frac{3}{4} = \frac{1}{2}$ dollar.

From the corresponding closure property of the natural numbers, it follows that the set of all fractions is closed with respect to multiplication.

DIVISION. For fractions, the definition of quotient is the same as it is for natural numbers: *If* p *and* q *are any two fractions, their quotient* $p \div q$ *is the fraction* x*, if there is one, such that* $q \cdot x = p$.

The student may recall that to divide a/b by c/d one may "invert the denominator and multiply"; *i.e.*, one may multiply a/b by d/c. We shall now prove that if a/b and c/d are two fractions, then

$$\frac{a}{b} \div \frac{c}{d} = \frac{a}{b} \cdot \frac{d}{c} = \frac{ad}{bc}.$$

Proof: We need only show that the result of multiplying c/d by ad/bc is a/b; thus

$$\frac{ad}{bc} \cdot \frac{c}{d} = \frac{adc}{bcd} = \frac{a}{b}.$$

This of course proves that the system is closed with respect to division. (Recall that the system of natural numbers is not thus closed.)

Example 1

$$\tfrac{3}{7} \div \tfrac{4}{21} = \tfrac{3}{7} \cdot \tfrac{21}{4} = \tfrac{9}{4}.$$

Example 2

If a and b are any natural numbers, then $a/(a + b)$ and $4/b$ are fractions. To divide the first by the second, we have

$$\frac{a}{a+b} \div \frac{4}{b} = \frac{a}{a+b} \cdot \frac{b}{4} = \frac{ab}{4(a+b)}.$$

We can prove now that if a and b are natural numbers, the quotient $a \div b$ is equal to the fraction a/b.

Proof: Since $a = \dfrac{a}{1}$ and $b = \dfrac{b}{1}$,

$$a \div b = \frac{a}{1} \div \frac{b}{1} = \frac{a}{1} \cdot \frac{1}{b} = \frac{a \cdot 1}{1 \cdot b} = \frac{a}{b}.$$

We may therefore use the symbols $a \div b$ and a/b interchangeably.

ADDITION. If two fractions have the *same denominator*, their sum is defined as the fraction which has this denominator for its denominator and has the sum of the two numerators for its numerator. Thus $\frac{2}{13} + \frac{5}{13} = \frac{7}{13}$.

If the fractions have different denominators, then one (or both) of the fractions is replaced by an equal fraction such that the new fractions do have equal denominators; these new fractions are then added as indicated above. (See also Prob. 21, page 12.)

Example

$$\frac{4}{7} + \frac{2}{5} = \frac{4}{7} \cdot \frac{\mathbf{5}}{\mathbf{5}} + \frac{2}{5} \cdot \frac{\mathbf{7}}{\mathbf{7}}$$
$$= \tfrac{20}{35} + \tfrac{14}{35} = \tfrac{34}{35}.$$

The procedure can be extended to the case of three or more fractions in an obvious way. In general it is convenient to use as a "common denominator" the L.C.M. of the denominators of the given fractions. Thus

$$\frac{1}{12} + \frac{7}{15} + \frac{11}{30} = \frac{1}{12} \cdot \frac{\mathbf{5}}{\mathbf{5}} + \frac{7}{15} \cdot \frac{\mathbf{4}}{\mathbf{4}} + \frac{11}{30} \cdot \frac{\mathbf{2}}{\mathbf{2}}$$
$$= \tfrac{5}{60} + \tfrac{28}{60} + \tfrac{22}{60}$$
$$= \tfrac{55}{60} = \tfrac{11}{12}.$$

Here, the L.C.M. of 12, 15, and 30 is 60, so we replaced each fraction by an equal fraction having the denominator 60.

SUBTRACTION. The definition is the same as it is for natural numbers: *If* p *and* q *are two fractions, the symbol* $p - q$ *denotes the fraction* x *such that* $q + x = p$, *if such a fraction exists.* The system is of course not closed with respect to subtraction; *i.e.*, for a given pair of fractions p and q, the difference $p - q$ may or may not exist in the system. When it does exist, we say that p *is greater than* q. The procedure used in finding the difference of two fractions is of course similar to that used in finding their sum.

Example

If a and b represent any natural numbers, then $a/(a+1)$ and $b/(b+1)$ are fractions. We may subtract the second from the first and express the difference as another fraction, by the following procedure in which we use $(a+1)(b+1)$ as a common denominator:

$$\frac{a}{a+1} - \frac{b}{b+1} = \frac{a}{a+1} \cdot \frac{\mathbf{b+1}}{\mathbf{b+1}} - \frac{b}{b+1} \cdot \frac{\mathbf{a+1}}{\mathbf{a+1}}$$
$$= \frac{a(b+1)}{(a+1)(b+1)} - \frac{b(a+1)}{(a+1)(b+1)}$$
$$= \frac{ab + a - ba - b}{(a+1)(b+1)}$$
$$= \frac{a-b}{(a+1)(b+1)}.$$

If $a > b$, all the above operations are possible within the system of fractions.

Finally, we give an example that involves a combination of the above defined operations.

Example

Assuming that a and b represent any numbers for which all the indicated operations can be performed, reduce the expression

$$\frac{\dfrac{a}{a+1} - \dfrac{b}{b+1}}{\dfrac{a+1}{b+1} - 1}$$

to an equivalent simple fraction.

Solution

$$\begin{aligned}
\frac{\dfrac{a}{a+1} - \dfrac{b}{b+1}}{\dfrac{a+1}{b+1} - 1} &= \frac{\dfrac{a(b+1)}{(a+1)(b+1)} - \dfrac{b(a+1)}{(b+1)(a+1)}}{\dfrac{a+1}{b+1} - \dfrac{b+1}{b+1}} \\
&= \frac{\dfrac{ab + a - ba - b}{(a+1)(b+1)}}{\dfrac{a+1-b-1}{b+1}} \\
&= \frac{\dfrac{a-b}{(a+1)(b+1)}}{\dfrac{a-b}{b+1}} \\
&= \frac{a-b}{(a+1)(b+1)} \cdot \frac{b+1}{a-b} \\
&= \frac{1}{a+1}.
\end{aligned}$$

If a and b represent any natural numbers or any fractions and if $a > b$, all these operations are possible within the system of fractions. When we define these same operations for the signed numbers in Sec. 5, we shall do so in such a way that all these steps remain valid even if a or b (or both) is a negative number, provided that neither is equal to -1.

PROBLEMS

Assuming that the letters represent natural numbers, carry out the indicated operations:

1. $\frac{4}{13} \cdot \frac{5}{4}; \frac{x}{y} \cdot \frac{z}{x}$.

2. $\frac{7}{18} \cdot \frac{12}{5}; \frac{x}{3a} \cdot \frac{2a}{b}$.

3. $\frac{\frac{3}{4}}{2}; \frac{a/b}{c}$.

4. $\frac{3}{\frac{4}{5}}; \frac{a}{b/c}$.

5. $\frac{\frac{3}{4}}{\frac{7}{8}}; \frac{x/a}{y/2a}$.

6. $\frac{2}{3} + \frac{5}{7}; \frac{a}{b} + \frac{c}{d}$.

7. $\frac{2}{5} + \frac{7}{10} + \frac{7}{15}; \frac{a}{b} + \frac{c}{2b} + \frac{c}{3b}$.

8. $\frac{1}{5} + \frac{1}{7} + \frac{1}{21}; \frac{1}{x} + \frac{1}{y} + \frac{1}{3y}$.

9. $\frac{\frac{3}{2} + \frac{9}{4}}{5}; \frac{(x/a) + (x/2a)}{5}$.

10. $\frac{3 + \frac{3}{5}}{\frac{2}{5}}; \frac{a + (a/b)}{c/b}$.

11. $\frac{\frac{1}{2} + \frac{1}{3}}{4}; \frac{(1/a) + (1/b)}{c}$.

12. $\frac{\frac{1}{4} + \frac{1}{5}}{\frac{1}{3}}; \frac{(1/a) + (1/b)}{1/c}$.

13. $\frac{\frac{2}{3} + \frac{5}{7}}{\frac{9}{11}}; \frac{(a/b) + (c/d)}{e/f}$.

14. $\frac{4 - \frac{4}{3}}{5}; \frac{x - (x/3)}{y}$.

15. $\frac{[(3 + 5)/7] - \frac{3}{14}}{\frac{5}{14}}; \frac{[(a + b)/x] - (a/2x)}{b/2x}$.

16. $\frac{[(2 + 11)/14] - \frac{1}{28}}{\frac{1}{2} - \frac{1}{3}}; \frac{[(a + b)/c] - (1/2c)}{(a/4) - (a/6)}$.

17. Is the set of all fractions less than 1 closed with respect to multiplication? With respect to addition?

18. If a and b are fractions, does there always exist a fraction x such that $ax = b$?

19. If a, b, and c are natural numbers, express as a single fraction the sum $(1/a) + (1/b) + (1/c)$.

20. If a and b are natural numbers, under what condition does the difference $(a/b) - (b/a)$ exist within the system of fractions?

21. We may regard fractions as pairs of natural numbers subject to the following definitions of product and sum: *If* $\mathbf{a}/\mathbf{b}$ *and* $\mathbf{c}/\mathbf{d}$ *are two such number pairs, then*

$$\frac{\mathbf{a}}{\mathbf{b}} \cdot \frac{\mathbf{c}}{\mathbf{d}} = \frac{\mathbf{ac}}{\mathbf{bd}} \quad \text{and} \quad \frac{\mathbf{a}}{\mathbf{b}} + \frac{\mathbf{c}}{\mathbf{d}} = \frac{\mathbf{ad} + \mathbf{bc}}{\mathbf{bd}}.$$

Show that this definition of sum is equivalent to that given in the text.

In each of the following problems, first perform the indicated operations and simplify the result assuming that the letters represent any natural numbers for which all these operations can be carried out. Then evaluate this final result for the given values of a, b, c, x, y, z. Check by substituting these numbers for the letters in the original expression and then performing the indicated arithmetic:

22. $\dfrac{[(a + b)/x] + (c/yz)}{(1/xy) + [(a + b)/z]}$; $a = 2, b = 1, c = 3,$
$x = 1, y = 4, z = 6.$

23. $\dfrac{[(x + 2y)/ab] - (xy/cb)}{(x/bc) + (y/a)}$; $a = 2, b = 4, c = 4,$
$x = 1, y = 3.$

24. $\dfrac{[(3x - 2y)/a] - (4/bc)}{[(x + y)/b] - (2/c)}$; $a = 4, b = 1, c = 3,$
$x = 6, y = 2.$

25. $\dfrac{[(a + 3b)/xy] + (b/yz)}{[(b - a)/x] + [(b + a)/x]}$; $a = 1, b = 3,$
$x = 2, y = 4, z = 1.$

26. $\dfrac{[(ax + y - 2az)/ab] + [(y - cx + 2cz)/bc]}{[(x + cy)/c] - [(ay - x)/a]}$; $a = 1, b = 2, c = 4,$
$x = 3, y = 12, z = 6.$

27. $\dfrac{[(3ax + 2y)/3ab] + [(3y - 2cx + 4z)/2bc]}{[(x + 3cy)/ac] - [(3by - x)/ab]}$; $a = b = c = 1,$
$x = 2, y = 4, z = 6.$

28. Show that if the letters represent any natural numbers for which all the indicated operations can be performed, then

$$\frac{[(a + b)/ab] + [(b - c)/bc]}{[(a + c)/ac] - [(b + c)/bc]} = \frac{ab + bc}{bc - ac}.$$

29. Show that if the letters represent any natural numbers for which all the indicated operations can be performed, then

$$\frac{a - [(ax + by)/(x + y)] + b}{(b/y) + (a/x)} = \frac{xy}{x + y}.$$

30. Show that if a and b represent any natural numbers for which all the indicated operations can be performed, then

$$\frac{[a/(a + 2)] - [b/(b + 3)]}{[(2a + 5)/(a + 2)] - 2} = \frac{3a - 2b}{b + 3}.$$

What condition must be satisfied by the natural numbers a and b in order that the indicated operations may be possible within the field of fractions?

31. Show that if a and b are natural numbers and if $a < b$, then

$$\frac{ab[(1/a) - (1/b)]}{(b - a)[(1/a) + (1/b)]} = \frac{ab}{a + b}.$$

32. Show that if the letters represent any natural numbers for which all the indicated operations can be carried out, then

$$\frac{[(3x + ay - 2az)/ac] - [(2x + by - 2bz)/bc]}{[(x - ay)/ac] + [(by + x)/cb]} = \frac{3b - 2a}{a + b}.$$

5. ***Signed numbers.*** We next invent a new and more useful number system in the following way:

Corresponding to each number a of the system studied in Sec. 4, we devise two new symbols, $+a$ and $-a$. The first is called a *positive number* and the second a *negative number*. We invent also a symbol 0, called *zero*, which is to be regarded as a number belonging to neither of these classes and having certain properties to be specified below.

Corresponding to the natural numbers we now have the *positive integers* and the *negative integers*. These, together with zero, constitute what we shall call the *integers* or *whole numbers*. Corresponding to the fractions of the preceding section we now have positive fractions and negative fractions.

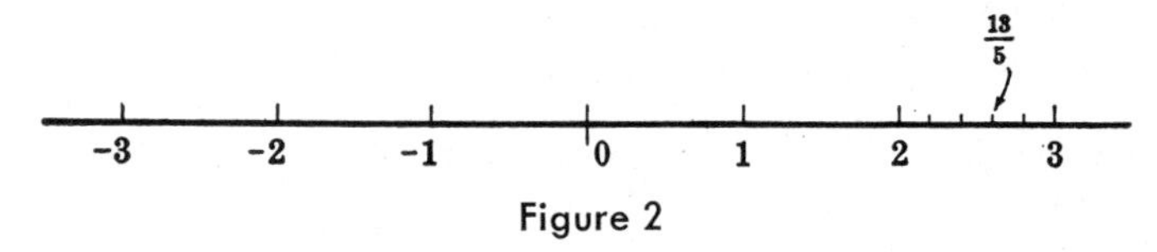

Figure 2

The numbers of this new system are called *signed numbers* or *directed numbers*. They can be given an interpretation involving the idea of direction in the following way:

On a straight line choose a point and associate with it the number zero. This point is then called the *origin*. Choose a unit of length, mark off this unit an indefinite number of times in both directions, and associate with the points so obtained the numbers $+1, +2, +3, \cdots$, and $-1, -2, -3, \cdots$, as indicated in Fig. 2. (It is purely a matter of convention that we put the positive numbers on the right and the negative ones on the left.) We now associate a definite point on the line with every number of our system in an obvious way; thus we get the point corresponding to $\frac{13}{5}$ or $2\frac{3}{5}$ by dividing the interval from 2 to 3 into five equal parts and locating the point at the third one of these marks (Fig. 2).

We shall now find it convenient to regard each number of the system studied in Sec. 4 as identical with the corresponding positive number of this new system. Thus the natural number 5 becomes identical with the signed number $+5$, and we may write 5 and $+5$ interchangeably. Our new number system then contains that of Sec. 4.

By the *absolute value* of a number a, we mean the number a itself if a is positive, the positive number $-a$ if a is negative, and 0 if a is zero. The symbol $|a|$ is used to denote the absolute value of the number a.

Examples

$$|-6| = 6; \qquad |\tfrac{2}{3}| = \tfrac{2}{3}; \qquad |0| = 0;$$
$$|a - b| = a - b \text{ if } a > b,\ b - a \text{ if } a < b,\ 0 \text{ if } a = b.^*$$

We must now agree upon definitions of sum, product, etc., for numbers of this system. The definitions should of course agree with those of Sec. 4 when applied to positive numbers and, if possible, they should be such that the basic laws (I to V, page 4) are preserved.

We shall first give the necessary modifications of our previous definitions of sum and product by stating the rules for addition and multiplication:

To add two numbers of like sign, add their absolute values and prefix this sign. To add two numbers of unlike sign, subtract the smaller absolute value from the larger and prefix the sign of the number having the larger absolute value. If in this latter case the absolute values are equal, the sum is *zero*. Thus $0 = a + (-a)$.

In terms of our representation of the signed numbers as points on a line (Fig. 2), we can interpret our definition of sum thus:

The sum $a + b$ is the number that lies b units to the right or left of a according as b is positive or negative.

Examples

$2 + 7 = 9$ (seven units to right of 2)
$2 + (-7) = -5$ (seven units to left of 2)
$-3 + (-4) = -7$ (four units to left of -3)
$-6 + 6 = 0$ (six units to right of -6)

To multiply two numbers of like sign, multiply their absolute values and prefix a plus sign. To multiply two numbers of unlike sign, multiply their absolute values and prefix a minus sign. Thus the product of two positive or two negative numbers is positive, and the product of a positive and a negative number is negative.

* The definition given below for $a - b$ is used in this example.

Examples

$$4 \cdot 5 = 20; \qquad (-4) \cdot (-5) = 20; \qquad 4(-5) = -20.$$

For the number zero we need the following additional definitions: *If a is any number of the system, then*

$$a + 0 = 0 + a = a; \qquad a \cdot 0 = 0 \cdot a = 0.$$

We have chosen these definitions because they are useful, because we want the signed numbers to behave as much like the unsigned numbers as possible, and because we want to preserve the previously mentioned basic laws. Thus we want $4(-3)$ to equal -12 because we wish to interpret it as the sum of four (-3)'s:

$$4 \cdot (-3) = (-3) + (-3) + (-3) + (-3) = -12.$$

Of course we want $(-3) \cdot 4$ to equal -12 also, because we wish to preserve the commutative law, even though we cannot interpret this as the sum of minus three 4's.

One of the reasons for setting up our definitions so that $(-4) \cdot (-2) = +8$ is our desire to preserve the distributive law. For example, $-4[2 + (-2)]$ should equal $-4 \cdot 0$ or zero. If the distributive law is to hold then

$$\begin{aligned} -4[2 + (-2)] &= -4(2) + (-4) \cdot (-2) \\ &= -8 + (-4) \cdot (-2). \end{aligned}$$

This is zero if and only if $(-4) \cdot (-2) = +8$.

For subtraction there is no need to modify our previous definition: *If a and b are any two numbers of the system, then the symbol $a - b$ denotes the number x such that $b + x = a$, if there is such a number in the system.* The only new fact is that the number x always exists; *i.e.*, the system of signed numbers is closed with respect to subtraction.

If $a - b$ is a positive number, we say that a is greater than b ($a > b$). Thus

$$5 > 3; \qquad 0 > -7; \qquad -2 > -5; \qquad 1 > -1.$$

For division we need a slight modification in the wording of our definition because of a difficulty created by the number zero:

If a *and* b *are any two numbers of the system, then* $a \div b$ *(or* a/b*) denotes the unique number* x *such that* $bx = a$*, if there is such a unique number in the system.**

Examples

$$\frac{-6}{2} = -3; \qquad \frac{0}{4} = 0; \qquad \frac{-27}{-6} = \frac{9}{2}.$$

The rule of signs is of course the same for division as for multiplication; *i.e.*, if a and b have like signs, the quotient a/b is positive; if they have unlike signs, it is negative.

It is easy to see that the system is closed with respect to division with the exception that *we cannot divide any number of the system by zero.* In order to make this last assertion clear, let us examine the symbol $4/0$. If it denotes any number of our system, then, in accordance with our definition, it must be a unique number x such that $x \cdot 0 = 4$. But there is no such number because $x \cdot 0 = 0$ for *every* number x of the system. Thus the symbol $4/0$ is certainly meaningless as far as our present definition of quotient is concerned. Consider next the symbol $0/0$. We again seek a unique number x such that $x \cdot 0 = 0$. The number does not exist because all numbers of the system have this last property. We agree then that the symbol $a/0$ where a is any number of the system is meaningless.

The number $1/a (a \neq 0)$ is called the *reciprocal* of the number a. Thus the reciprocal of $\frac{2}{3}$ is $1/\frac{2}{3}$ or $\frac{3}{2}$; the reciprocal of -4 is $-1/4$. Division by a is equivalent to multiplication by the reciprocal of a.

By the *negative* of a number a we mean the number that must be added to a in order to obtain the sum *zero.* Thus the negative of $+6$ is -6 and the negative of -6 is $+6$. The negative of a is denoted by $-a$. It can be proved that

$$a - b = a + (-b) \qquad \text{and} \qquad a - (-b) = a + b.$$

This means that subtracting a directed number is equivalent to adding its negative. Proofs will be left to the exercises.

6. *The system of rational numbers.* All the numbers of Sec. 5—the positive and negative fractions and zero, including of course

* That is, if there is one *and only one* such number.

all the integers—constitute the system of *rational numbers.* The word rational comes from the idea of ratio—the ratio of two integers.

This system is closed with respect to addition, subtraction, multiplication, and division, with the single exception that division by zero is impossible. The definitions that we have given for sum, product, etc., are such that the basic commutative, associative, and distributive laws apply to the above-mentioned fundamental operations within this number system.

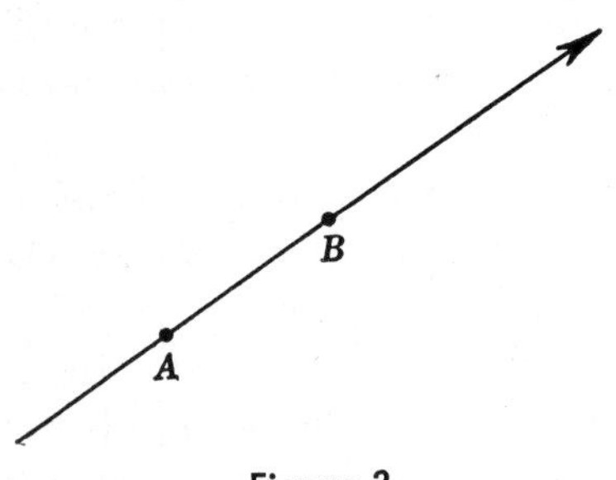

Figure 3

7. *Directed lines.* A *directed line* is a line on which a positive direction or sense has been designated. Segments on such a line are regarded as positive or negative in length according to the way in which they are read. Thus if the positive direction on the line shown in Fig. 3 is upward to the right as indicated by the arrowhead, then the segment AB is positive and BA is negative. If the length of the segment is 2 units, then $AB = +2$ and $BA = -2$; thus $BA = -AB$.

Consider now the line shown in Fig. 4, and let it be assumed that with every rational number we have associated a point of

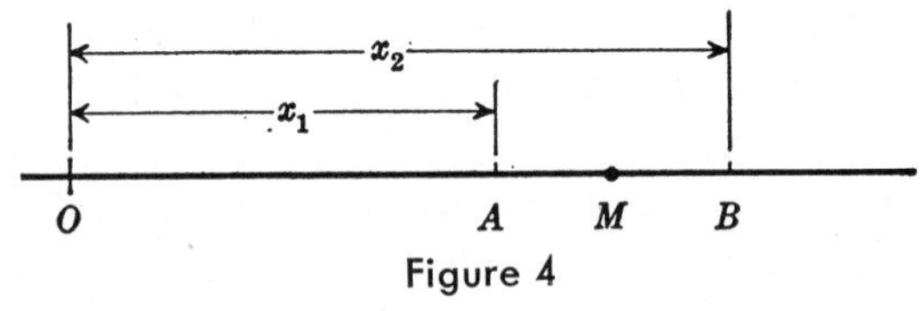

Figure 4

the line, in the manner described in Sec. 5. Let A and B be the points associated with the numbers x_1 and x_2, respectively. Then, if O is the origin,

$$OB = x_2; \qquad OA = x_1; \qquad \text{and} \qquad AB = x_2 - x_1.$$

Furthermore, if M is the mid-point of the segment AB,

$$\begin{aligned} OM &= OA + \tfrac{1}{2}AB \\ &= x_1 + \tfrac{1}{2}(x_2 - x_1) \\ &= \tfrac{1}{2}(x_1 + x_2). \end{aligned}$$

We may now show that the points on the line that correspond to the rational numbers lie very close together—so close in fact that between any two of them, however close these two may be, there are as many millions of others as we please. Thus, for example, even though the points corresponding to the numbers $\frac{121}{125}$ and $\frac{122}{125}$ are very close, there are infinitely many points between them that correspond to other rational numbers. We may show this as follows: Let P and Q be the points corresponding to the numbers $\frac{121}{125}$ and $\frac{122}{125}$, respectively. Then the mid-point M_1 of PQ, which certainly lies between P and Q, is associated with the rational number $\frac{1}{2}(\frac{121}{125} + \frac{122}{125})$. Next, the mid-point M_2 of PM_1 is associated with the rational number

$$\frac{1}{2}\left[\frac{121}{125} + \frac{1}{2}\left(\frac{121}{125} + \frac{122}{125}\right)\right].$$

Continuing in this way, we can obviously find as many points as we please between P and Q that are associated with rational numbers.

One might be tempted to conclude from this that *every* point on the line is associated with some rational number. It turns out, however, that this is not true. We have shown above that there are an unlimited number of points that lie between P and Q and *do* correspond to rational numbers. There are also an unlimited number of points on the line that lie between P and Q and *do not* correspond to rational numbers. We must have a more extensive system of numbers—one that includes numbers that are not rational numbers—if we wish to have a number associated with every point of the line. This extension will be discussed briefly in Sec. 11.

PROBLEMS

Compute each of the following:

1. $-6 - (-8)$.

2. $-4 + (-7) - (+3)$.

3. $+\frac{7}{16} - (-\frac{5}{12})$.

4. $\frac{7}{3} - (-\frac{2}{9}) - (\frac{4}{6} - \frac{1}{3})$.

5. $-\left(\frac{-8}{+2} + \frac{+20}{-4}\right)$.

6. $\frac{64}{-4} - \left(\frac{16}{-1}\right)$.

7. $(-\frac{2}{3}) \div (-\frac{4}{9})$.

8. $\frac{6}{5} \div (-\frac{21}{10})$.

9. $\frac{(-4) + (-3)}{10 - (-4)} \div \frac{\frac{1}{2} - (-\frac{1}{6})}{3(-2)}$.

10. $\frac{4}{3}(\frac{5}{8}) - (-\frac{3}{7})(-\frac{28}{3}) + (-\frac{1}{3})(+\frac{15}{6})$.

11. $(\frac{2}{3})(-\frac{3}{5})(-\frac{1}{6})(-\frac{12}{5})$.

12. $\dfrac{8 - (\frac{2}{3})(-\frac{1}{2})}{5 + (\frac{3}{8})(\frac{1}{2})(-\frac{5}{3})}$.

13. Is there a smallest positive rational number? If so, what is it?

14. Consider the set of all positive rational numbers less than 1. Is there a largest number in this set?

In each of the following problems let A be the point associated with the first given number and B be the point associated with the second. Find the length of the directed segment AB Find also the number associated with the mid-point M of AB:

15. $1\frac{1}{2}$; $7\frac{1}{2}$.
16. $-3\frac{1}{2}$; $6\frac{1}{2}$.
17. $4\frac{1}{3}$; $-\frac{4}{3}$.
18. $-\frac{8}{3}$; $-\frac{26}{3}$.
19. $-1\frac{3}{8}$; $2\frac{1}{4}$.
20. $3\frac{3}{4}$; $-7\frac{1}{2}$.

21. Let A and B be the points associated with the numbers $4\frac{1}{3}$ and $12\frac{2}{3}$, respectively. What number is associated with the point P that is two-thirds of the way from A to B?

22. If a and b are rational numbers, does there always exist a rational number x such that $ax = b$? If an additional restriction on a or b is necessary, what is this restriction?

In each of the following problems simplify the given expression as much as possible and then find its value for the given values of x and a. Check your result by substituting the given values for x and a directly into the given expression and then combining the resulting numbers:

23. $\dfrac{x + 3a}{x - a} + \dfrac{x}{a - x}$; $x = 4$, $a = 2$.

24. $\dfrac{x(4 - a)}{3x - 2a} - \dfrac{ax}{2a - 3x}$; $x = \frac{5}{8}$, $a = -\frac{5}{8}$.

25. $x\left(\dfrac{2}{x - a}\right) + 2a\left(\dfrac{1}{a - x}\right) + \dfrac{a}{x - a}$; $x = -\frac{4}{3}$, $a = 2$.

26. $\dfrac{(a - 2x)/(x - a)}{2/(a - x)} + \frac{1}{2}(2x - a)$; $x = \frac{7}{6}$, $a = -\frac{2}{3}$.

27. $\left(\dfrac{2x + 5a}{x - 3a}\right)\left(\dfrac{6a - 2x}{4x - 10a}\right) - \dfrac{5a}{5a - 2x}$; $x = -2$, $a = \frac{2}{5}$.

28. $\left(\dfrac{x + a}{x - 2a}\right)\left(\dfrac{x - a}{x}\right)\left(\dfrac{2a - x}{a - x}\right) - \dfrac{a}{2x}$; $x = \frac{5}{2}$, $a = -10$.

29. $\dfrac{(4x + 2a)/x}{(x - a)/3x} + \dfrac{8ax/(a - x)}{2a/3}$; $x = -\frac{1}{6}$, $a = -\frac{2}{3}$.

30. $\dfrac{(x - a)/(x - 2a)}{(2a - 2x)/(x + a)} - \dfrac{a(x + a)}{2a - x} \cdot \dfrac{2}{3x + 3a}$; $x = \frac{2}{3}$, $a = \frac{1}{2}$.

31. If a and b are rational numbers show that

$$a - b = a + (-b) \qquad \text{and} \qquad a - (-b) = a + b.$$

32. Show that if a, b, and c are rational numbers, then, except for certain special cases,

$$\frac{[(3a + bc)/(3a - b)] + c}{c + [(a - bc)/(a + b)]} = \frac{3(a + b)}{3a - b}.$$

The relation is obviously not true if $b = 3a$ or if $b = -a$. Are there any other exceptions? HINT: Try $a = 0$, $b = 1$, $c = 2$.

33. Show that if a, b, c, and d are rational numbers, then, except for certain cases,

$$\frac{[(d - 2a)/(d + 2a)] - [(b - c)/(b + c)]}{[b/(b + c)] - [d/(d + 2a)]} = -2.$$

The relation is obviously not true if $b = -c$ or if $d = -2a$. Are there any other exceptions? HINT: Try $a = 3$, $b = 1$, $c = 2$, $d = 3$.

34. Show that if a, b, c, and d are rational numbers, then, except for certain special cases,

$$\frac{[(3a + 3b)/(a + 3b)] + [(c - d)/(c + d)]}{[3d/(a + 3b)] - d/b} = -\frac{2b(3bc + 2ac + ad)}{ad(c + d)}.$$

List all the special cases in which the relation is not true.

35. Show that if the letters represent any rational numbers for which all the indicated operations can be performed, then

$$\frac{[(2ax + 3y)/2ac] - [(3bx + 2y)/3bc]}{[(2y - bx)/b] + [(3y + 2cx)/2c]} = \frac{9b - 4a}{3a(4c + 3b)}.$$

It is obvious that a, b, and c must be different from zero. Are any additional restrictions necessary? HINT: Try $a = 1$, $b = 4$, $c = -3$, $x = -1$, $y = 2$.

8. *Positive integral powers.* Let n be a positive integer and a be any rational number. We define the symbol a^n to mean $a \cdot a \cdot a \cdots a$ where there are n of the a's. In particular, $a^2 = a \cdot a$ is the *square* of a, and $a^3 = a \cdot a \cdot a$ is the *cube* of a. In general a^n is called the nth *power* of a, and n is called the *exponent* of the power.

The following laws concerning the use of such exponents follow immediately from the above definition:

(I) $$a^m \cdot a^n = a^{m+n}.$$

Example: $$2^3 \cdot 2^5 = 2^8.$$

(II) $$\frac{a^m}{a^n} = a^{m-n} \quad \text{if } m > n \text{ and } a \neq 0;$$

$$= \frac{1}{a^{n-m}} \quad \text{if } n > m \text{ and } a \neq 0;$$

$$= 1 \quad \text{if } n = m \text{ and } a \neq 0.*$$

* If $m = n$, we have $(a^m/a^n) = (a^n/a^n) = 1$ if $a \neq 0$. If the law $(a^m/a^n) = a^{m-n}$ is to hold for this case, then we have $(a^n/a^n) = a^{n-n} = a^0$. It therefore appears desirable to define the symbol a^0 to have the value 1 if a is any number different from zero. This is done in Chap. VI.

Example: $$\frac{2^8}{2^5} = 2^3; \frac{2^3}{2^7} = \frac{1}{2^4}.$$

(III) $$(a^m)^n = a^{mn}.$$

Example: $$(3^2)^4 = 3^8.$$

(IV) $$(a \cdot b)^n = a^n \cdot b^n.$$

Example: $$(2x)^3 = 2^3 \cdot x^3 = 8x^3.$$

(V) $$\left(\frac{a}{b}\right)^n = \frac{a^n}{b^n} \qquad \text{if } b \neq 0.$$

Example: $$\left(\frac{3}{4}\right)^2 = \frac{3^2}{4^2} = \frac{9}{16}.$$

The proof of (**I**) is as follows:

$$\begin{aligned} a^m &= a \cdot a \cdot a \cdots \text{ to } m \text{ factors (by definition).} \\ a^n &= a \cdot a \cdot a \cdots \text{ to } n \text{ factors (by definition).} \\ a^m \cdot a^n &= (a \cdot a \cdot a \cdots \text{ to } m \text{ factors}) \cdot (a \cdot a \cdot a \cdots \text{ to } n \text{ factors}) \\ &= (a \cdot a \cdot a \cdots \text{ to } m + n \text{ factors}) \\ &= a^{m+n}. \end{aligned}$$

Proofs of the others will be left to the exercises.

9. *Positive integral roots.* If there exists a number x such that $x^n = a$, where n is an integer greater than 1, then x is called an nth *root* of a. In particular if $x^2 = a$, then x is a *square root* of a; if $x^3 = a$, then x is a *cube root* of a.

Examples

Both 4 and -4 are square roots of 16 because $4^2 = 16$ and $(-4)^2 = 16$. A cube root of -8 is -2 since $(-2)^3 = -8$.

It will be seen later that within the system of complex numbers (this includes the so-called imaginary numbers) every number except zero has exactly n distinct nth roots—two square roots, three cube roots, etc.

If a is positive, there is one nth root of a that is also positive. If a is negative, there is no nth root of a that is positive, but if n is odd, there is a negative one. In order to avoid ambiguity, we shall define the symbol $\sqrt[n]{a}$* for these two cases as follows:

* The symbol $\sqrt[n]{a}$ is called a *radical.* The integer n is the *index* of the radical and the number a is the *radicand.*

If a *is positive,* $\sqrt[n]{a}$ *denotes the positive* n*th root. If* a *is negative and* n *is odd,* $\sqrt[n]{a}$ *denotes the negative* n*th root.* In each of these cases this particular root is called the *principal* nth root.

Examples

$$\sqrt{25} = 5; \qquad \sqrt[3]{-8} = -2; \qquad \sqrt{(-6)^2} = +6 \ (not\ -6).$$

$$\begin{aligned} \sqrt{(a-b)^2} &= a - b \quad \text{if } a > b \\ &= b - a \quad \text{if } b > a \\ &= 0 \quad \text{if } a = b. \end{aligned}$$

It is thus incorrect to write $\sqrt{16} = \pm 4$; instead, we have $\sqrt{16} = 4$, $-\sqrt{16} = -4$, and $\pm\sqrt{16} = \pm 4$.

10. *The square root of* 2. It is easy to show that there is no rational number whose square is 2. In the proof that follows we use the fact (page 5) that a natural number p can be expressed as a product of prime numbers in one and only one way. We also use the rather obvious fact that, if p^2 is expressed as a product of primes, the result is the same primes obtained for p, each occurring twice as many times. This of course means that each prime factor of p^2 occurs an *even* number of times. Thus

$$60 = 2 \cdot 2 \cdot 3 \cdot 5 \quad \text{and} \quad \begin{aligned} 60^2 &= 2^2 \cdot 2^2 \cdot 3^2 \cdot 5^2 \\ &= 2 \cdot 2 \cdot 2 \cdot 2 \cdot 3 \cdot 3 \cdot 5 \cdot 5. \end{aligned}$$

The proof of our theorem is as follows: Assume that there is a rational number $\frac{p}{q}$ whose square is 2; *i.e.*, assume that

$$\left(\frac{p}{q}\right)^2 = 2 \tag{1}$$

where p and q are natural numbers. If this is true, then

$$\frac{p^2}{q^2} = 2$$

and

$$p^2 = 2q^2. \tag{2}$$

Now q^2 must contain 2 as a factor an even number of times if at all; consequently, $2q^2$ *must contain* 2 *as a factor an odd number of times.* It cannot therefore be equal to p^2 because p^2 contains 2 as a factor an *even* number of times if at all. We thus have the following state of affairs:

(*a*) If our assumption (1) is true, then (2) must be true.

(*b*) (2) cannot be true.

We therefore conclude that our assumption (1) must be false; *i.e.*, there *does not* exist any pair of natural numbers p and q such that $(p/q)^2 = 2$.

11. ***Irrational numbers.*** We have shown that there is no rational number whose square is 2. Similarly, there is no rational number which is a square root of 3, or 5, or 7, or any integer that is not a "perfect square" (*i.e.*, the square of an integer). The situation is the same in regard to cube roots of integers that are not perfect cubes, etc. This means of course that the system of rational numbers is not closed with respect to the operation of taking roots. It means that there is no number in the system which describes exactly the length of the diagonal of a square whose sides are each one unit long. It means that, after we have associated a point on the line of Fig. 2 with each rational number, there are still points on the line that have not been associated with any number; for example, the point whose distance from the origin is $\sqrt{2}$. For these and other reasons, the system of rational numbers is inadequate for our purposes. We therefore supplement it by the addition of new numbers called *irrational* numbers.

We cannot at this point give any adequate definition of irrational number.* We must therefore content ourselves with a brief discussion of some of its important properties.

The first essential fact is that there is an irrational number associated with every point of the line (Fig. 2) that does not have a rational number associated with it. Thus there is a number, rational or irrational, associated with *every* point.

A second fact is that, just as in the case of rational numbers, the points associated with the irrational numbers are very close together. Thus if A and B are the points associated with two irrational numbers (say, $\sqrt{2}$ and $\sqrt{3}$) then there are infinitely many points between A and B that also correspond to irrational numbers [for example, $\frac{1}{2}(\sqrt{2} + \sqrt{3})$]. It can be shown, incidentally, that the sum of a rational and an irrational number is an irrational number, and that the product of a rational and an irra-

* The student will find a fairly elementary discussion of this subject in G. H. Hardy, "A Course of Pure Mathematics," New York, The Macmillan Company, 1938.

tional number is irrational. Thus $\frac{2}{3} + \sqrt{5}$ and $-\frac{7}{4}\sqrt{2}$ are irrational.

Another important property of irrational numbers concerns their representation in decimal form. In the first place we have the following:

Theorem. *Any terminating decimal represents a rational number.* For example, 3.42 is just another way of writing $\frac{342}{100}$. It is, therefore, a rational number.

The converse of this theorem is not true; *i.e.*, it is not true that every rational number has a decimal representation that is terminating. Thus we have $\frac{2}{3} = 0.6666$ which is a nonterminating decimal. However we can prove the following:

Theorem. *The decimal representation of a rational number either terminates or is periodic.* By a periodic decimal we mean one like $7.6333 \cdots$ in which the 3 is repeated indefinitely, or $2.68454545 \cdots$ in which the numbers 45 continue. (A convenient notation is $7.6\dot{3}$ and $2.68\dot{4}\dot{5}$, the dots being placed over the numbers that form the repeating block.)

We shall indicate the proof of our theorem by means of an example in which we obtain the decimal representation of the number $\frac{24}{55}$:

```
       .436
 55|24.000
    22 0
     2 00
     1 65
       350
       330
        20
```

Observe that in the process of division the only remainders that can occur are 0, 1, 2, $\cdots$, up to 54. If 0 occurs, the decimal terminates. If 0 does not occur, then after not more than 54 steps one of the remainders must occur a second time—then the whole process must repeat itself. In our example the first remainder is 20, the second is 35, and in the third step the remainder 20 appears again. Hence

$$\frac{24}{55} = 0.4363636 \cdots \quad \text{or} \quad 0.4\dot{3}\dot{6}.$$

Later on, after we have studied geometric series, we shall be able to prove the converse of this theorem; *i.e., every periodic decimal represents a rational number.*

It can be proved that every irrational number also has a decimal representation that approximates it to any desired degree of accuracy, and it follows from the above theorems that this decimal is neither terminating nor periodic. In the case of $\sqrt{2}$ we know that since $1^2 = 1$ and $2^2 = 4$,

$$1 < \sqrt{2} < 2.$$

By squaring the numbers 1.1, 1.2, etc., we find that, since $1.4^2 = 1.96$ and $1.5^2 = 2.25$,

$$1.4 < \sqrt{2} < 1.5.$$

Proceeding in this fashion we can show that, correct to three decimal places, $\sqrt{2} = 1.41$; to five places, it is 1.41421; etc. We know that however far we carry the decimal approximation it will not terminate or become periodic.

In regard to computations involving irrational numbers it can be stated that the sum, product, or quotient of two irrational numbers can be approximated as closely as desired by using sufficiently close approximations to the individual numbers; the nth root of an irrational number a is approximated with any desired degree of accuracy by the nth root of a sufficiently close approximation to a, etc.

12. ***Summary. Real numbers.*** The positive and negative rational numbers, the positive and negative irrational numbers, and the number zero, constitute the system of *real numbers.* In this chapter we have sketched in an informal way the evolution of this number system, starting with the natural numbers. First, we observed that the fundamental operations with the natural numbers obeyed certain basic laws, and when the inadequacy of our system for certain operations led us to enlarge it in successive stages by introducing new kinds of numbers, we let the requirement of consistency, and our desire to preserve the basic laws, guide us in setting up new definitions of sum, product, etc. For example, our desire that the distributive law $a(b + c) = ab + ac$ should hold for signed numbers led us to define the product of two negative numbers as a positive number.

One further enlargement will be necessary in order to obtain a system that is adequate for all our needs, but for the present we shall confine our attention to real numbers. This system is closed with respect to addition, subtraction, multiplication, and division, except for division by zero. It also is subject to the following basic laws:

I. *Commutative law for addition.* $a + b = b + a$.
II. *Commutative law for multiplication.* $a \cdot b = b \cdot a$.
III. *Associative law for addition.* $a + (b + c) = (a + b) + c$.
IV. *Associative law for multiplication.* $a(b \cdot c) = (a \cdot b)c$.
V. *Distributive law.* $a(b + c) = ab + ac$.

In addition, we assume the following axioms concerning equality:

VI. $a = a$.
VII. *If* $a = b$, *then* $b = a$.
VIII. *If* $a = b$ *and* $b = c$, *then* $a = c$.
IX. *If* $a = b$ *and* $c = d$, *then* $a \pm c = b \pm d$, $a \cdot c = b \cdot d$, *and* $\frac{a}{c} = \frac{b}{d}$ *if* c *and* d *are* $\neq 0$.

Various definitions and theorems that have arisen in this chapter in connection with the natural numbers or rational numbers may now be extended in an obvious way to the general case of real numbers. Thus, in our definition of a^n, and in the corresponding laws of exponents, the base a may be any real number. In our definition that $a > b$ if $a - b$ is positive, a and b may be any real numbers. We used the word fraction originally to mean what we would now call a positive rational number. The word fraction is usually used to denote any symbol of the form p/q and thus refers to the way in which the number is written rather than to the kind of number. Thus $\frac{3 - 4\sqrt{17}}{2}$ is called a fraction.

PROBLEMS

Prove each of the following theorems assuming that m and n are positive integers and a and b are real numbers:

1. $\frac{a^m}{a^n} = a^{m-n}$ if $m > n$, $a \neq 0$.

$= \frac{1}{a^{n-m}}$ if $n > m$, $a \neq 0$.

$= 1$ if $n = m$, $a \neq 0$.

2. $(a^m)^n = a^{mn}$. **3.** $(a \cdot b)^n = a^n \cdot b^n$.

4. $\left(\dfrac{a}{b}\right)^n = \dfrac{a^n}{b^n}$ if $b \neq 0$.

5. Which, if any, of the following statements are true, a and b denoting any two positive rational numbers?

$\sqrt{a^2 - b^2} = a - b$; $\sqrt{a^2 + b^2} = a + b$;
$\sqrt{(a + b)^2} = a + b$; $\sqrt{a + b} = \sqrt{a} + \sqrt{b}$.

Evaluate each of the following:

6. $\sqrt[3]{-64} + \sqrt{\frac{4}{25}}$. **7.** $-\frac{1}{2}\sqrt[5]{-32} + 4\sqrt[3]{8}$.

8. $\sqrt[3]{1{,}000} - 2\sqrt{10{,}000}$. **9.** $\frac{3}{4}\sqrt[6]{64} - \frac{2}{3}\sqrt[3]{-\frac{1}{8}}$.

10. $-\sqrt{\frac{1}{36}} + \sqrt{(-\frac{4}{9})^2}$. **11.** $\dfrac{1}{2} + \sqrt{1 - \left(\dfrac{\sqrt{3}}{2}\right)^2}$.

12. Simplify the expression $y + \sqrt{(x - y)^2}$ for the case in which $x > y$ and then for the case in which $x < y$. As a check, evaluate the expression for $x = 6$ and $y = 2$ and then for $x = 2$ and $y = 6$.

13. Prove that there is no rational number whose square is 5.

14. Prove that there is no rational number whose cube is 2.

In each of the following cases find the terminating or periodic decimal that represents the given rational number:

15. $\dfrac{7}{9}$. **16.** $\dfrac{41}{50}$. **17.** $\dfrac{2}{13}$. **18.** $\dfrac{50}{14}$.

19. $\dfrac{68}{21}$. **20.** $\dfrac{23}{11}$. **21.** $\dfrac{500}{666}$. **22.** $\dfrac{2{,}293}{9{,}900}$.

23. Prove that the sum of a rational number R and an irrational number I is irrational. HINT: Let $S = R + I$. S is a real number and is therefore either rational or irrational. From $S = R + I$ we may infer $I = S - R$. Now if S is rational, $S - R$ is rational and hence I is rational, which is contrary to the assumption that I is irrational.

24. Show by means of an example that the sum of two irrational numbers may be rational. HINT: Consider the numbers $3 + \sqrt{2}$ and $3 - \sqrt{2}$.

In each of the following cases the letters represent any real numbers different from zero. Perform the indicated operations and express your result in a simple form:

25. $(2a)^3 \cdot \left(\dfrac{b}{2a}\right)^4$. **26.** $(2a^3) \cdot \left(\dfrac{b}{2}\right)^3 \cdot \left(\dfrac{8}{a^2b^2}\right)$.

27. $\left(\dfrac{x^4}{5}\right)^2 \cdot \left(\dfrac{y^2}{x^3}\right)^2 \cdot \left(\dfrac{50}{xy}\right)$. **28.** $\left(\dfrac{xy}{z}\right)^2 \cdot \left(\dfrac{zx}{y}\right)^3 \cdot \left(\dfrac{yz}{x}\right)^4$.

29. $(2ab^2)^3 \cdot \left(\dfrac{1}{4a^2b}\right)^3 \cdot \left(\dfrac{2a}{b}\right)^4$. **30.** $\dfrac{(4/a)^2}{(2/x)^4} \cdot \left(\dfrac{a}{x}\right)^3$.

31. $\left[\left(\frac{a}{x}\right)^2 + \left(\frac{x}{a}\right)^2\right] \cdot \frac{ax^2}{a^4 + x^4}$.

32. $\frac{x+y}{2(2xy)^2} \cdot \left[(2xy)^3 + \left(\frac{xy}{2}\right)^3\right]$.

33. $\left[\left(\frac{x}{a}\right)^2 + \left(\frac{x}{2a}\right)^2\right] \cdot \left[\left(\frac{2a^2}{5x}\right)^2 + \left(\frac{a^2}{5x}\right)^2\right]$.

34. $\frac{4x + (1/x)(4/y)^2}{x^2y^2 + 4} \cdot \left(\frac{xy}{2}\right)^2$.

35. Simplify the expression

$$x + \sqrt{(3x - 2y)^2} + \sqrt{(x - 2y)^2}$$

given that x and y are positive numbers and $\frac{2}{3}y < x < 2y$. Simplify it also assuming that $x > 2y > 0$. Evaluate the expression for $x = 4$, $y = 5$, and for $x = 7$, $y = 3$.

36. Given that x is a positive number and y is a negative number, simplify the expression

$$3y - \sqrt{(x - 2y)^2} + 2\sqrt{(y - x)^2}.$$

If x and y are both positive, under what additional condition does your result remain true? Evaluate the expression for $x = 4$, $y = -2$, for $x = 3$, $y = 1$, and for $x = 3$, $y = 2$.

37. Given that x is a positive number and y is a negative number, simplify the expression

$$2x + 3\sqrt{(3x - y)^2} - 3\sqrt{(y - 2x)^2}.$$

Simplify it also assuming that x and y are positive numbers such that $\frac{1}{3}y < x < \frac{1}{2}y$. Evaluate the expression for $x = 2$, $y = -1$, and for $x = 5$, $y = 12$.

CHAPTER II

OPERATIONS WITH POLYNOMIALS AND RATIONAL FRACTIONAL EXPRESSIONS

13. ***Constants and variables.*** In this chapter, and in fact in much of our subsequent work, the letters that we use to denote numbers will be understood to stand for *real* numbers. We may frequently use the word *number* to mean *real number*. The word *constant* will signify a particular number or a letter that is understood to denote some particular number, even though the number has not been specified. The word *variable* will signify a letter that denotes any one of a specified set of numbers.

Example

The total surface area of a right circular cylinder of radius x in. and height y in. is $2\pi x^2 + 2\pi xy$ sq. in. The number 2 and the Greek letter π are *constants*. The letters x and y may denote any positive numbers, and hence may be regarded as *variables* in the above sense. If we wish to think of y as having some fixed value, then x is the only variable in the expression, and it then gives the surface areas of cylinders having various radii, but having this fixed height.

We shall frequently employ the later letters of the alphabet, such as x, y, z, u, v, for variables, and the early letters such as a, b, c, d, for constants. We shall also use symbols such as a_0, a_1, a_2, a_3, $\cdots$, to denote constants. These are read "a sub-zero," "a sub-one," etc.

14. ***Algebraic expressions. Polynomials and rational fractional expressions.*** By an *algebraic expression* we shall mean a symbol that is built up from numbers and letters representing numbers, and involving a finite number of the operations studied in Chap. I—addition, subtraction, multiplication, and division, and the operations of taking integral powers and roots. Thus $x^2 - 4y^2$ and $7x + y\sqrt{a^2 + x^2}$ are algebraic expressions. We may regard x and y in the first expression as variables in the sense described above. If the letter a in the second expression is thought of as denoting some particular number, then this expression involves also the two variables x and y.

An algebraic expression is said to be *rational* in its variables if it can be written in a form in which none of the variables occur under radical signs. The first of the above expressions is rational in x and y, but the second is not.

A *polynomial* or *rational integral expression* in one variable, say x, is an expression that can be written as the sum of a finite number of terms each of which is either of the form ax^n, where a is a constant and n is a positive integer, or is a constant. (In a term of the form ax^n, a is called the *coefficient* of x^n.)

Examples

$4x^3 - \sqrt{5}\,x^2 + 3$ and $x^4 - 8$ are polynomials in x. The expressions $2/x$ and $x\sqrt{x} + 5x$ are not polynomials.

Any polynomial in x can be written in the form

$$a_0x^n + a_1x^{n-1} + a_2x^{n-2} + \cdots + a_{n-1}x + a_n,$$

where the a's are constants (some of which may be zero) and n is a positive integer. If $a_0 \neq 0$, the polynomial is said to be of *degree n*.

A polynomial in two variables, x and y, is an expression that can be written as the sum of a finite number of terms each of which either has one of the forms ax^n, ay^n, ax^my^n, where a is a constant and m and n are positive integers, or is a constant. The degree of a term is the sum of the exponents of x and y in that term, and the degree of the polynomial is that of the term of highest degree.

Example

$7x^2y^3 - \frac{1}{3}xy^2 + 3y + 2$ is a polynomial in x and y of degree **5**. The expression $(3x^2 + y^2)/(x - y)$ is not a polynomial.

A polynomial in any number of variables is defined in an obviously similar way.

An expression that can be written as the quotient of two polynomials is called a *rational fractional expression.* The second expression in the above example is of this type.

15. ***Fundamental operations with polynomials.*** One may add two or more polynomials or subtract one polynomial from another in an obvious way. Thus

$$(4x^2 - 2xy + 6y) + (x^2 + xy - y^2 - x) = 5x^2 - xy - y^2 + 6y - x.$$
$$(4x^2 - 2xy + 6y) - (x^2 + xy - y^2 - x) = 3x^2 - 3xy + y^2 + 6y + x.$$

Multiplication of one polynomial by another can be carried out in the manner indicated by the following example:

$$\begin{aligned}(x^2 - xy + y^2) \cdot (x + y) &= (x^2 - xy + y^2) \cdot x + (x^2 - xy + y^2) \cdot y \\ &= x(x^2 - xy + y^2) + y(x^2 - xy + y^2) \\ &= x^3 - x^2y + xy^2 + yx^2 - xy^2 + y^3 \\ &= x^3 + y^3.\end{aligned}$$

In the first step we used the distributive law (V, page 5). What law is used in the second step? It is sometimes convenient to arrange the work as follows:

$$\begin{array}{llll} x^2 & - xy & + y^2 & \\ x & + y & & \\ \hline x^3 & - x^2y & + xy^2 & \\ & + x^2y & - xy^2 & + y^3 \\ \hline x^3 & & & + y^3 \end{array} \qquad \textit{(product)}$$

For dividing one polynomial in x by another, one may use the process illustrated below in which we divide $x^3 + 5x^2 - 2x - 12$ by $x + 2$. We assume that the student has had previous experience with this and therefore we omit the details:

$$\begin{array}{r|l} & x^2 + 3x \quad - 8 \qquad \textit{(quotient)} \\ x + 2 & x^3 + 5x^2 - 2x - 12 \\ & x^3 + 2x^2 \\ & \qquad 3x^2 - 2x \\ & \qquad 3x^2 + 6x \\ & \qquad\qquad - 8x - 12 \\ & \qquad\qquad - 8x - 16 \\ & \qquad\qquad\qquad + 4 \qquad \textit{(remainder)} \end{array}$$

The result enables us to say that

$$\frac{x^3 + 5x^2 - 2x - 12}{x + 2} = x^2 + 3x - 8 + \frac{4}{x + 2}.$$

It will be observed that the word *quotient* as used here does not mean quite the same thing as in Chap. I. This should not cause any confusion because the student is familiar with both usages. If one divides 375 by 13, he gets $28\frac{11}{13}$ as the quotient in the sense of Chap. I. In elementary arithmetic it is common practice to call 28 the quotient and 11 the remainder. This corresponds to the way in which the word quotient is used here.

It is worth while to observe that when one uses "long division" to divide 375 by 13 he is, in effect, *finding the number of times that* 13 *can be subtracted from* 375, and what remainder is then left over:

$$\begin{array}{r|l l}
 & 28 & \\
13 & 375 & \\
 & 260 & \textit{subtract } 20 \cdot (13) \\
 & 115 & \\
 & 104 & \textit{subtract } 8 \cdot (13) \\
 & 11 & \textit{remainder}
\end{array}$$

He thus regards 375 as being composed of twenty-eight 13's plus 11:

$$375 = 28(13) + 11.$$

In a precisely similar way, we have above first subtracted $x^2(x + 2)$, then $3x(x + 2)$, then $-8(x + 2)$, from $x^3 + 5x^2 - 2x - 12$, leaving a remainder of 4. The corresponding result is

$$x^3 + 5x^2 - 2x - 12 = (x^2 + 3x - 8)(x + 2) + 4.$$

Our interpretation of the division process makes it clear that this last equation holds for *all* values of x, even for $x = -2$, for which the divisor is zero. This fact will be of fundamental importance in connection with the remainder theorem in Chap. XIV (page 195).

16. *Special products.* By using the rules for multiplication one can easily verify the following "formulas" for certain special products. The letters of course represent any (real) numbers and can be replaced by other letters or expressions that represent numbers:

(I) $$(x + y)^2 = x^2 + 2xy + y^2.$$
(II) $$(x - y)^2 = x^2 - 2xy + y^2.$$
(III) $$(x + y)^3 = x^3 + 3x^2y + 3xy^2 + y^3.$$
(IV) $$(x - y)^3 = x^3 - 3x^2y + 3xy^2 - y^3.$$
(V) $$(x + y)(x - y) = x^2 - y^2.$$
(VI) $$(x + b)(x + d) = x^2 + (b + d)x + bd.$$
(VII) $$(ax + b)(cx + d) = acx^2 + (ad + bc)x + bd.$$
(VIII) $$(x + y)(x^2 - xy + y^2) = x^3 + y^3.$$
(IX) $$(x - y)(x^2 + xy + y^2) = x^3 - y^3.$$
(X) $$(x + y + z)^2 = x^2 + y^2 + z^2 + 2xy + 2xz + 2yz.$$

The proofs are left to the student. An example of the use of formula X is as follows:

$$\begin{aligned}(x + 2y - 1)^2 &= x^2 + (2y)^2 + (-1)^2 + 2(x)(2y) + 2(x)(-1) \\ &\qquad + 2(2y)(-1) \\ &= x^2 + 4y^2 + 1 + 4xy - 2x - 4y \\ &= x^2 + 4xy + 4y^2 - 2x - 4y + 1.\end{aligned}$$

17. ***Factoring of certain kinds of polynomials.*** We have seen (page 5) that a natural number can be expressed in one and only one way as a product of prime numbers. A generalization of this is the fact that any positive or negative integer can be expressed as the product of primes and the negatives of primes in one and only one way, except for trivial variations in the choice of signs. Thus, $-70 = (-7)(5)(2)$ or $(7)(-5)(2)$, etc.

A problem that corresponds closely to the above is that of factoring a given polynomial into its prime factors. We shall consider here only polynomials with integral coefficients, and require that the factors shall also be polynomials with integral coefficients. A polynomial will be called *prime* if it has no factors of this kind except plus or minus itself and plus or minus one.

Examples

(1) $$\begin{aligned}4x^3 - xy^2 &= x(4x^2 - y^2) \\ &= x(2x + y)(2x - y).\end{aligned}$$

(2) $$\begin{aligned}x^4 - y^4 &= (x^2 + y^2)(x^2 - y^2) \\ &= (x^2 + y^2)(x + y)(x - y).\end{aligned}$$

(3) $$x^2 - 7x + 10 = (x - 2)(x - 5).$$

(4) $$x^3 + 8y^3 = (x + 2y)(x^2 - 2xy + 4y^2).$$

The polynomials $2x + y$, $x^2 + y^2$, and $x^2 - 2xy + 4y^2$, which appear in the above examples, are prime in accordance with our definition. If we allowed fractional coefficients, then we could of course "factor" $2x + y$ by writing it, for example, as $2(x + \frac{1}{2}y)$. Similarly, if we allowed irrational coefficients, we could factor $x^2 - 5$ into $(x + \sqrt{5})(x - \sqrt{5})$, and if we allowed factors that were not polynomials, we could factor $x - y$ as $(\sqrt{x} + \sqrt{y})(\sqrt{x} - \sqrt{y})$. With the restrictions that we have made, there is a unique representation of a given polynomial as a product of prime factors, except for trivial changes in sign, just as in the case of an integer. For example,

$$2x^2 - 11xy + 15y^2 = (2x - 5y)(x - 3y)$$

or

$$(-2x + 5y)(-x + 3y).$$

In the simple cases that will be considered here, the factors can usually be found by using the special product formulas of the preceding section. Thus, one factors $4x^2 - 25y^2$ by using the fact that

$$u^2 - v^2 = (u + v)(u - v).$$

If we let $u = 2x$ and $v = 5y$, we have

$$\begin{aligned} 4x^2 - 25y^2 &= (2x)^2 - (5y)^2 \\ &= (2x + 5y)(2x - 5y). \end{aligned}$$

The polynomial $2x^2 + 3x - 14$ is factored by observing that, if it can be factored, we must have

$$2x^2 + 3x - 14 = (ax + b)(cx + d)$$

where a, b, c, d are integers to be determined so that the product of these factors is equal to $2x^2 + 3x - 14$. This means that we must have $ac = 2$, $bd = -14$, and $bc + ad = 3$. After trying a few combinations, one finds that

$$2x^2 + 3x - 14 = (2x + 7)(x - 2).$$

In a similar way one discovers that the polynomial $x^2 - 2xy + 4y^2$ is prime. For, if we set

$$x^2 - 2xy + 4y^2 = (ax + by)(cx + dy),$$

we find that there are no integers a, b, c, d, such that the product of the factors on the right is equal to the polynomial on the left.

PROBLEMS

1. Which of the following are polynomials in x and what in each case is the degree of the polynomial?

(a) $2x^3 - 7$. (b) $\frac{x^3 + 4}{x}$. (c) $4x^3 + 7x^2 - 3\sqrt{x} + 5$.

(d) $\frac{4}{x + 1}$. (e) $\sqrt{x^2 + 4}$. (f) $5x^3 + x^4 + 6x^7$.

2. Which of the following are polynomials in x and y, and what in each case is the degree of the polynomial?

(a) $xy + 4$. (b) $3x^2y^2 + y^2 + x$. (c) $xy + x^2\sqrt{y} + 4x$.
(d) $5xy^3 + 6xy + 2$. (e) $x + y^2 + 2$. (f) $\sqrt{x} + y^2\sqrt{y}$.

3. Prove formulas I, VII, and IX of Sec. 16.
4. Prove formulas II, IV, and X of Sec. 16.
5. Show that $(x - y)(x^3 + x^2y + xy^2 + y^3) = x^4 - y^4$.
6. Show that $(x + y)(x^4 - x^3y + x^2y^2 - xy^3 + y^4) = x^5 + y^5$.
7. Show that $(x - y)(x^4 + x^3y + x^2y^2 + xy^3 + y^4) = x^5 - y^5$.

Carry out the following multiplications:

8. $(x + 4)(x - 3)$; $(2x^2 + 7)(3x^2 - 5)$.
9. $(x + 2y)(3x - y)$; $(4x^2 + 5y)(x^2 - 7y)$.
10. $(2xy + 5y)(3xy - 2y)$; $(x^2y^2 + 1)(x^2y^2 - 1)$.
11. $(x^2 + 3y)^3$; $(3a - 4b)^3$.
12. $(3x - 2y^2)^3$; $(x^2 + 2y^2)^3$.
13. $(5a^2b^2 - 4k)(5a^2b^2 + 4k)$; $(3a^2b + k)(k - 2a^2b)$.
14. $(x + y + 1)^2$; $(2u^2 - v + y)^2$.
15. $(a - b + c)(a + b - c)$; $(3 - y + 4z)(3 + y - 4z)$.
16. $(a + b + c)(a + b - c)$; $(3x + 4y + 2)(3x + 4y - 2)$.
17. $[(a + b) - 2]^3$; $(2x + y + 1)^3$.
18. $[(u + v) - 1]^3$; $(2u - v + 1)^3$.

Factor each of the following polynomials into its prime factors:

19. $x^2 + 7x + 6$; $x^3 + 6x^2 - 7x$.
20. $a^2 - 8a + 15$; $a^4 + 2a^3 - 15a^2$.
21. $x^4 + x^2 - 20$; $x^8 - 12x^4 - 64$.
22. $4x^2 + 3x - 1$; $4u^4 + 3u^2 - 1$.
23. $15x^2 + 3x - 12$; $4x^3 + 6x^2 - 4x$.
24. $4x^2 + 40x + 100$; $3x^5 + 6x^3 - 24x$.
25. $2x^2 - 9xy - 18y^2$; $18x^3 + 21x^2y - 6xy^2$.
26. $24u^2 + 20uv - 24v^2$; $12x^2 + 2xy - 70y^2$.
27. $u^2 - 4v^2$; $9a^2 - 4b^2$.
28. $(2x - 3y)^2 - 4$; $(x + 5)^2 - 4y^2$.
29. $16x^4 - y^4$; $9x^2y^2 - 4y^4$.

30. $x^3y - xy^3$; $8x^3 - 18xy^2$.
31. $3x^6 - 48x^2y^4$; $x^5y - 16xy^5$.
32. $3x^5 - 18x^3y^2 + 27xy^4$; $x^4 - 6x^2y + 9y^2$.
33. $x^3 - 1$; $x^3 + 64$.
34. $8x^3 - y^3$; $u^3 + 27v^3$.
35. $h^3 - 8$; $x^5 - 8x^2$.
36. $(x^2)^3 - 2^3$; $x^6 + 27$.
37. $3xy^3 + 192x$; $16x^4y - 2xy$.
38. $(a + b)^3 - (a - b)^3$; $(x + y)^3 - 1$.
39. $x^3(x + y) - y^3(x + y)$; $(x^2 - y^2)x^2 - (x^2 - y^2)y^2$.
40. $4(2x - y) + x(y - 2x)$; $x^2(3x - 4y) + y^2(4y - 3x)$.

41. By dividing $x^5 - 32$ by $x - 2$, show that $x - 2$ is a factor of $x^5 - 32$ and find another factor.

42. Show that $x^2 + x + 5$ is a factor of $x^4 - 2x^3 + 4x^2 - 13x + 10$, and find two other factors.

43. Show that $x^2 + 4$ is a factor of $x^4 - 4x^3 + 9x^2 - 16x + 20$, and find the other factor.

44. Show that $x^2 + 2$ is a factor of $x^5 + 2x^3 - 27x^2 - 54$, and find two other factors.

45. Show that $x^2 + x - 6$ is a factor of $2x^4 + 5x^3 - 11x^2 - 20x + 12$, and then express this polynomial of fourth degree as a product of four polynomials of first degree.

46. Show that $x^2 - 4$ is a factor of $6x^4 + 7x^3 - 27x^2 - 28x + 12$, and then express this polynomial of fourth degree as a product of four polynomials of first degree.

In each of the following problems carry out the indicated division and express the result in the form

$$\frac{\textit{Dividend}}{\textit{divisor}} = \textit{quotient} + \frac{\textit{remainder}}{\textit{divisor}}.$$

Then check your result by reducing the form on the right to that on the left:

47. $x^3 + 7x^2 - 2x + 15 \div x + 2$.
48. $x^4 - 2x^3 + 5x - 4 \div x + 1$.
49. $6x^3 - 14x^2 - 27x - 12 \div 3x + 2$.
50. $3x^3 - 7x^2 + 15x - 30 \div 3x - 7$.
51. $2x^4 + 13x^3 + 15x^2 - 4x - 9 \div 2x + 3$.
52. $4x^3 - 8x^2 + 7 \div 2x - 3$.
53. $2x^4 + 5x^3 - 6x^2 + 11x + 2 \div 2x + 5$.
54. $x^4 - x^3 + 6x^2 + 8x + 4 \div x^2 + 4$.
55. $2x^5 + 5x^3 + 3x^2 - 12x + 5 \div x^2 - 2$.
56. $x^4 - 2x^3 + 6x^2 + 5x + 12 \div x^2 + x + 2$.
57. $2x^5 + x^4 + x^3 - 13x^2 + 7 \div 2x^2 + x - 1$.
58. $2x^5 - x^4 + x^3 + 5x^2 - 2x + 3 \div x^2 - x + 1$.

18. ***Operations with rational fractional expressions.*** In Chap. I we gave definitions of the sum, product, etc., of two fractions and reviewed the various elementary operations with fractions. In this section we shall consider some slightly more difficult cases—those in which it is necessary or convenient to use what we have just learned about factoring in order to carry out the specified operations and express the result in a simple form. For example, when we encounter the fraction $\frac{63}{84}$ in arithmetic, we simplify it by expressing numerator and denominator as a product of prime factors and canceling out the common factors. Thus

$$\frac{63}{84} = \frac{3 \cdot 3 \cdot 7}{2 \cdot 2 \cdot 3 \cdot 7} = \frac{3}{4}.$$

A fraction is said to be in *lowest terms* if its numerator and denominator have no common factor other than ± 1.

We employ precisely the above procedure in reducing a rational fractional expression in algebra to lowest terms. Thus

$$\frac{x^2 - 7x + 10}{4x^2 - 16} = \frac{(x - 2)(x - 5)}{4(x - 2)(x + 2)} = \frac{x - 5}{4(x + 2)}.$$

In canceling the factor $x - 2$, we have divided numerator and denominator of the original fraction by $x - 2$. This operation is permissible provided x is any number for which $x - 2 \neq 0$; *i.e.*, if $x \neq 2$. The original expression is meaningless for $x = \pm 2$. Hence we can say that

$$\frac{x^2 - 7x + 10}{4x^2 - 16} = \frac{x - 5}{4(x + 2)} \qquad \text{if } x \neq \pm 2.$$

When we multiply two or more fractions, we may use the above procedure in order to express the result in lowest terms. Thus

$$\frac{4x}{x^2 - 4y^2} \cdot \frac{x + 2y}{3x^2 + xy} = \frac{4(x)(x + 2y)}{(x - 2y)(x + 2y)(x)(3x + y)}$$
$$= \frac{4}{(x - 2y)(3x + y)}.$$

Again we must exclude all values of x and y that would make any denominator zero.

In arithmetic we carry out the operation $\frac{5}{16} - \frac{7}{12} + \frac{5}{3}$ by replacing each fraction by an equivalent fraction having the denominator 96. Thus

$$\begin{aligned}\frac{5}{16} - \frac{7}{12} + \frac{5}{3} &= \frac{5}{16}\left(\frac{3}{3}\right) - \frac{7}{12}\left(\frac{4}{4}\right) + \frac{5}{3}\left(\frac{16}{16}\right)\\ &= \frac{15}{48} - \frac{28}{48} + \frac{80}{48}\\ &= \frac{15 - 28 + 80}{48} = \frac{67}{48}.\end{aligned}$$

Here, 48 is the lowest common multiple (see page 6) of the denominators 16, 12, and 3 and is the most convenient number to use for a common denominator. In adding rational fractional expressions in algebra, we follow the same procedure. Thus

$$\begin{aligned}\frac{x}{4x^2 - y^2} - \frac{3}{x} + \frac{x + y}{2x^2 + xy} &= \frac{x}{4x^2 - y^2}\left(\frac{x}{x}\right) - \frac{3}{x}\left(\frac{4x^2 - y^2}{4x^2 - y^2}\right)\\ &\qquad + \frac{x + y}{2x^2 + xy}\left(\frac{2x - y}{2x - y}\right)\\ &= \frac{x^2}{x(4x^2 - y^2)} - \frac{3(4x^2 - y^2)}{x(4x^2 - y^2)}\\ &\qquad + \frac{(x + y)(2x - y)}{x(4x^2 - y^2)}\\ &= \frac{x^2 - 12x^2 + 3y^2 + 2x^2 + xy - y^2}{x(4x^2 - y^2)}\\ &= \frac{2y^2 + xy - 9x^2}{x(4x^2 - y^2)}.\end{aligned}$$

In this case the factors of the first, second, and third denominators are, respectively, $(2x + y)(2x - y)$, x, and $x(2x + y)$. Their lowest common multiple (page 6) is then the product $x(2x + y)(2x - y)$ or $x(4x^2 - y^2)$. We used this for a common denominator and carried out precisely the same operations that were used in the preceding example from arithmetic. Our final result means that

$$\frac{x}{4x^2 - y^2} - \frac{3}{x} + \frac{x + y}{2x^2 + xy} = \frac{2y^2 + xy - 9x^2}{x(4x^2 - y^2)},$$

where x and y are any (real) numbers for which none of the denominators is zero.

PROBLEMS

In each of the following cases reduce the given fraction to lowest terms and note any values of the variables that must be excluded:

1. $\frac{81}{123}$; $\frac{3x^2 - 3y^2}{x + y}$; $\frac{5xy + 5x}{4 + 4y}$.
2. $\frac{75}{105}$; $\frac{6y - 6x}{x^2 - y^2}$; $\frac{x^2 + 2x - 8}{8 + 2x}$.
3. $\frac{44}{76}$; $\frac{2x^3 - x^2 - 15x}{4x + 10}$; $\frac{x^3 - y^3}{x^2 - y^2}$.
4. $\frac{144}{174}$; $\frac{x^4 - 16y^4}{(2x^2 + 8y^2)(x + 2y)}$; $\frac{x^2 + 5xy + 6y^2}{2x^2 - 18y^2}$.

In each of the following problems perform the indicated operations and express the result in lowest terms:

5. $\frac{x + y}{x + 3y} \cdot \frac{x^2 - 9y^2}{y^2 - x^2}$.
6. $\frac{xy}{x^2 - 4} \cdot \frac{2x^2 + 3x - 14}{4xy + 14y}$.
7. $\frac{4x^2 - 25}{3x + 1} \cdot \frac{9x^2 - 1}{4x^2 + 10x}$.
8. $\frac{3x^2}{x^2 + x + 1} \cdot \frac{4x^4 - 4x}{x^2 + 4}$.
9. $\left(1 - \frac{2xy}{x^2 + y^2}\right)\left(\frac{x^2 + y^2}{x^2 - y^2}\right)$.
10. $\left(x + \frac{2xy}{x - 2y}\right)\left(\frac{x^2 - 4y^2}{x}\right)$.
11. $\left(\frac{x^2}{y^2} - 4\right)\left(\frac{y}{x + 2y}\right)\left(2 - \frac{x}{y}\right)$.
12. $\left(\frac{3xy - y^2}{x^2 + 1}\right) \div \frac{y^2 - 9x^2}{4x^3 + 4x}$.
13. $\left(\frac{2x^2 - 9x - 35}{6x + 4}\right) \div \left(\frac{49 - x^2}{9x^2 + 12x + 4}\right)$.
14. $\frac{1}{x + y} + \frac{1}{x - y} + \frac{2y}{x^2 - y^2}$.
15. $\frac{x}{x - y} - \frac{y}{x + y} - \frac{y^2}{x^2 - y^2}$.
16. $\frac{x - y}{xz} - \frac{2y + z}{2yz} + \frac{x + z}{2xy}$.
17. $2x + \frac{xy}{x - y} + \frac{x^2}{y - x}$.
18. $1 - x + x^2 - \frac{x^3}{1 + x}$.
19. $\left(\frac{1}{4x - 5} - \frac{1}{4x + 5}\right)(4x^2 + 9x + 5)$.
20. $\left(\frac{x}{x - y} - \frac{y}{x + y}\right)\left(\frac{x + y}{x^2 + y^2}\right)$.

21. $\left(x + \frac{1}{x} + 2\right)\left(\frac{x^2}{x^2 - 1}\right)$.

22. $\frac{(1/x) - (1/y)}{(1/x) + (1/y)}$.

23. $\frac{x + (x/y)}{1 + (1/y)}$.

24. $\frac{x - (12/x) - 1}{x + (12/x) + 7}$.

25. $\frac{1/(x - y) + 1/(x + y)}{1/(x + y) - 1/(x - y)}$.

26. $\frac{4/(x^2 - 4) + 2/(2 - x)}{1 + 2/(x - 2)}$.

27. $\frac{y^2 - (x^3/y)}{y + x + (x^2/y)}$.

28. $\frac{(x + 3)/(x - 3) - (x - 3)/(x + 3)}{x/(x + 3) - x/(x - 3)}$.

29. $\frac{x - (x - 1)/(x + 2)}{x^3 - 1} + \frac{1 - 2x/(x + 2)}{x^2 - 4}$.

30. $\frac{[(x + 3y)^2/(x^2 - x)][1 - 3y/(x + 3y)]}{1 - \{[(9y^2 + x)/(x^2 - 1)] - [1/(x - 1)]\}}$.

31. Show that if x is a real number not equal to 0 or ± 1, then

$$\frac{1/(1 + x) + 1/(1 - x)}{1/(1 - x) - 1/(1 + x)} = \frac{1}{x}.$$

32. Show that if a and b are nonzero real numbers such that $a^2 \neq b^2$, then

$$\frac{(a + b)/(a - b) + (a - b)/(a + b)}{(a + b)/(a - b) - (a - b)/(a + b)} = \frac{a^2 + b^2}{2ab}.$$

33. Show that if a and b are real numbers, not both zero and such that $a^2 \neq b^2$, then

$$\frac{a/(a + b) + b/(a - b)}{b/(a + b) - a/(a - b)} = -1.$$

34. Show that if a is a real number not equal to 0 or 1, then

$$1 - \frac{1}{1 - 1/(1 - a)} = \frac{1}{a}.$$

35. Show that if a is a real number different from 1, then

$$1 - \frac{1}{1 + a/(1 - a)} = a.$$

36. Show that if a and b are real numbers, not both zero and such that $a^2 \neq b^2$, then

$$\frac{1/(a^2 + b^2) + 1/(a^2 - b^2)}{a/(a - b) - b/(a + b)} = \frac{2a^2}{(a^2 + b^2)^2}.$$

37. Show that if a and b are real numbers neither of which is zero, and such that $a^2 \neq b^2$, then

$$\frac{(a^2 + b^2)/(a^2 - b^2) + 1}{(a + b)/(a - b) - (a - b)/(a + b)} = \frac{a}{2b}.$$

38. Show that with certain restrictions on the real numbers a and b,

$$\frac{[2a - 3a^2/(3a + b)][3a + (b^2 - 6ab)/(3a + 2b)]}{3a - b + 2b^2/(3a + b)} = a.$$

What restrictions on the values of a and b are necessary?

CHAPTER III

EQUATIONS

19. ***Introduction.*** We have used the word *equation* to mean a statement that two numbers, or two expressions representing numbers, are equal. If the statement is true for all permissible values of the variables, it is called an *identity*. Thus the equations

$$(x + y)^2 = x^2 + 2xy + y^2 \qquad \text{and} \qquad \frac{x^2 - 9}{x - 3} = x + 3$$

are identities. By permissible values of the variables, we mean here any numbers for which the two expressions both have values. In the first of the above equations, x and y may be any numbers; in the second, x may be any number except 3. We sometimes use the symbol $\equiv$ instead of $=$ to indicate an identity. Thus we might write $(x - 1)^2 \equiv x^2 - 2x + 1$. In Chap. II we were concerned with equalities of this kind, but we did not use the symbol $\equiv$ because it is not common practice to do so.

In this chapter we shall be concerned with equations that are *not* identities. Such equations are often called *conditional equations*, but we shall refer to them simply as equations. We shall confine our attention for the present to some simple types of equations in which only one variable or *unknown* is involved. Our problem will be to discover those values of this unknown for which the equation is true. In accordance with the definition given in Chap. I, we shall call these values of the unknown the *roots* or *solutions* of the equation, and we shall refer to the process of finding them as *solving* the equation.

Examples

The equation $2x - 12 = 0$ is true if and only if $x = 6$. It has then only this one root. The equation $x^2 + 4 = 0$ is not true for any real value of x and therefore has no (real) root. The equation $x^2 = 5x$ is true if $x = 5$ because $5^2 = 5 \cdot 5$; it is true also if $x = 0$, for $0^2 = 5 \cdot 0$. It is not true for any other value of x, although the student may not see immediately how this could be proved. The equation has then the two roots 5 and 0.

In dealing with equations we shall refer to the expression on the right-hand side of the equality sign as the *right side* or *right member* of the equation, and that on the left as the *left side* or *left member*.

20. *Solving an equation.* In some simple cases, solutions of an equation can be found by inspection or by trial. Usually, however, we shall find solutions by performing various operations upon the two members of the equation, and we must determine what operations are permissible. For example, to solve the equation

$$4x + 3 = 21, \tag{1}$$

we first subtract 3 from both members. We then have

$$4x = 18. \tag{2}$$

Next, we divide both members of (2) by 4. The result is

$$x = 4\tfrac{1}{2}. \tag{3}$$

It is essential that we follow through the reasoning that is involved here. We first assume that x is a number for which (1) is true. Then, if (1) is true, (2) must be true because of the axiom, "if equals be subtracted from equals the remainders are equal." Finally, if (2) is true, then (3) must be true because of the axiom, "If equals be divided by equals (zero excepted), the quotients are equal." We thus know that if (1) is to be true, (3) must be true; *i.e.*, if there is a number x such that $4x + 3 = 21$, that number must be $4\frac{1}{2}$.

Does it follow automatically that if $x = 4\frac{1}{2}$, then $4x + 3 = 21$? The answer is of course NO! The converse of a true proposition may very well be false. In this particular case, the converse is true. This can be proved by starting with the hypothesis that $x = 4\frac{1}{2}$ and showing that the equation $4x + 3 = 21$ is a consequence by reversing the above steps. Thus, if $x = 4\frac{1}{2}$, then $4x = 18$; if $4x = 18$, then $4x + 3 = 21$. An equivalent procedure would be simply to find the value of $4x + 3$ under the hypothesis that $x = 4\frac{1}{2}$. An example in which the converse is not true will soon be encountered (page 46).

21. ***Some essentials of logic.*** It seems desirable at this point to review briefly certain matters of logic and define certain terms that will be used in our subsequent work.

Let A and B denote two statements so related to each other that B *must be* true *if* A is true. We say then that B is a *logical consequence* of A, or that B *follows from* A, or that A *implies* B. We call A the *hypothesis* and B the *conclusion* of the proposition or theorem, "If A is true, then B is true." If the truth of the conclusion is actually an inescapable result of the truth of the hypothesis, then the proposition is called *valid*, or is said to be a *valid argument*. Thus the proposition, "If $x = 2$ and $y = 5$, then $x + y = 7$," is valid. The following is not a valid argument: "I own a black car and this car is black, therefore I own this car."

The *converse* of a given theorem is the proposition that results when the hypothesis and conclusion of the given theorem are interchanged. The converse of a valid theorem may or may not be valid.

Example

Theorem: If $x = 2$ and $y = 5$, then $x + y = 7$.
Converse: If $x + y = 7$, then $x = 2$ and $y = 5$.

In this case the converse is not valid—the truth of the conclusion is not an inescapable consequence of the truth of the hypothesis.

In the case of a valid proposition, the truth of the hypothesis of course ensures the truth of the conclusion. The truth of the conclusion does not, however, ensure that of the hypothesis. Consider, for example, the following proposition: *If* $\boldsymbol{a} = -\boldsymbol{a}$, *then* $\boldsymbol{a}^2 = (-\boldsymbol{a})^2$. The proposition is valid because, if two numbers are equal, their squares must be equal. Furthermore, the conclusion is true because, for any number a of our system, $a^2 = (-a)^2$. The hypothesis is of course false if $a \neq 0$.

If A implies B, we say that A is a *sufficient* condition for B, and that B is a *necessary* condition for A; *i.e.*, the truth of A is sufficient to ensure that of B, and B is necessarily true if A is true. If A implies B and B implies A, we say that either A or B is a *necessary and sufficient condition* for the other. We also describe this situation by saying that B is true *if and only if* A is true, or by saying that the statements A and B are *equivalent.*

Examples

$x = 2$ and $y = 5$ are sufficient, but not necessary, for $x + y = 7$. $x = 4$ is necessary and sufficient for $2x = 8$, and the statements $x = 4$ and $2x = 8$ are equivalent.

22. *Equivalent equations.* In order to solve the equation

$$\frac{4x + 9}{2} = \frac{21}{2} + \frac{2x}{3}, \tag{1}$$

we may first remove all the denominators by multiplying both sides by their lowest common multiple which is 6. The result is

$$12x + 27 = 63 + 4x. \tag{2}$$

Next, we may subtract $4x + 27$ from both members of (2) to obtain the equation

$$8x = 36. \tag{3}$$

Finally, if we divide both sides of (3) by 8 we have

$$x = 4\tfrac{1}{2}. \tag{4}$$

As in the similar example on page 43, we may reason that (1) implies (4); *i.e.*, if x is a number such that (1) is true, then (4) must be true, so that $4\frac{1}{2}$ *is the only possible root of* (1). If we start with equation (4) as our hypothesis and reverse the steps, we can similarly show that (4) implies (1), and thus we can be assured that (1) is true if and only if $x = 4\frac{1}{2}$.

Each of the equations that we obtained in the successive steps of the above example is *equivalent* to the original equation in the sense defined in Sec. 21. For example, the truth of (1) implies that of (3), and the truth of (3) implies that of (1). When two equations are equivalent in this sense, any root of one of them is a root of the other.

If an equation B can be obtained from an equation A by employing only the operations of substituting equals for equals, adding the same number or expression representing a number to both members,* and multiplying or dividing both members by the same nonzero constant, then B is equivalent to A.

Certain other operations, such as multiplying or dividing both members by the same expression involving the unknown, or squaring both members, may yield an equation that is not equivalent to the original one. Let us try, for example, to find all roots of the equation

$$\frac{3}{x-3} + 4 = \frac{x}{x-3}. \tag{1}$$

We may first remove the denominators by multiplying both members by $x - 3$. The result is

$$3 + 4(x - 3) = x. \tag{2}$$

If we replace the left member of (2) by $4x - 9$ (substituting equals for equals), we have

$$4x - 9 = x. \tag{3}$$

From this we get immediately

$$x = 3. \tag{4}$$

We started with the hypothesis that x is a number such that (1) is true and found that equation (4) is a logical consequence. This means that (1) cannot be true unless $x = 3$, but it does not mean that (1) *must* be true if $x = 3$. In fact if we substitute 3 for x in (1), we see that 3 is not a root, and we can conclude that (1) has no root.

It is important to observe that while (1) implies (4), (4) does not imply (1). If we start with (4) as our hypothesis and try to reverse the above steps, we see that (4) implies (3) and (3) implies (2). Now, in order to get from (2) to (1), we would have to divide both sides of (2) by $x - 3$, *but our hypothesis is that* x *equals* 3 *so this would require division by zero which is impossible.* We therefore cannot say that (1) must be true if (4) is true. (4) is a *necessary* but not a *sufficient* condition for (1).

* Observe that the equations $x^2 = 2x$ and $x^2 + (1/x) = 2x + (1/x)$ are not equivalent. When we add $1/x$ to both members of the equation $x^2 = 2x$, we are adding the same number to both members only if $x \neq 0$. We can say only that any root of the equation $x^2 = 2x$, *other than zero*, is a root of the new equation.

The student should study the following examples very carefully and supply the details of the reasoning that is involved.

Example 1

Find all roots of the equation

$$\frac{1}{2x + 1} = \frac{2x}{4x^2 - 3}. \tag{1}$$

Solution

First, multiply both sides of (1) by $(2x + 1)(4x^2 - 3)$. The result is

$$4x^2 - 3 = 2x(2x + 1) \tag{2}$$

which is equivalent to

$$2x = -3 \qquad \text{or} \qquad x = -\tfrac{3}{2}. \tag{3}$$

(1) implies (3) so that (1) can have no root except that given by (3); but because of the operation used in getting from (1) to (2), a root of (2) or (3) might not satisfy (1). Therefore we substitute $-\frac{3}{2}$ for x in (1), and we find that it is a root. Hence (1) is true if and only if $x = -\frac{3}{2}$.

Example 2

Solve the equation

$$25 + 2\sqrt{x^2 + 1} = 2x. \tag{1}$$

Solution

Squaring both sides of (1) as it stands would not remove the radical so we first subtract 25 from both sides. We then have

$$2\sqrt{x^2 + 1} = 2x - 25. \tag{2}$$

Now we square both sides of (2). The result is

$$4(x^2 + 1) = 4x^2 - 100x + 625 \tag{3}$$

which is equivalent to

$$100x = 621 \qquad \text{or} \qquad x = 6.21. \tag{4}$$

We can reason that 6.21 is the only possible root of (1), but it may not be a root. Direct substitution shows that it is not. We can conclude that (1) has no root.

PROBLEMS

1. Which of the following are identities, and which are conditional equations?

(a) $(x - 2)(x + 2) = x^2 - 4$. (b) $x(x - 3) = x^2 - 6$.

(c) $\dfrac{x^2 - 4x + 3}{x - 1} = x - 3$. (d) $\dfrac{x^2 - 4}{x - 2} = 3x$.

In each of the following problems write down the hypothesis and the conclusion of the given proposition and state whether or not the proposition is valid. Then write down the converse and state whether or not it is valid:

2. If $2x = 7$ and $y = 3$, then $2x + y = 10$.
3. If two numbers are equal, their squares are equal.
4. If x is an even number, then $2x$ is an even number.
5. If x and y are positive numbers, their sum is a positive number.
6. If $x = 2$, then $x^2 - 4 = 0$.
7. If $x(x + 4) = x^2 + 20$, then $x = 5$.
8. If $x - 2 = 0$, then $x(x - 2) = 0$.
9. If $x^2 = 6x$, then $x = 6$.
10. If two triangles are congruent, their corresponding angles are equal.

11. If we substitute 5 for x in the equation $x^2 + 10 = 7x$, we get $35 = 35$. Which of the following propositions have we proved?
(a) If $x^2 + 10 = 7x$, then $x = 5$.
(b) If $x = 5$, then $x^2 + 10 = 7x$.

12. If we multiply both members of the equation

$$\frac{x}{x - 2} = \frac{2}{x - 2}$$

by $x - 2$, we get $x = 2$. Which of the following propositions have we proved?

(a) If $\dfrac{x}{x - 2} = \dfrac{2}{x - 2}$, then $x = 2$.

(b) If $x = 2$, then $\dfrac{x}{x - 2} = \dfrac{2}{x - 2}$.

13. Prove that the equations $\frac{2}{3}x + 5 = x - 7$ and $x - 36 = 0$ are equivalent. (Show that the truth of either implies that of the other.)

14. Show that if $x = -3$, then $(2x^2 - 7)/x = (7 + 6x)/3$. Show that the converse is true and hence conclude that the equation has no other root.

15. Show that if $(2/x) + 3/(x - 3) = 5/(x - 2)$, then $x = 12$. Is the converse true?

16. Show that if $x/(x - 5) = (3x - 10)/(x - 5)$, then $x = 5$. Is the converse true?

17. Is the condition $x = -1$ necessary or sufficient or neither or both for $\sqrt{4x^2 - 3} = 2x + 1$?

18. Is the condition $x = 3$ necessary or sufficient or neither or both for

$$(2x - 3)(2x + 3) = x(4x - 3)?$$

Find all roots of each of the following equations:

19. $x(x - 5) = (x - 3)(x - 4)$.

20. $3(x^2 - 2) = (x + 6)(3x - 4)$.

21. $(4z + 3)(z - 2) = (2z - 1)(2z + 1)$.

22. $4 + x(6x - 1) = (3x + 4)(2x - 1)$.

23. $\dfrac{x}{2} + \dfrac{3x + 1}{5} = \dfrac{x + 3}{10}$.

24. $\dfrac{5 - 6x}{2} + \dfrac{2x - 1}{4} = 7$.

25. $\dfrac{2w + 3}{4} - \dfrac{3w + 2}{12} = 1$.

26. $\dfrac{7x + 5}{3} + \dfrac{1}{6} = 2x$.

27. $\dfrac{1}{x} - \dfrac{x}{5} = \dfrac{2 - 3x}{15}$.

28. $\dfrac{1}{x - 4} + \dfrac{2}{x - 2} = \dfrac{3}{x}$.

29. $\dfrac{2x}{2x + 3} = 5 - \dfrac{3}{3 + 2x}$.

30. $\dfrac{5x^2}{x^2 - 4} = 4 + \dfrac{x}{x + 2}$.

31. $\dfrac{4}{x} + \dfrac{3}{x + 4} = \dfrac{7x - 5}{x^2 + 4x}$.

32. $\dfrac{3}{x - 3} - \dfrac{1}{x - 1} = \dfrac{x}{x^2 - 4x + 3}$.

33. $2 + \sqrt{x^2 + 6} = x$.

34. $x + \sqrt{x^2 - 4} = 4$.

35. $3\sqrt{x^2 - 8} = 3x - 2$.

36. $\sqrt{x - 28} = \sqrt{x} - 2$.

37. $\sqrt{2x + 2} = \sqrt{3x + 12}$.

38. $\sqrt{12 + 2x} = \sqrt{2x} + 3$.

39. $\sqrt{2x} + 1 = \sqrt{2x - 7}$.

40. $\sqrt{9 - x} + \sqrt{21 - x} = 6$.

41. Prove that there is no number x satisfying the equation

$$\frac{2x + 1}{x(2x + 3)} + \frac{1}{2x} = \frac{3}{2x + 3}.$$

42. Prove that if $\dfrac{4}{x - 2} + \dfrac{5}{3x + 2} = \dfrac{2x + 7}{3x^2 - 4x - 4}$, then $x = 0.6$. Prove also the converse of this statement.

43. Prove that if $\dfrac{3}{x} + \dfrac{4x - 3}{x^2 + 3x} = \dfrac{5}{x + 3}$, then $x = -3$. Prove that the converse is not true, and hence infer that the given equation has no solution.

44. Prove that if $\dfrac{12}{2x - 5} = \dfrac{13 - 8x}{2x^2 - 3x - 5} + \dfrac{3}{x + 1}$, then $x = -1$. Show that the converse is not true, and hence infer that the given equation has no solution.

45. Investigate the truth of each of the following statements:

(*a*) If $\dfrac{x}{x + 1} - \dfrac{3}{x} = \dfrac{6 - 3x}{x^2 + x}$, then $x = 3$.

(*b*) If $x = 3$, then $\dfrac{x}{x + 1} - \dfrac{3}{x} = \dfrac{6 - 3x}{x^2 + x}$.

23. ***Rational integral equations.*** A rational integral equation in x of degree n is an equation that can be written in the form

$$a_0x^n + a_1x^{n-1} + \cdots + a_{n-1}x + a_n = 0 \qquad (a_0 \neq 0),$$

in which the a's are constants and n is a positive integer. If the degree is *one*, it is called a *linear* equation; if the degree is *two*, it is called a *quadratic* equation. The names *cubic* and *quartic* are usually associated with rational integral equations of degree *three* and *four*, respectively.

Any linear equation in x can be written in the form

$$a_0x + a_1 = 0 \qquad \text{or} \qquad ax + b = 0$$

in which the coefficient of x is different from zero. Many of the equations that we have solved in this chapter have been linear equations or have led to linear equations, and we have seen that the problem of solving such an equation is almost trivial. It is easy to prove that, if a and b are real numbers and $a \neq 0$, then the equation $ax + b = 0$ has the unique solution $x = -b/a$.

Before proceeding to the study of quadratic equations, we prove the following theorem in which u and v denote, for the present, any real numbers.

Theorem. *If either* $\mathbf{u} = 0$ *or* $\mathbf{v} = 0$, *then* $\mathbf{u} \cdot \mathbf{v} = 0$, *and conversely if* $\mathbf{u} \cdot \mathbf{v} = 0$, *then either* $\mathbf{u} = 0$ *or* $\mathbf{v} = 0$ (*or both*).

The first part of the theorem is an immediate consequence of the definition on page 16. The proof of the second part is as follows: Assume that

$$u \cdot v = 0 \tag{1}$$

and $u \neq 0$; then we can divide both sides of (1) by u and have

$$\frac{u \cdot v}{u} = \frac{0}{u} \qquad \text{or} \qquad v = 0. \tag{2}$$

Thus if $u \neq 0$, then $v = 0$. Similarly, if $v \neq 0$, we can divide by v and show that in this case $u = 0$. This completes the proof. The extension to the case of more than two factors is obvious.

Example

The product $(2x + 5)(x - 3)$ is equal to zero if and only if x is a number such that either $2x + 5 = 0$ or $x - 3 = 0$. This means that the equation

$$(2x + 5)(x - 3) = 0$$

has the two roots $x = -\frac{5}{2}$ and $x = 3$, *and no others.*

24. ***Quadratic equations. Solution by factoring.*** Any quadratic equation in x can be written in the form

$$ax^2 + bx + c = 0$$

where a, b, and c are constants and $a \neq 0$. If the left member can be factored into two factors of the type allowed in Chap. II, the solution is immediate.

Example

Determine whether or not there is a real number x such that

$$3x^2 = x + 14. \tag{1}$$

Solution

The given equation is equivalent to

$$3x^2 - x - 14 = 0, \tag{2}$$

which in turn is equivalent to

$$(3x - 7)(x + 2) = 0. \tag{3}$$

Equation (3) is satisfied if and only if either $3x - 7 = 0$ or $x + 2 = 0$. Hence there are exactly two numbers having the required property, namely

$$x = \frac{7}{3} \qquad \text{and} \qquad x = -2.$$

For our present purposes it is desirable to modify the requirements that we set up in Chap. II in regard to the factoring of polynomials. In particular, we replace the requirement that the coefficients in the factors must be integers by the stipulation that they may be any real numbers. Then, for example, we may solve the equation

$$x^2 - 5 = 0$$

by writing it in the form

$$(x + \sqrt{5})(x - \sqrt{5}) = 0.$$

This gives immediately the two solutions $x = \pm\sqrt{5}$. This procedure may be extended as indicated in the following:

Example

Find a number x whose square is equal to 31 more than six times x.

Solution

We require a solution of the equation

$$(1) \qquad x^2 = 6x + 31$$

or

$$(2) \qquad x^2 - 6x - 31 = 0.$$

We cannot factor the left member of (2) immediately, but it will take the form $u^2 - v^2$ if we add and subtract 9. Thus (2) is equivalent to

$$(3) \qquad (x^2 - 6x + 9) - 40 = 0$$

or

$$(4) \qquad (x - 3)^2 - 40 = 0.$$

We may now factor (4) as the difference of two squares. Thus we have

$$(5) \qquad [(x - 3) + \sqrt{40}] \cdot [(x - 3) - \sqrt{40}] = 0.$$

By equating each factor separately to zero, we obtain the two solutions:

$$(6) \qquad \begin{aligned} x - 3 - \sqrt{40} &= 0 \quad \text{or} \quad x = 3 + \sqrt{40}; \\ x - 3 + \sqrt{40} &= 0 \quad \text{or} \quad x = 3 - \sqrt{40}. \end{aligned}$$

There are then exactly two real numbers, both irrational, having the required property. In decimal form they are approximately 9.32 and -3.32. It should be observed that our operations are all reversible so that not only does (1) imply (6), but either of the conditions given by (6) implies (1). The student should also check the result by direct substitution.

25. ***Solution by completing the square.*** A procedure which differs only slightly from that followed in the last example above makes use of the following:

Theorem. *If* $u^2 = v^2$, *then either* $u = v$ *or* $u = -v$, *and conversely if* $u = \pm v$ *then* $u^2 = v^2$.

The proof of the first part is as follows: If $u^2 = v^2$, then $u^2 - v^2 = 0$ or $(u + v)(u - v) = 0$. This in turn implies that either $u - v = 0$ or $u + v = 0$, and hence that either $u = v$ or $u = -v$. The proof of the second part is quite obvious.

We shall illustrate the procedure by solving the equation

$$x^2 + 3x + 1 = 0. \tag{1}$$

First, we subtract 1 from both members, thus obtaining the equation

$$x^2 + 3x = -1. \tag{2}$$

Next we add to both members of (2) whatever number is needed in order that the left side may be a "perfect square." In this case, the number needed is $\frac{9}{4}$. Thus we get

$$x^2 + 3x + \tfrac{9}{4} = -1 + \tfrac{9}{4} \tag{3}$$

or

$$(x + \tfrac{3}{2})^2 = \tfrac{5}{4}. \tag{4}$$

Now, in accordance with the above theorem, (4) is true if and only if

$$x + \frac{3}{2} = \pm \frac{\sqrt{5}}{2}. \tag{5}$$

Finally, if we subtract $\frac{3}{2}$ from both sides of (5), we have the solutions

$$x = -\frac{3}{2} \pm \frac{\sqrt{5}}{2}. \tag{6}$$

The result means that (1) is true if and only if

$$x = \frac{-3 + \sqrt{5}}{2} \quad \text{or} \quad \frac{-3 - \sqrt{5}}{2}.$$

In decimal form, these roots are approximately -0.38 and -2.62.

In using this method, we observe that the constant needed to complete the square when we have $x^2 + px$ is $(p/2)^2$, *the square of one-half the coefficient of* x. This results from the fact that

$$\left(x + \frac{p}{2}\right)^2 = x^2 + px + \left(\frac{p}{2}\right)^2.$$

If the coefficient of x^2 is not 1, then it is advisable to divide both members of the equation by this coefficient in order that the simple procedure given above for completing the square may be applicable.

Example

Solve the equation

$$(1) \qquad 2x^2 - 15x + 7 = 0.$$

Solution

We first divide both sides by 2 and then move the constant term over to the right side. The result is

$$(2) \qquad x^2 - \tfrac{15}{2}x = -\tfrac{7}{2}.$$

Next we add the square of one-half the coefficient of x to both members in order that the left member may be a perfect square:

$$(3) \qquad x^2 - \tfrac{15}{2}x + \tfrac{225}{16} = -\tfrac{7}{2} + \tfrac{225}{16}$$

or

$$(4) \qquad (x - \tfrac{15}{4})^2 = \tfrac{169}{16}.$$

We have then

$$(5) \qquad x - \tfrac{15}{4} = \pm\tfrac{13}{4}$$

or

$$(6) \qquad x = \tfrac{15}{4} \pm \tfrac{13}{4} = 7 \quad \text{or} \quad \tfrac{1}{2}.$$

We have used only reversible operations so we know that (1) will be true if x is either 7 or $\frac{1}{2}$. However, the student should check these results by direct substitution.

It should be remarked here that a given quadratic equation may not have any solution within the field of real numbers. Consider, for example, the equation

$$(1) \qquad x^2 + 2x + 5 = 0.$$

If we complete the square, we obtain the equation

$$(2) \qquad (x + 1)^2 = -4.$$

(1) is true if and only if x is a number such that (2) is true. But if x is any real number, $(x + 1)^2$ is positive and cannot be equal to -4. Therefore, there is no real number x which satisfies (1). In the next section, we shall extend our number system so that it will include numbers whose squares are negative numbers. Within this system, equation (1) will have two solutions.

PROBLEMS

1. Which of the following are rational integral equations, and what in each case is the degree?

(a) $x^3 = 4(x^2 + 3)^2$.

(b) $x^2 + 2\sqrt{x + 4} = 6$.

(c) $(x + 1)^3 = x(x^2 + 5)$.

(d) $x + \sqrt{x} = 12$.

2. Let a and b denote real numbers and assume $a \neq 0$. Prove that

(a) If $ax + b = 0$, then $x = -b/a$.

(b) If $x = -b/a$, then $ax + b = 0$.

Solve each of the following equations by factoring:

3. $x^2 + x - 6 = 0$.

4. $2x^2 + 7x - 15 = 0$.

5. $7x^2 - 31x + 12 = 0$. check

6. $4x^2 + 13x + 9 = 0$.

7. $8x^2 + 34x + 15 = 0$.

8. $2x^2 - 29x + 77 = 0$.

9. $2x^2 = 5x + 12$.

10. $x(5x + 24) = 5$.

11. $3x(x + 4) = 4x - 5$. check

12. $2x(x + 2) = 3x + 1$.

13. $(2x + 3)(x + 1) = 15$.

14. $(x + 4)(3x - 7) + 14 = x(8 + x)$.

15. $(3x + 4)(x + 9) = 2(x^2 + 11x + 36)$.

16. $x^2(x + 4) = x^3 + 4x + 24$.

Solve each of the following equations by factoring it as illustrated in the example on page 52:

17. $x^2 - 10x + 21 = 0$.

18. $4x^2 + 12x + 2 = 0$.

19. $x^2 - 4x - 1 = 0$.

20. $x^2 + 6x + 7 = 0$.

21. $4x^2 + 12x + 3 = 0$.

22. $9x^2 - 12x - 2 = 0$.

23. $16x^2 = 8x + 19$.

24. $25x^2 = 6(5x - 1)$.

Solve each of the following equations by using the method of Sec. 25:

25. $x^2 - 14x + 24 = 0$.

26. $x^2 - 2x = 5$.

27. $x^2 + 6x + 5 = 0$.

28. $x^2 + 22x + 96 = 0$.

29. $4x^2 = 6x + 7$.

30. $3x^2 + 4x = 15$.

31. $9x^2 - 6x - 4 = 0$.

32. $x^2 - 7x = 22$.

33. $x^2 = 6(x + 1)$.

34. $9x^2 = 8(3x - 1)$.

35. $x(x + 6) = x - 5$.

36. $45 + 4(x + 1)(x - 1) = 28x$.

Solve each of the following equations:

37. $\dfrac{3x}{x+2} + \dfrac{2}{x} = \dfrac{5}{2}$.

38. $\dfrac{4x}{3x-2} + \dfrac{5}{x} = \dfrac{7}{3}$.

39. $\dfrac{4x}{2x-7} + x = 11$.

40. $\dfrac{6x}{3x+8} + 2x = 12$.

41. $\dfrac{x}{x-2} + \dfrac{7}{3} = \dfrac{35}{x-4}$.

42. $\dfrac{2x}{2x-5} - \dfrac{15}{2x+3} = 2$.

43. $\dfrac{x+5}{x-3} - \dfrac{x+6}{2x} = \dfrac{1}{6}$.

44. $x - 9 = \sqrt{2x - 10}$.

45. $\sqrt{12x + 10} = 2(2x - 5)$.

46. $x + 2\sqrt{x - 1} = 25$.

47. $\sqrt{2x + 2} = 2 + \sqrt{3x + 15}$.

48. $\sqrt{3 - x} + \sqrt{2x + 16} = 5$.

49. $\sqrt{3x - 6} + \sqrt{2x + 4} = 2\sqrt{x}$.

50. $\sqrt{\dfrac{7x+1}{4}} = \dfrac{5x-7}{6}$.

51. $\sqrt{\dfrac{x+41}{x-7}} = \dfrac{x+6}{x-6}$.

52. $2 + \sqrt{\dfrac{x}{x-10}} = \dfrac{x-4}{x-14}$.

53. Show that if x is a number such that

$$\frac{5(5-4x)}{2x+5} = 5(1-x) + \frac{12x^2}{2x+5},$$

then $2x^2 + 5x = 0$, and thus prove that the first equation can have no roots other than 0 and $-\frac{5}{2}$. Are these numbers roots?

54. Show that if x is a number such that

$$\frac{3(x^2+4)}{x-3} = 3 + 2x + \frac{13x}{x-3},$$

then $x^2 - 10x + 21 = 0$, and thus prove that the first equation can have no roots other than 3 and 7. Are these numbers roots?

55. Show that if x is a number such that

$$\sqrt{3x} = 1 + \sqrt{1 + 4x},$$

then $x^2 - 12x = 0$, and thus prove that the first equation can have no roots other than 0 and 12. Then, by showing that neither of these numbers is a root, complete the proof of the fact that the equation has no root.

56. Show that $-\frac{44}{81}$ is a root of the equation

$$2\sqrt{1 - 2x} + \sqrt{x + 5} = 5,$$

and prove that the equation has no other root. HINT: Show that if the given equation is to be satisfied, then x must satisfy the equation

$$81x^2 + 368x + 176 = 0.$$

From this infer that -4 and $-\frac{44}{81}$ are the only numbers that could be roots and show that -4 is not a root.

57. Determine whether each of the following statements is true or false:

(*a*) If $\sqrt{4 - x} + \sqrt{3x + 16} = 2$, then $x = 0$ or -5.

(*b*) If $x = 0$ or -5, then $\sqrt{4 - x} + \sqrt{3x + 16} = 2$.

26. ***Complex numbers.*** There is no real number x such that $x^2 + 4 = 0$, for this equation is equivalent to $x^2 = -4$ and there is no real number whose square is a negative number. Similarly, the equation

$$x^2 + 6x + 14 = 0$$

has no solution among the real numbers because it is equivalent to

$$(x + 3)^2 = -5,$$

and this cannot be true for any real number x.

Our position here is similar to that of a person who is attempting to solve the equation $3x = 16$ and has at his disposal a number system consisting of only the natural numbers. He must either say that the equation has no solution or invent new numbers; and just as he can overcome his difficulty by inventing fractions, we can overcome ours by inventing numbers whose squares are negative numbers. We accomplish this in the following way:

We adjoin to the system of real numbers one additional symbol. This symbol is to be regarded as a new number—*a number whose square is* -1. It is customary to use the letter i as this new symbol. Then

$$i^2 = -1,$$

and we can regard i as representing a square root of -1.

We specify that this new symbol can be used along with the real numbers and letters representing real numbers, in the operations of addition, subtraction, multiplication, and division, precisely as if it represented another real number, and in accordance with the rules that have been prescribed for real numbers. It is subject to the one additional rule that i^2 is equivalent to -1. Thus, just as $(3a) \cdot (3a) = 9a^2$, we have

$$(3i) \cdot (3i) = 9i^2;$$

and since $i^2 = -1$ this means that $(3i) \cdot (3i) = -9$, and that $3i$ is a square root of -9.

In regard to the positive integral powers of i, we obtain immediately the following results:

$$
\begin{aligned}
i &= i; \\
i^2 &= -1; \\
i^3 &= i^2 \cdot i = (-1) \cdot i = -i; \\
i^4 &= i^3 \cdot i = (-i) \cdot i = -i^2 = +1; \\
i^5 &= i^4 \cdot i = (1) \cdot i = i, \text{ etc.}
\end{aligned}
$$

We thus see that any positive integral power of i is equal to i, -1, $-i$, or $+1$. Furthermore,

$$\frac{1}{i} = \frac{i}{i^2} = \frac{i}{-1} = -i,$$

so that the reciprocal of i is equal to the negative of i.

The symbol

$$a + bi,$$

where a and b are any real numbers, is called a *complex number*. If $b = 0$, it is the real number a. If $b \neq 0$, it is called an *imaginary number*. If $a = 0$ and $b \neq 0$, it is called a *pure imaginary number*.

Examples

5 or $5 + 0i$ is a real number; $3 - 2i$ and $0 + 6i$ are imaginary numbers, the second being a pure imaginary number. All are complex numbers.

The names real number and imaginary number are unfortunate because of course neither is any more real or any more imaginary, within the usual meanings of these words, than the other. One must not think of the words real and imaginary as having their ordinary meanings at all when he speaks of a *real number* or an *imaginary number*. There is nothing any more "imaginary" about a number whose square is -5 than about one whose square is $+5$.

We define equality for complex numbers as follows:

$$a + bi = c + di \quad \textit{if and only if} \quad a = c \text{ and } b = d.$$

We call a the *real part* and b the *imaginary part* of the complex number $a + bi$, and the above definition states that two complex numbers are equal if and only if their real parts are equal and their imaginary parts are equal. It follows that

$$a + bi = 0 \qquad \textit{if and only if} \qquad a = 0 \text{ and } b = 0.$$

The fundamental operations have been defined above by the requirement that they be carried out precisely as if i represented a real number, and subject to the additional provision that $i^2 = -1$; thus, for example,

$$\begin{aligned}(2 + 3i) + (5 - 2i) &= 7 + i.\\ (2 + 3i) \cdot (5 - 2i) &= 10 + 11i - 6i^2\\ &= 10 + 11i + 6\\ &= 16 + 11i.\end{aligned}$$

The system of complex numbers is closed with respect to addition, subtraction, multiplication, and division, with the exception that division by zero is impossible. In fact,

$$\begin{aligned}(a + bi) + (c + di) &= (a + c) + (b + d)i.\\ (a + bi) \cdot (c + di) &= ac + (ad + bc)i + bdi^2\\ &= (ac - bd) + (ad + bc)i.\end{aligned}$$

In order to see how the quotient $a + bi$ divided by $c + di$ can be expressed in the form $p + qi$, we shall first divide $16 + 11i$ by $5 - 2i$:

$$\begin{aligned}\frac{16 + 11i}{5 - 2i} &= \frac{16 + 11i}{5 - 2i} \cdot \frac{5 + 2i}{5 + 2i}\\ &= \frac{80 + 87i + 22i^2}{25 - 4i^2}\\ &= \frac{58 + 87i}{29}\\ &= 2 + 3i.\end{aligned}$$

We had already seen above that

$$(2 + 3i)(5 - 2i) = 16 + 11i,$$

and consequently the result of the present division was to be expected.

By the *conjugate* of a complex number $m + ni$, we mean the number $m - ni$. It is clear that the product of a complex number and its conjugate is a real number:

$$(m + ni)(m - ni) = m^2 - n^2i^2 = m^2 + n^2.$$

We can always express the quotient, $a + bi$ divided by $c + di$, in the form $p + qi$ by multiplying numerator and denominator by the conjugate of the denominator:

$$\frac{a + bi}{c + di} = \frac{a + bi}{c + di} \cdot \frac{c - di}{c - di}$$
$$= \frac{(ac + bd) + (bc - ad)i}{c^2 + d^2}$$
$$= \frac{ac + bd}{c^2 + d^2} + \frac{bc - ad}{c^2 + d^2} i.$$

It will be recalled that, if a and b are two real numbers, then $a > b$ if $a - b$ is a positive number. We do not set up any corresponding definition for the case of imaginary numbers. Thus we do not speak of $6 + 4i$ as being either greater than or less than $3 + i$.

Imaginary numbers are used extensively in both pure and applied mathematics. Without them we would be laboring under as serious a handicap as if we had no negative numbers or no fractions. Our immediate use for them, as indicated in the beginning of this section, is in connection with quadratic equations.

***Example* 1**

Solve the equation $x^2 + 5 = 0$ by factoring.

Solution

We may replace $+5$ by $-5i^2$ and the equation becomes

$$x^2 - 5i^2 = 0,$$

or

$$(x + \sqrt{5}\, i)(x - \sqrt{5}\, i) = 0.$$

This last equation is satisfied if and only if

$$x = +\sqrt{5}\, i \quad \text{or} \quad -\sqrt{5}\, i.$$

If we replace 5 throughout the above example by k, where k represents any positive number, we have the proof of the fact that the equation $x^2 = -k (k > 0)$ has the two roots $x = +\sqrt{k}\, i$ and $x = -\sqrt{k}\, i$, and no others. We use this result in getting from (3) to (4) in the next example.

Example 2

By completing the square, find the roots of the equation

$$(1) \qquad x^2 - 6x + 13 = 0.$$

Solution

If we subtract 4 from each side of (1), we have

$$(2) \qquad x^2 - 6x + 9 = -4$$

or

$$(3) \qquad (x - 3)^2 = -4.$$

Equation (3) is satisfied if and only if

$$(4) \qquad x - 3 = \pm 2i$$

or

$$(5) \qquad x = 3 \pm 2i.$$

The operations that led from (1) to (5) are reversible so we know that (1) has the two roots $3 + 2i$ and $3 - 2i$. Furthermore there can be no other roots. Why? The student should check the results by direct substitution.

PROBLEMS

Perform the indicated operations:

1. $(2 + 3i) + (7 - i) - (5 - 2i)$.
2. $3(4 + i) - 4(2 + 3i)$.
3. $2i(1 + i) + 4(3 - 2i)$.
4. $(3 + 7i)i + 4(8 - i)$.
5. $(2 + 3i)(4 - i) + 2$.
6. $2i + 4(i + 1)(2i - 1)$.
7. $(4 + 3i) + (2 - i)^2 + 3$.
8. $(2 + i)^2 - 4(2 + i) + 5$.
9. $4(2 + 3i)^2 - 8(2 + 3i) + 13$.
10. $(3 + 2i)^2 - 4(1 + i)^2$.

11. Show that $5 + i$ and $5 - i$ are roots of the equation $x^2 + 26 = 10x$.
12. Show that $1 + 4i$ and $1 - 4i$ are roots of the equation $x^2 - 2x + 17 = 0$.
13. Show that $2 + i$ is a root of the equation $x(x^2 - 7x + 17) = 15$.
14. Show that $1 - 3i$ is a root of the equation $x^2(x + 1) + 2(2x + 15) = 0$. Show also that -3 is a root.
15. Show that $\sqrt{5}$, $-\sqrt{5}$, $2i$, and $-2i$ are roots of the equation $x^4 = x^2 + 20$.
16. Show that the cube of each of the following numbers is 1, and hence that each is a cube root of 1: $-\frac{1}{2} + \frac{1}{2}\sqrt{3}i$; $-\frac{1}{2} - \frac{1}{2}\sqrt{3}i$; 1.

17. Show that the cube of each of the following numbers is -1, and hence that each is a cube root of -1: $\frac{1}{2} + \frac{1}{2}\sqrt{3}i$; $\frac{1}{2} - \frac{1}{2}\sqrt{3}i$; -1.

18. Show that $(1 + i)^4 = -4$ and that $(1 + i)^3 = -2 + 2i$.

Solve each of the following equations:

19. $x^2 + 2x + 2 = 0$.

20. $x^2 + 6x + 13 = 0$.

21. $x^2 = 8x - 25$.

22. $x^2 = 14x - 65$.

23. $4x^2 - 4x + 17 = 0$.

24. $9x^2 + 12x + 5 = 0$.

25. $x^2 + 6x + 4 = 0$.

26. $x^2 + 2x = 9$.

27. $4(x^2 + 9x) = 19$.

28. $x^2 + x = -1$.

29. $4x(x - 1) = -7$.

30. $3x^2 = 3x - 2$.

31. Is every imaginary number a complex number? Is every complex number an imaginary number? Is every real number a complex number?

Reduce each of the following complex numbers to the form $a + bi$:

32. $\frac{6}{i} + 4i^3$.

33. $\frac{4}{i^2} + \frac{3}{i}$.

34. $(2i)^3 + (3i)^2$.

35. $(i + 1)^3 - 2i^5$.

36. $\frac{4i^3 + 6i^2}{i^3}$.

37. $4 + (1 + i)^5$.

38. Show that $x = 1 + i$ is a solution of the equation

$$x^4 - 6x^2 + 12x = 8.$$

39. Show that $x = 2 + 3i$ is a solution of the equation

$$2x^3 = 7x^2 - 22x - 13.$$

40. Show that $x = -2 + i$ is a solution of the equation

$$x^3 + 4ix^2 + 8ix = 7i + 6.$$

In each of the following problems perform the indicated operations and express the result in the form $a + bi$:

41. $\frac{3 + 7i}{1 - i}$.

42. $\frac{2 + 6i}{3 + i}$.

43. $\frac{0.9 + 0.1i}{2 + i}$.

44. $\frac{-4 - 9i}{-2 + 6i}$.

45. $\frac{1 + i}{2 - i} + \frac{3 + i}{(1 + i)^2}$.

46. $\frac{2i}{1 + 4i} + \frac{4}{(1 + i)^2}$.

47. $\frac{i}{(2 + i)^2} + \frac{4}{1 - i}$.

48. $\frac{(2 + 6i)(-3 + 4i)}{(2 + i)(2 - i)}$.

49. $\frac{(3 - i)(1 + 5i)}{(2 + i)(3 - 2i)}$.

50. $\frac{(3 + i)^2}{1 + i} - \frac{2}{(1 + 3i)^2}$.

27. ***The quadratic formula.*** Consider the quadratic equation $ax^2 + bx + c = 0$ in which a, b, and c denote real numbers and $a \neq 0$. By completing the square, we may solve this equation and obtain the roots in terms of a, b, and c. We can then use this result as a formula and solve all quadratic equations that we meet in the future merely by substituting the proper numbers for a, b, and c, in our formula. The steps are as follows:

Starting with the equation

$$ax^2 + bx + c = 0, \tag{1}$$

first subtract c from both sides and then divide both sides by a. The result is

$$x^2 + \frac{b}{a}x = -\frac{c}{a}. \tag{2}$$

Now add $b^2/4a^2$ to both sides of (2) in order to make the left side a perfect square. We then have

$$x^2 + \frac{b}{a}x + \frac{b^2}{4a^2} = \frac{b^2}{4a^2} - \frac{c}{a} \tag{3}$$

or

$$\left(x + \frac{b}{2a}\right)^2 = \frac{b^2 - 4ac}{4a^2}. \tag{4}$$

Equation (4) is satisfied if and only if

$$x + \frac{b}{2a} = \pm \frac{\sqrt{b^2 - 4ac}}{2a}. \tag{5}$$

We can conclude that, if (1) is to be true, then x must have one or the other of the following two values:

$$x = \frac{-b + \sqrt{b^2 - 4ac}}{2a} \quad \text{or} \quad x = \frac{-b - \sqrt{b^2 - 4ac}}{2a}. \tag{6}$$

Direct substitution, or a study of the reversibility of the operations used in getting from (1) to (6), shows that if x has either of the values given by (6) then (1) is true, so (1) has the two roots given by (6) and no others.

The formula

$$\boldsymbol{x = \frac{-b \pm \sqrt{b^2 - 4ac}}{2a},}$$

MEMORIZE

which gives both roots, is called the *quadratic formula.*

Example 1

The equation $2x^2 - 9x - 5 = 0$ in which $a = 2$, $b = -9$, and $c = -5$, has the two solutions:

$$x = \frac{-(-9) \pm \sqrt{(-9)^2 - 4(2)(-5)}}{2(2)}$$
$$= \frac{9 \pm \sqrt{121}}{4}$$
$$= \frac{9 \pm 11}{4} = 5 \quad \text{or} \quad -\frac{1}{2}.$$

Example 2

The equation $x^2 + x + 1 = 0$ in which $a = 1$, $b = 1$, and $c = 1$, has the two solutions:

$$x = \frac{-1 \pm \sqrt{(1)^2 - 4(1)(1)}}{2(1)}$$
$$= \frac{-1 \pm \sqrt{-3}}{2}$$
$$= -\frac{1}{2} + \frac{\sqrt{3}}{2}i \quad \text{or} \quad -\frac{1}{2} - \frac{\sqrt{3}}{2}i.$$

Example 3

For what value or values of w is

$$6 + w(w + 3) = 5(w + 2)?$$

Solution

This is a quadratic equation in w, but it is not in the standard form $aw^2 + bw + c = 0$. In order to reduce it to this form, we perform the indicated multiplications and then subtract $5w + 10$ from both sides. The result is

$$w^2 - 2w - 4 = 0,$$

so that $a = 1$, $b = -2$, and $c = -4$. The solutions are

$$w = \frac{-(-2) \pm \sqrt{(-2)^2 - 4(1)(-4)}}{2(1)}$$
$$= \frac{2 \pm \sqrt{20}}{2}$$
$$= 1 \pm \sqrt{5}.$$

In decimal form the two roots are, approximately,

$$w_1 = 3.236 \quad \text{and} \quad w_2 = -1.236.$$

28. ***The character of the roots. Sum and product of the roots.*** The two roots of the quadratic equation $ax^2 + bx + c = 0$ are the numbers x_1 and x_2 where

$$x_1 = \frac{-b + \sqrt{b^2 - 4ac}}{2a} \quad \text{and} \quad x_2 = \frac{-b - \sqrt{b^2 - 4ac}}{2a}.$$

If a, b, and c are real numbers, these roots are obviously imaginary numbers if and only if the number $b^2 - 4ac$, which appears under the radical sign, is a negative number. The situation is, in fact, as follows:

If $b^2 - 4ac > 0$, *the roots are real and unequal.*
If $b^2 - 4ac = 0$, *the roots are real and equal.*
If $b^2 - 4ac < 0$, *the roots are conjugate imaginary numbers.*

Additional conclusions concerning the roots can be reached by studying the above expressions for x_1 and x_2. For example, if a, b, and c are rational numbers, then the roots are rational numbers if and only if $b^2 - 4ac$ is a perfect square.

The number $b^2 - 4ac$ which thus plays an important part in the determination of the character of the roots is called the *discriminant* of the quadratic equation.

As stated above, if the discriminant is zero, the two roots x_1 and x_2 are equal. In fact we have in this case

$$x_1 = -\frac{b}{2a} \quad \text{and} \quad x_2 = -\frac{b}{2a}.$$

We prefer to say that the equation has two *equal* roots rather than that it has only one root. We may then say that *every* quadratic equation with real coefficients has *two* roots and that these roots are both real numbers unless the discriminant is negative, in which case they are conjugate imaginary numbers. Thus, in this latter case, if one of the roots is $2 + 3i$ the other *must be* $2 - 3i$. The numbers 4 and $2 + i$, or the numbers $2 + 3i$ and $5 - 2i$, could not be the roots of any quadratic equation whose coefficients are real numbers.

Example

For what value or values of k will the equation

$$(1) \qquad k(x + 1) = x(2 - kx)$$

have two equal roots?

Solution

When equation (1) is reduced to the form $ax^2 + bx + c = 0$, it becomes

$$kx^2 + (k - 2)x + k = 0,$$

so that $a = k$, $b = k - 2$, and $c = k$. The two roots will be equal if and only if we select the number k so that $b^2 - 4ac = 0$. This means that we must choose for k a number that satisfies the equation

$$(k - 2)^2 - 4 \cdot k \cdot k = 0.$$

This equation reduces to

$$3k^2 + 4k - 4 = 0$$

or

$$(3k - 2)(k + 2) = 0.$$

The roots are $k = \frac{2}{3}$ and $k = -2$. If we use either of these numbers for k, equation (1) will have two equal roots. The student should verify this by actually finding the roots of (1) for the case in which $k = \frac{2}{3}$ and for the case in which $k = -2$.

By adding the expressions given at the beginning of this section for the two roots x_1 and x_2 of the equation $ax^2 + bx + c = 0$, and by multiplying them, the student can prove the following results:

$$x_1 + x_2 = -\frac{b}{a};$$

$$x_1 \cdot x_2 = \frac{c}{a}.$$

The sum of the two roots is equal to $-b/a$ *and the product of the two roots is equal to* c/a.

Example

Without actually finding the two roots of the equation

$$4x^2 + 28x + 53 = 0$$

we can state that the sum of the two roots is -7, and that the product of the two roots is $\frac{53}{4}$.

29. ***Equations that lead to quadratic equations.*** The problem of solving a given equation involving fractions or radicals may of course reduce to that of solving a quadratic equation. We give two examples which the student should study carefully.

Example 1

Solve the equation

$$\frac{2x+1}{x-3} - \frac{x+4}{x} = \frac{8}{5}. \tag{1}$$

Solution

First, in order to remove the denominators, we multiply both sides of (1) by $5x(x-3)$. The result is

$$5x(2x+1) - 5(x-3)(x+4) = 8x(x-3). \tag{2}$$

This reduces to

$$x^2 - 8x - 20 = 0. \tag{3}$$

Now (3) has the two roots

$$x = -2 \quad \text{and} \quad x = 10. \tag{4}$$

We can reason that if (1) is to be true, then (2) must be true, etc., and finally assert that if (1) is to be true, x must be either -2 or 10 so that no other number could be a root. It does not follow automatically that if $x = -2$, or if $x = 10$, (1) must be true. In order to investigate this question, we start with the hypothesis that $x = -2$, or that $x = 10$, and try to show that (1) is a consequence by reversing the steps, or we may use direct substitution. It turns out in this case that both -2 and 10 are roots of (1).

Example 2

Solve the equation

$$\sqrt{2x+1} - \sqrt{4x+6} = 2. \tag{1}$$

Solution

We first isolate one of the radicals and then remove it by squaring both sides:

$$\sqrt{2x+1} = 2 + \sqrt{4x+6}; \tag{2}$$

$$2x+1 = 4 + 4\sqrt{4x+6} + 4x + 6; \tag{3}$$

$$-2x - 9 = 4\sqrt{4x+6}. \tag{4}$$

Now we square both sides again in order to remove the remaining radical:

(5) $4x^2 + 36x + 81 = 16(4x + 6);$

(6) $4x^2 - 28x - 15 = 0;$

(7) $(2x - 15)(2x + 1) = 0;$

(8) $x = \frac{15}{2}, \text{ or } -\frac{1}{2}.$

We can assert that if (1) is true, then (8) must be true so that the only possible roots of (1) are those given by (8). *Direct substitution shows that neither of these numbers is a root, hence* (1) *has no roots.*

PROBLEMS

1. Show by direct substitution that the two numbers given by the quadratic formula are roots of the equation $ax^2 + bx + c = 0$.

2. Show that one root of the equation $ax^2 + bx + c = 0$ is equal to zero if and only if $c = 0$. Show that both roots are zero if and only if $b = 0$ and $c = 0$.

3. Show that the two roots of the equation $ax^2 + bx + c = 0$ are equal in absolute value but opposite in sign if and only if $b = 0$ and $c \neq 0$.

4. Show that if $x = 2 - 3i$ then $x^2 = 4x - 13$. Is the converse true?

5. Show that if $x = \frac{1}{2}(3 + i)$ then $2x^2 = 6x - 5$. Is the converse true?

6. Show that both $4 - i$ and $3 + i$ are roots of the equation

$$x^2 - 7x + 13 + i = 0.$$

not a rational integral equation

7. Show that $2 + i$ and $3(1 - 2i)$ are roots of the equation

$$x^2 + 5(i - 1)x + 3(4 - 3i) = 0.$$

Why is this not a contradiction of our statement in regard to the roots being conjugate complex numbers?

8. In the equation $ax^2 + bx + c = 0$, let a, b, and c be rational numbers. Is it possible that there may be one rational and one irrational root?

Solve each of the following equations using the quadratic formula:

9. $x^2 - 4x + 20 = 0$.
10. $x^2 - 2x - 4 = 0$.
11. $9x^2 - 3x = 20$.
12. $4w^2 + 12w = 11$.
13. $4y^2 + 13 = 16y$.
14. $x^2 + 34 = 10x$.
15. $36x(1 - x) = 25$.
16. $(3x - 4)(x - 2) = 16$.
17. $(6u + 2)(u - 1) = 14$.
18. $x(3x + 4) + 2x(x - 1) = 17$.

Solve each of the following equations for x in terms of y:

19. $x^2y + 4x - 3y = 0$.
20. $x^2y = 2xy - 1$.
21. $x^2 + 2xy + y^2 - 4 = 0$.
22. $x^2 + 4xy + 3y^2 - 2 = 0$.
23. $x^2 + 2xy + 2y^2 = 4y$.
24. $x^2 - y^2 + 8x - 4y + 16 = 0$.

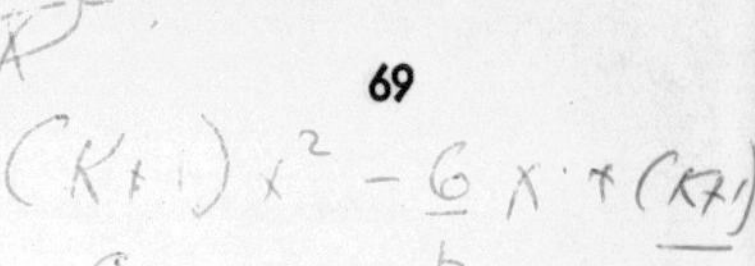

Solve each of the following equations for y in terms of x:

25. $y^2 - 2xy + 2x^2 = 16.$

26. $y^2 - 4x^2 + 3xy - 5x = 1.$

27. $x^2 - 2xy + y^2 + 2y - 3x - 3 = 0.$

28. $y^2 + xy + 6y = 2x^2 - 3x - 3.$

29. For what value or values of k does the quadratic equation

$$3x^2 + kx + 2x = 1 - k$$

have two equal roots? Are the roots real for all real values of k?

In each of the following cases determine the value or values of k, if there are any, for which the given equation has equal roots:

30. $kx^2 + 4kx = x - 4k.$

31. $kx^2 = 6x - x^2 - k - 1.$

32. $(kx + 1)(x - 2) + k = 0.$

33. $\dfrac{3x + k}{2x + 3} = x - k.$

34. $\dfrac{3x + 2k}{3x + 6} = x + k.$

35. Discuss the character of the roots of the equation

$$2k(x^2 + x) + x^2 + 5x + 4 = 0,$$

in which k is a real number.

36. Discuss the character of the roots of the equation

$$x^2 + 2k^2 + k(3x + 2) + x + 1 = 0,$$

in which k is a real number.

37. Show that there is no real number k such that the equation

$$(kx + 1)(x - 2) = k$$

has equal roots. Is there a real value of k for which one root is equal to zero? Is there a real value of k for which the two roots are equal in absolute value but opposite in sign?

38. For what value of k does the equation $(kx + 2)(x - 1) = 2k$ have roots that are equal in absolute value but opposite in sign? For what value of k is one root equal to zero? Is there a real value of k for which the roots are equal?

Solve each of the following equations:

39. $\dfrac{3x + 10}{x} + \dfrac{2x - 1}{3} = 8.$

40. $\dfrac{2(x + 9)}{5} + \dfrac{3}{x} = \dfrac{35}{6}.$

41. $\dfrac{x^2 + 2}{x - 2} + \dfrac{9x - 2}{2} = \dfrac{12x - 18}{x - 2}.$

42. $\dfrac{x + 3}{(x - 2)(x + 1)} = \dfrac{2x}{x - 2}.$

43. $10 + \dfrac{x(x - 6)}{x + 1} = \dfrac{2(2 - x)}{3} + \dfrac{7}{x + 1}.$

44. $\sqrt{4 - 3x} - x = 12.$ **45.** $3x + \sqrt{6x + 37} + 11 = 0.$

46. $2\sqrt{4x + 7} = 7 + \sqrt{2x}.$ **47.** $\sqrt{2x - 3} - \sqrt{x + 2} = 1.$

48. $\sqrt{3x + 8} + \sqrt{9x + 25} = 1.$ **49.** $\sqrt{2x - 2} - \sqrt{x - 3} = \sqrt{x - 15}.$

50. $4 + \sqrt{\dfrac{x}{x - 9}} = 3\left(\dfrac{x - 2}{x - 7}\right).$ **51.** $2 - \sqrt{\dfrac{x}{x - 10}} = \dfrac{x - 4}{x - 14}.$

52. $3\left(\dfrac{x + 2}{3x - 8}\right) = \sqrt{\dfrac{x + 4}{x - 4}}.$ **53.** $\sqrt{\frac{7}{4}x + \frac{1}{4}} + \frac{5}{6}x = \frac{7}{6}.$

54. Find two consecutive odd numbers whose product is 483.

55. Find two consecutive even numbers whose product is 1,088.

56. The difference between two numbers is 3 and their product is 238. Find the numbers.

57. How long is each side of a square if a diagonal is 10 in. longer than a side?

58. The length of a rectangular sheet of tin is equal to four times its width. By cutting a 2-in. square from each corner and turning up the sides, an open rectangular box with volume 224 cu. in. can be made. What are the dimensions of the sheet?

59. The sum of the first n consecutive positive integers, $1, 2, 3, \cdots, n$, is given by the formula $s = \frac{1}{2}n(n + 1)$. What number n of these consecutive integers has a sum of 2,211?

60. The sum of two numbers is 20 and the sum of their squares is 218. What are the numbers?

61. The length of the longer leg of a right triangle is 12 in. If the shorter leg were lengthened by 4 in. the hypotenuse would be lengthened by 2 in. Find the length of the shorter leg.

62. The numerator of a fraction is x and the denominator is $\sqrt{x + 14}$. If the numerator were decreased by 18 and the denominator doubled, the value of the fraction would be decreased by $3\frac{1}{3}$. What is the value of x?

63. Two pipes are used for running water into a tank. When both pipes are used, the tank can be filled in 48 min. The larger pipe alone will fill the tank in 40 min. less time than the smaller one. Find the number of minutes required for each pipe separately to fill the tank.

30. ***Equations of higher degree.*** We have seen that the rational integral equation of first degree, $ax + b = 0$, has precisely one root, namely $x = -b/a$. We have seen that the corresponding equation of second degree, $ax^2 + bx + c = 0$, has the two roots given by the quadratic formula. Furthermore, we have seen that in many cases equations involving fractions or radicals, or both, may lead to equations of one of these types and hence may be solved.

The next step would naturally be to examine the equation

$$ax^3 + bx^2 + cx + d = 0.$$

In view of the results that we obtained for the linear and quadratic equations, we might suspect that this equation has three roots and we might hope to obtain a formula for these roots in terms of the coefficients a, b, c, and d—a formula that would correspond to the quadratic formula for rational integral equations of degree two. On the other hand we might reason as follows:

We found that if a and b are real numbers ($a \neq 0$), then the equation $ax + b = 0$ has a real root; *i.e.*, it has a root within the system of real numbers if its coefficients belong to that system. This is not true for the quadratic equation. In fact we found that if a, b, and c are real numbers ($a \neq 0$), then the equation $ax^2 + bx + c = 0$ may not have any root among the real numbers. We had to invent a new kind of number and extend our number system to include this new kind of number, before we could say that the equation has a root at all. We found that it has precisely two roots within the system of complex numbers. We might then wonder whether or not it may be necessary to invent a still different kind of number in order that the equation

$$ax^3 + bx^2 + cx + d = 0$$

may have a root. The answer turns out to be *no*. The German mathematician K. F. Gauss proved in 1799 the following:

Fundamental theorem of algebra: *Every rational integral equation of degree one or higher, whose coefficients are any complex numbers (real or imaginary), has a root among the complex numbers.*

The proof of this important theorem is not elementary and will not be given here.* The theorem assures us, for example, that the equation

$$2x^3 + (4 + 3i)x^2 - 2x + 7 = 0$$

is satisfied by some number of the form $p + qi$ where p and q are real numbers.

It can be proved, using the above fundamental theorem, that every rational integral equation of degree n has precisely n roots (page 199). Thus, just as every linear equation has one root and every quadratic has two roots, every cubic equation has three

* See, for example, Birkhoff and MacLane, "A Survey of Modern Algebra," New York, The Macmillan Company, 1941, or Uspensky, "Theory of Equations," New York, McGraw-Hill Book Company, Inc., 1948.

roots, etc. It is possible to obtain formulas for the three roots of the equation $ax^3 + bx^2 + cx + d = 0$ in terms of the coefficients a, b, c, and d, but the formulas are complicated and are seldom used. The same is true in the case of the corresponding equation of degree four. It was proved by the Norwegian mathematician N. H. Abel in 1824 that no such formulas can exist for equations of degree higher than four.

There are various methods of finding the roots of a given equation with numerical coefficients, and one of these is usually more convenient than solution by formulas, even for those cases in which the formulas do exist. Some of these methods will be discussed in Chap. XIV.

In this section we shall consider only equations of higher degree that can be solved by simple factoring and the use of the quadratic formula.

Example 1

Find all roots of the equation

$$x^3 + 8 = 0. \tag{1}$$

Solution

We can factor the left side of (1) by using formula VIII on page 34; thus (1) is equivalent to

$$(x + 2)(x^2 - 2x + 4) = 0. \tag{2}$$

This equation is satisfied if and only if we have either

$$x + 2 = 0 \quad \text{or} \quad x^2 - 2x + 4 = 0. \tag{3}$$

The solutions of the two equations in (3) are

$$x = -2 \quad \text{and} \quad x = \frac{2 \pm \sqrt{4 - 16}}{2} = 1 \pm \sqrt{3}\, i.$$

Our given equation (1) then has the three roots:

$$x = -2; \quad x = 1 + \sqrt{3}\, i; \quad x = 1 - \sqrt{3}\, i.$$

The student should check each of these roots by direct substitution in (1). Observe that each of these numbers can be regarded as a cube root of -8. Why?

Example 2

Find all roots of the equation

$$4x^4 + 23x^2 - 72 = 0.$$

Solution

We may regard this as a quadratic equation in which x^2 is the unknown, and we may first solve for x^2 by using the quadratic formula:

$$4(x^2)^2 + 23(x^2) - 72 = 0.$$

$$x^2 = \frac{-23 \pm \sqrt{23^2 - 4(4)(-72)}}{8}$$

$$= \frac{-23 \pm 41}{8} = \frac{9}{4} \quad \text{or} \quad -8.$$

Corresponding to $x^2 = \frac{9}{4}$ we have $x = \pm\frac{3}{2}$, and corresponding to $x^2 = -8$ we have $x = \pm 2\sqrt{2}\,i$. The given equation then has the following four roots: $\frac{3}{2}$, $-\frac{3}{2}$, $2\sqrt{2}\,i$, $-2\sqrt{2}\,i$.

We need not continue to repeat in every case the obvious requirement that the converse theorem must be verified either by substitution or by making sure that every step used is reversible.

The product of several polynomials in x is equal to zero for those values of x, and only those values, for which one or more of the factors is equal to zero. Thus the equation

$$8(x - 1)^2(x^2 - 4)(x + 5)(x - 6) = 0$$

is satisfied if and only if we have $x - 1 = 0$, or $x^2 - 4 = 0$, or $x + 5 = 0$, or $x - 6 = 0$. It has therefore the roots 1, 2, -2, -5, and 6, and no others. The equation is of degree six and in accordance with the statement on page 71 it should have six roots. It is understood in connection with this statement that, whenever a factor appears k times, the corresponding root is to be regarded as k equal roots. In this case 1 is a double root, and the six roots are 1, 1, 2, -2, -5, and 6.

Example 3

Find all roots of the equation

$$3(x^2 + 4)(2x^2 - x - 15) = 0.$$

Solution

The given equation is satisfied if and only if x is a number such that either

$$x^2 + 4 = 0 \qquad \text{or} \qquad 2x^2 - x - 15 = 0.$$

We thus find immediately the four roots $2i$, $-2i$, 3, $-\frac{5}{2}$.

PROBLEMS

1. Solve the equation $x^3 - 1 = 0$ and thus find the three cube roots of 1. Check by cubing each of these numbers.

2. Solve the equation $x^3 + 1 = 0$ and thus find the three cube roots of -1. Check by cubing each of these numbers.

3. Solve the equation $x^4 - 1 = 0$ and thus find the four fourth roots of 1. Check by computing the fourth power of each of these numbers.

Solve each of the following equations:

4. $x^4 - 16 = 0$.
5. $x^4 - 4 = 0$.
6. $z^3 - 8 = 0$.
7. $x^3 + 27 = 0$.
8. $u^3 + 5u = 0$.
9. $x^3 + x^2 + x = 0$.

10. $x^3 - x^2 - 5x + 5 = 0$.
11. $x^3 + x^2 = 9x + 9$.
12. $2x^3 - 3x^2 + 8x - 12 = 0$.
13. $4y^3 - 12y^2 = 27 - 9y$.
14. $\dfrac{x^3 - 1}{x + 1} = 0$.
15. $\dfrac{x^3 + 8}{x + 2} = 0$.

16. Show that -2 is a root of the equation

$$x^3 - 3x^2 - (3 - i)x + 14 + 2i = 0.$$

Show also that $2 + i$ and $3 - i$ are roots.

17. Solve the equation $x^3 + x^2 - x + 15 = 0$, given that $x + 3$ is a factor of the left side.

18. Solve the equation $2x^3 - 7x^2 + 6x + 5 = 0$, given that $2x + 1$ is a factor of the left side.

19. Solve the equation $x^4 - 2x^3 + 4x - 4 = 0$, given that $x^2 - 2$ is a factor of the left side.

20. Solve the equation $2x^4 - 4x^3 + 11x^2 + 18x - 90 = 0$, given that $2x^2 - 9$ is a factor of the left side.

Solve each of the following equations by regarding it as a quadratic equation in x^2:

21. $x^4 + 21x^2 - 100 = 0$.
22. $x^4 + 8x^2 - 9 = 0$.
23. $x^4 - 3x^2 - 108 = 0$.
24. $x^4 + 20x^2 + 96 = 0$.
25. $4x^4 - 35x^2 - 9 = 0$.
26. $4x^4 + 13x^2 - 75 = 0$.

27. The fourth power of a certain number is equal to twice the sum of its square and cube. Find all numbers that satisfy this condition.

Find all roots of each of the following equations:

28. $(3x - 1)(3x^2 - 9)(x^2 + 1) = 0$.
29. $2(x^2 - 5)(x^2 + 4) = 0$.
30. $(x + 4)^2(2x^2 - x - 3) = 0$.
31. $4(x - 1)^3(x^2 + 9)^2 = 0$.
32. $(x^2 + 1)(x^2 - 4x + 13) = 0$.
33. $(3x^2 - 5x - 2)(x^2 - 7x + 10) = 0$.
34. $x^2(x^2 - 4x + 5)(x^3 - 8) = 0$.
35. $x(x^2 - 2x + 2)(x^2 - 6x + 10) = 0$.
36. $2x^2(x^4 - 16)(3x - x^2) = 0$.
37. $9(x - 3)(3x + 2)^2 = 16(3x + 2)(x - 3)^2$.
38. $2(x + 5)(2x + 7)^2 = 7(x + 5)^2(2x + 7)$.
39. $13(2x + 3)(x - 4)^2 = 7(2x + 3)(x - 4)^2$.
40. $(6 - x)(4x + 5)^3 = 6(6 - x)^2(4x + 5)^2$.

CHAPTER IV

FUNCTIONS AND GRAPHS

31. ***Rectangular coordinates.*** In Chap. I we defined directed lines and discussed the correspondence between the real numbers and the points of such a line. In this section we wish to extend this idea.

Let $x'x$ and $y'y$ be two mutually perpendicular directed lines intersecting at a point O (Fig. 5). On $x'x$ the positive direction

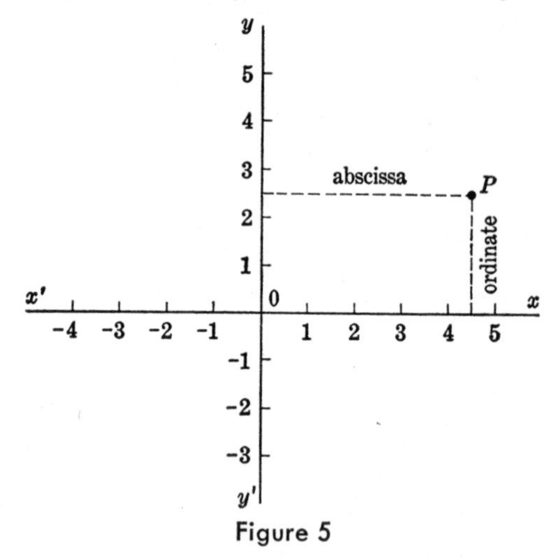

Figure 5

is chosen to the right; on $y'y$ it is upward. This is of course an arbitrary agreement. On each line we choose a unit of measurement and lay off a scale with the zero point at O, as indicated in the figure. The lines $x'x$ and $y'y$ are called the *coordinate axes*, $x'x$ being the x-axis and $y'y$ the y-axis. Point O is called the *origin*.

Consider now any point P in the plane of the coordinate axes. Its directed distance from the y-axis is called its *abscissa* or *x-coordinate*, this being positive if P is to the right of the y-axis and negative if to the left. Its directed distance from the x-axis is called its *ordinate* or *y-coordinate,* this being positive if P is above the x-axis and negative if below. The abscissa and ordinate together are called the *rectangular coordinates* (or simply *coordinates*) of P. In writing down the coordinates of a point, we put them in parentheses, putting the abscissa first. Thus the coordinates of P and Q in Fig. 6 are written, respectively, as $P(7,4)$ and $Q(-4,-5)$.

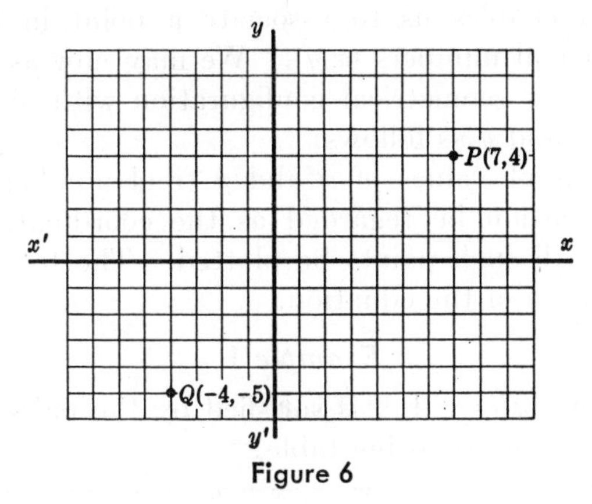

Figure 6

The process of locating and marking a point whose coordinates are given is called *plotting* the point. Plotting is facilitated by the use of paper that is ruled into small squares as in Fig. 6. Such paper is called *rectangular coordinate paper.*

We may assume that to every point of the plane there corresponds one and only one pair of real numbers, and that to every pair of real numbers, taken in a given order, there corresponds a single point. This is stated briefly by saying that there is a one-to-one correspondence between the points of the plane and the pairs of real numbers (x,y).

y
II
I
(−, +)
(+, +)
x'
x
(−, −)
(+, −)
III
IV
y'

Figure 7

The coordinate axes divide the plane into four portions called quadrants. These quadrants are numbered as indicated in Fig. 7.

The points in quadrant I are those whose coordinates have the signs $(+,+)$. Quadrant II consists of the points whose coordinates have the signs $(-,+)$, etc.

32. ***Graph of an equation.*** The coordinate system that we have just discussed enables us to associate a point in the xy-plane with a pair of real numbers (x,y). We may now associate a line or curve or other geometrical configuration with an equation in the variables x and y as follows:

Let each pair of values of x and y (real numbers only) that satisfy the equation be regarded as the coordinates (x,y) of a point, and let all such points be plotted. The totality of these is called the *graph* of the equation.

Example 1

The equation $2y = x + 2$ is satisfied by the pairs of values of x and y given in the following table:

x	-4	-3	-2	-1	0	1	2	3	4	5
y	-1	$-\frac{1}{2}$	0	$\frac{1}{2}$	1	$1\frac{1}{2}$	2	$2\frac{1}{2}$	3	$3\frac{1}{2}$

When these points are plotted, they appear to lie on a straight line (Fig. 8). We may suspect that the coordinates of every point

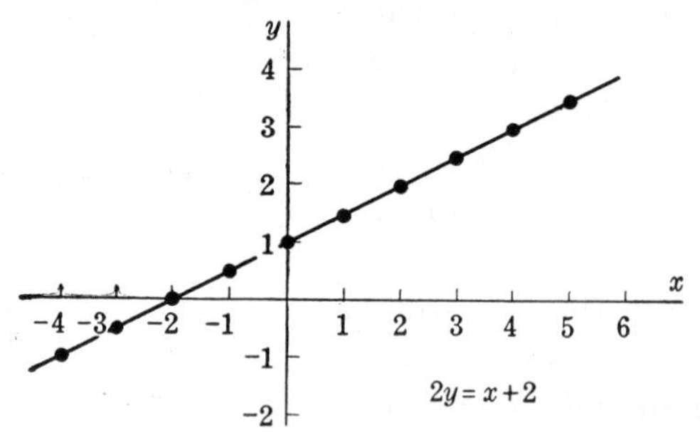

Figure 8

on this line satisfy the equation and that the equation is not satisfied by the coordinates of any point not on the line. This is true, but we shall not give the proof here.

Example 2

The equation $x^2 + y^2 = 25$ is satisfied by the pairs of values of x and y given in the following table:

x	-5	-4	-3	-2	-1	0	1	2	3	4	5
y	0	± 3	± 4	$\pm\sqrt{21}$	$\pm\sqrt{24}$	± 5	$\pm\sqrt{24}$	$\pm\sqrt{21}$	± 4	± 3	0

When these points are plotted, they appear to lie on a circle with center at the origin and radius 5 (Fig. 9). It is easy to prove that

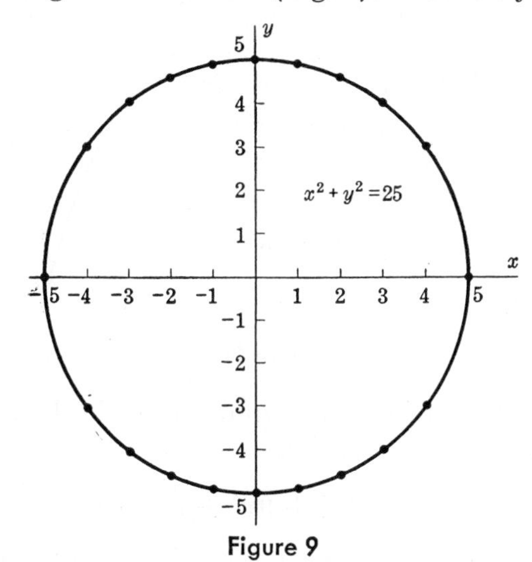

Figure 9

the equation is satisfied by the coordinates of every point on this circle and that it is not satisfied by the coordinates of any point not on the circle. The proof is as follows:

The distance from the origin to any point (x,y) in the plane is equal to $\sqrt{x^2 + y^2}$, as is apparent from Fig. 10. Hence for any point on the circle we have $\sqrt{x^2 + y^2} = 5$ or $x^2 + y^2 = 25$. If (x,y) is a point of the plane that does not lie on this circle, the value of $\sqrt{x^2 + y^2}$ is more or less than 5, and hence $x^2 + y^2$ is not equal to 25.

Figure 10

In making the graph of a given equation, we of course cannot actually plot all the points whose coordinates satisfy the equation. We actually plot as many as are deemed necessary to indicate the graph with sufficient accuracy for the purpose at hand, and then draw a curve through these points. This curve is in general an

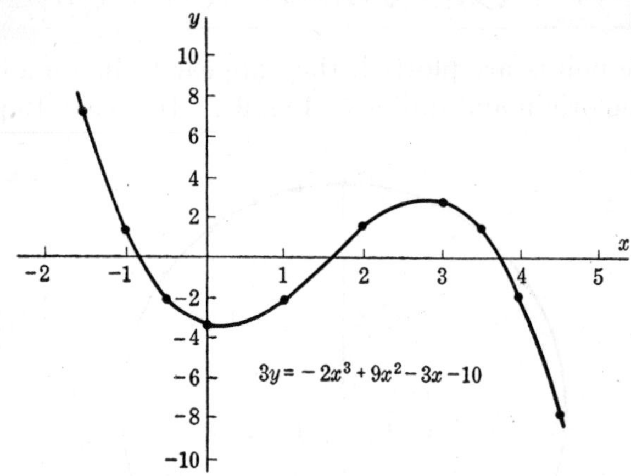

Figure 11

approximation to the actual graph. Thus in Fig. 11 we have drawn the graph of the equation

$$3y = -2x^3 + 9x^2 - 3x - 10$$

in the interval from $x = -1.5$ to $+4.5$ by making the following table:

x	-2	-1.5	-1	-0.5	0	1	2	3	3.5	4	4.5
y	16	7.2	1.3	-2	-3.3	-2	1.3	2.7	1.3	-2	-7.8

After plotting the corresponding points, we have drawn a smooth curve through them. In taking this curve as an approximation to the graph of the equation, we are assuming that if additional points were plotted in this interval they would lie on or near this curve. This is actually true, but we cannot give the proof here.

It is of course possible that a given equation may have no graph, or that the graph may consist of one or more isolated points. Thus the equation $x^2 + y^2 = 0$ is not satisfied by any pair of real numbers except (0,0). The graph then consists of this one point. The equation $2x^2 + 3y^2 + 4 = 0$ has no graph because it is not satisfied by any pair of real numbers.

PROBLEMS

1. Plot each of the following points and compute its distance from the origin: $A(-3,4)$; $B(4,-4)$; $C(-12,-5)$; $D(5,5)$.

2. Plot each of the following points and compute its distance from the origin: $A(-2,-4)$; $B(0,-3)$; $C(3,-4)$; $D(-6,6)$.

3. Draw the triangle whose vertices are the points $A(-2,-5)$; $B(6,-5)$; $C(6,1)$. Compute its area.

4. Draw the triangle whose vertices are the points $A(-3,0)$; $B(6,0)$; $C(8,3)$. Compute its area.

5. The line segment joining $A(-2,0)$ and $B(7,0)$ is one side of a parallelogram and the segment from $A(-2,0)$ to $C(0,4)$ is another side. Find the coordinates of the fourth vertex. Compute the area of the parallelogram.

6. A line is drawn from the origin to the point $P(-7,-4)$. What are the coordinates of its mid-point?

7. For what points is the ordinate equal to 3? Write down an equation that has these points for its graph.

8. For what points is the abscissa equal to -5? Write down an equation that has these points for its graph.

9. For what points is the abscissa equal to the ordinate? Write down an equation that has these points for its graph.

10. Prove that the graph of the equation $xy = 0$ consists of all points on the x-axis and all points on the y-axis.

11. Make the graph of each of the following equations:

(a) $x^2 + xy = 0$; (b) $(x^2 + y^2)(x^2 - y^2) = 0$.

12. Make the graph of each of the following equations:

(a) $x^2 + 4y^2 = 0$; (b) $x^2 - 4y^2 = 0$.

13. What difference exists between the graphs of the two equations

$$x^2 + y^2 - 25 = 0$$

and $(x^2 + y^2)(x^2 + y^2 - 25) = 0$?

In each of the following cases make a table of corresponding values of the two variables, plot the points, and draw a smooth curve through them. Use the specified interval on the x-axis:

14. $x + 2y = 1$; $x = -3$ to $+4$.

15. $3y = 2x + 3$; $x = -3$ to $+4$.

16. $x - 3y = 6$; $x = -2$ to $+8$.

17. $x + 4y = 4$; $x = -2$ to $+6$.

18. $y = x^2 - 4x - 5$; $x = -2$ to $+6$.

19. $2y = 2x^2 + 3x - 14$; $x = -4$ to $+3$.

20. $4y = 8x - x^2$; $x = -1$ to $+9$.

21. $3y + (x - 3)^2 = 0$; $x = -1$ to $+7$.

22. $y + 2x^2 = 5x + 12$; $x = -3$ to $+6$.

23. $y = \frac{1}{4}x^2(6 - x)$; $x = -2$ to $+7$.

24. $6y = x(x + 5)^2$; $x = -7$ to $+2$.

25. $2y = 2x^3 - 3x^2 - 9x + 10$; $x = -3$ to $+3$.

26. $8y = x^3 - 6x^2 + 4x - 24$; $x = -3$ to $+7$.
27. $6y = x^2(x^2 - 16)$; $x = -4.5$ to $+4.5$.
28. $y = \dfrac{x^2(x-6)^2}{8}$; $x = -1$ to $+7$.
29. $y = \dfrac{24}{x^2+4}$; $x = -5$ to $+5$.
30. $y = \dfrac{16x}{x^2+4}$; $x = -6$ to $+6$.

In each of the following cases draw the graph of the given equation after choosing a suitable interval on the x-axis and making a table of corresponding values of x and y:

31. $x^2 + y^2 = 36$.
32. $x^2 + 4y^2 = 16$.
33. $9x^2 + 16y^2 = 144$.
34. $y^2 = 4x$.
35. $y^2 = 8 - 4x$.
36. $x^2 - y^2 = 1$.
37. $x^4 + y^4 = 64$.
38. $x^2 + y^2 = 8x$.
39. $9x^2 + 16(y-3)^2 = 144$.
40. $y^2 - 8y + 8x = 0$.

33. *Functions.* *A variable* **y** *is said to be a* **function** *of a second variable* **x** *if a relation exists between them such that to each of a certain set of values of* **x** *there corresponds one or more values of* **y**. If there is just *one* value of y for each admissible value of x, y is said to be a *single-valued* function of x. The set of all admissible values of x is called the *range of definition* of the function.

For example, if we let x denote any real number and specify that y shall depend upon x in accordance with the relation

$$y = \frac{8x}{x^2+4},$$

then we have defined y as a function of x. It is a single-valued function because to each value of x there corresponds just one value of y. The student is familiar with many examples of such dependence of one variable upon another. Thus the volume of a sphere is a function of its radius, the relation being $V = \frac{4}{3}\pi r^3$. For each (positive) value of r there is a definite value of V. The amount of one's monthly electric bill is a function of the number of kilowatt-hours used. The relation in this case may be fairly complicated, but when the number of kilowatt-hours has been determined by reading the meter, there is a way of arriving at the amount of the bill. This is all we mean by saying that the amount of the bill is a function of the number of kilowatt-hours used.

The equation $3x - 4y = 12$ yields a definite value of y for each value assigned to x, and we therefore may say that it defines y as a function of x. It also, of course, defines x as a function of y; *i.e.*, we may assign values to y and the equation determines corresponding values of x. The variable to which we assign values is called the *independent variable;* the other is called the *dependent variable.*

The statement, "y is a function of x," is abbreviated by writing $y = f(x)$, which is read "y equals f of x." We also use symbols such as $f(x)$, $g(x)$, and $\varphi(x)$ to denote specific functions of x. Thus if we are concerned in a particular problem with the two functions $x^3 - 7x^2 + 16x + 2$ and $5x^2 + 4$, we may find it convenient to let $f(x)$ denote the first and $g(x)$ the second of these functions. We can then refer to the first function, for example, as "the function f" or "the function $f(x)$." This will be simpler than saying "the function $x^3 - 7x^2 + 16x + 2$."

If $f(x)$ denotes a certain function of x, then $f(a)$ denotes the value of $f(x)$ when x has the value a; it is found by substituting a for x in the expression defining $f(x)$. Thus if

$$f(x) = x^2 - 4x + 2,$$

then

$$\begin{aligned} f(0) &= 0^2 - 4(0) + 2 = 2, \\ f(1) &= 1^2 - 4(1) + 2 = -1, \\ f(-2) &= (-2)^2 - 4(-2) + 2 = 14, \\ f(h) &= h^2 - 4h + 2. \end{aligned}$$

34. *Graph of a function.* By the graph of a function $f(x)$ we mean the graph of the equation $y = f(x)$. Thus we may speak of the graph of the function $x^2 + 5$ and of the graph of the equation $y = x^2 + 5$ interchangeably.

It will be recalled that a polynomial in x is an expression, or function of x, that can be written in the form

$$a_0x^n + a_1x^{n-1} + a_2x^{n-2} + \cdots + a_{n-1}x + a_n$$

where the a's are constants, which we assume for the present to be real numbers. The polynomial is of degree n if $a_0 \neq 0$.

The polynomial in x of degree *one* is called a *linear function of* x, and the polynomial of degree *two* is called a *quadratic function.* These functions can be written in the form $ax + b$ and $ax^2 + bx + c$, respectively. It can be proved that the graph of a linear function is a straight line. Figure 12 is an example. The graph of a quadratic function is a curve called a parabola. An example is shown in Fig. 13. In the case of a polynomial of degree three (cubic), the graph may have the general form shown in Fig. 11,

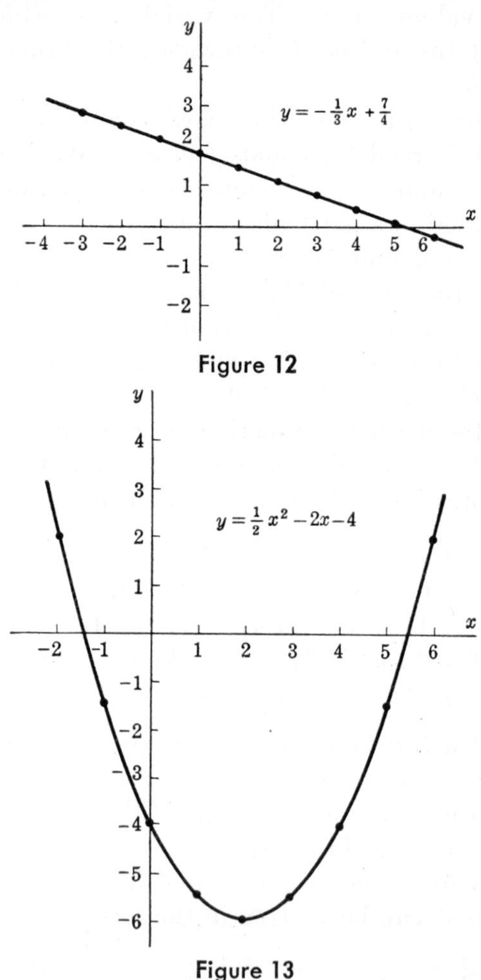

Figure 12

Figure 13

but there are several variations that may occur, depending upon the values of the coefficients. Figure 14 shows the graph of a particular polynomial of degree four.

It can be proved that the graph of a polynomial is a smooth continuous curve without any breaks or sharp corners. The examples shown here are typical.

35. *Graphical solution of an equation in one unknown.* In order to find or locate approximately the real roots of the equation $f(x) = 0$, one may make the graph of the equation $y = f(x)$ and then locate the points on the graph for which $y = 0$. These are of course the points at which the graph crosses, or touches, the x-axis. Thus in Fig. 13 it is evident that we have $y = 0$ for

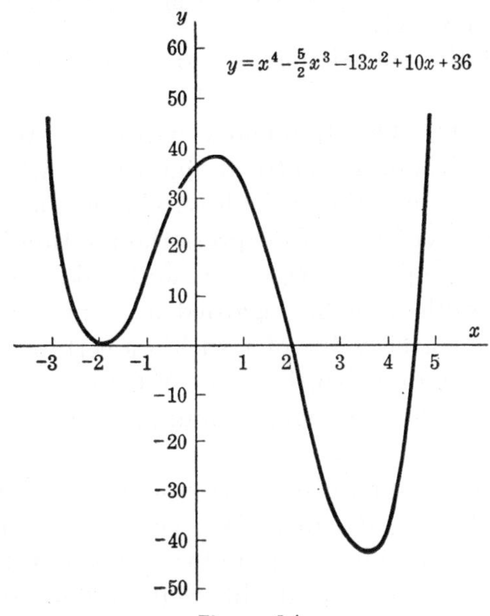

Figure 14

a value of x between -1 and -2 and for another value of x between 5 and 6. This means that the equation

$$\tfrac{1}{2}x^2 - 2x - 4 = 0$$

has two real roots, one of them being between -1 and -2, and the other between 5 and 6. By using the quadratic formula, one can show that the roots are actually $2 + 2\sqrt{3}$ or 5.46, and $2 - 2\sqrt{3}$ or -1.46.

Figure 14 exhibits the fact that the equation

$$x^4 - \tfrac{5}{2}x^3 - 13x^2 + 10x + 36 = 0$$

has both $+2$ and -2 as roots and has another real root between 4 and 5. In this case it happens that the left member can be factored. In fact the equation can be written in the form

$$\tfrac{1}{2}(x + 2)(x + 2)(x - 2)(2x - 9) = 0.$$

From this it follows that the roots are -2, -2, $+2$, and $4\frac{1}{2}$. We call -2 a "double root." The behavior of the graph at such a root is that shown in the figure; *i.e.*, the graph touches but does not cross the x-axis. A method of proving this is indicated in Probs. 7 and 8 of the next set.

The roots of the equation $f(x) = 0$ are called the *zeros* of the function $f(x)$.

36. *Applications.* One frequently wishes to study the way in which some physical or geometrical quantity Q, which is known to be a function of a certain variable x, changes as the value of x changes. One procedure is to express the relation between Q and x in the form of an equation, $Q = f(x)$, and make the corresponding graph. This graph may be regarded as a picture which shows the value of Q for each value of x within a certain range; it shows pictorially the way in which the value of Q increases and decreases as x increases. The following example should make the idea clear:

A manufacturer of an antifreeze compound packages his product in 1-gal. cylindrical tin cans. He recognizes that such a can may be relatively small in radius and rather tall, or larger in radius and shorter. He knows that different combinations of radius and height yield cans requiring different amounts of tin, and he suspects that there should be a particular radius that would result in a can having the smallest amount of surface area and therefore requiring the smallest amount of tin. He knows that when the radius is specified the height is fixed by the requirement that the volume must be 1 gal. (231 cu. in.). Consequently, the surface area S is determined if the radius x is specified; in other words S is a function of x.

In order to get the relation between S and x, we observe that for any radius x(inches) there will be a bottom and a top each having an area of πx^2 sq. in. In addition there is a lateral area of $2\pi xh$ sq. in. where the height h is related to x by the requirement that $\pi x^2 h = 231$ cu. in. Thus we have

$$S = 2\pi x^2 + 2\pi xh, \qquad \text{where } h = \frac{231}{\pi x^2}.$$

After substituting for h its value in terms of x, we have the relation

$$S = 2\pi x^2 + \frac{462}{x}.$$

For any number x of inches in the radius, this equation gives the number S of square inches in the surface area of the 1-gal. can.

The graph of this equation is shown in Fig. 15. It shows pictorially that S is large if x is small, and that S decreases as x increases, until x reaches a value around $3\frac{1}{4}$ in. where S appears

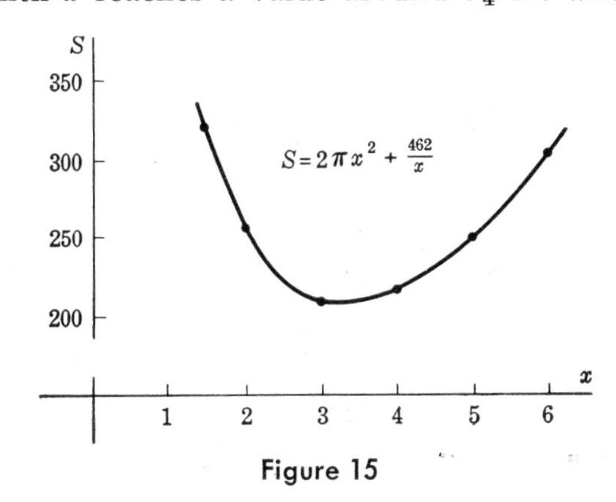

Figure 15

to be smallest. For x greater than this, the surface area is larger again.

Using methods of calculus, one can show that the lowest point on the curve is actually at $x = \sqrt[3]{231/2\pi}$ or 3.325 in.

PROBLEMS

1. Sketch a curve to represent the graph of a single-valued function $f(x)$. What directed distances on the graph are equal to $f(0)$, $f(1)$, $f(2)$, etc.?

2. If $f(x) = 8x/(x^2 + 4)$, find the value of $f(4) - f(2)$. Does the value of $f(x)$ increase or decrease, and by how much, when the value of x changes from 2 to 4?

3. Does the equation $x^2 + y^2 = 25$ define y as a single- or double-valued function of x? To what range of values must x be restricted if y is to be real?

4. The volume and surface area of a sphere are expressed as functions of its radius by the equations $V = \frac{4}{3}\pi r^3$ and $S = 4\pi r^2$, respectively. Express the volume as a function of the surface area.

5. Express the volume of a cube as a function of its surface area.

6. Solve the equations $x^2 - 6x + 5 = 0$, $x^2 - 6x + 9 = 0$, and

$$x^2 - 6x + 10 = 0.$$

Draw the graphs of the left members. Use the identities $x^2 - 6x + 9 = (x - 3)^2$ and $x^2 - 6x + 10 = (x - 3)^2 + 1$ to prove that the graphs of these functions do not cross the x-axis.

7. Sketch the graph of the equation

$$y = (x - 1)(x - 2)(x - 3)^2.$$

By considering the signs of the three factors on the right for $x < 1$, for x between 1 and 2, etc., prove that the graph crosses the x-axis at $x = 1$ and 2 but does not cross it at $x = 3$.

8. Sketch the graph of the equation

$$y = \tfrac{1}{4}(x - 1)(x - 2)^2(x - 3)^3.$$

By considering the signs of the factors on the right for $x < 1$, for x between 1 and 2, etc., show that the graph crosses the x-axis at $x = 1$ and 3 but does not cross it at $x = 2$. In what way does the appearance of the graph near $x = 3$ differ from that near $x = 1$?

In each of the following cases make a graph of the given function $f(x)$, finding as many of its real zeros as you can and locating others between consecutive integers:

9. $2x^2 + 5x - 7$.

10. $-4x^2 + 20x - 25$.

11. $2x^2 - 8x + 6$.

12. $20x - 0.4x^2$.

13. $0.2x^2 + 0.5x - 4$.

14. $\dfrac{x^2}{8} + \dfrac{x}{2} + \dfrac{1}{2}$.

15. $11x^2 - 2x^3$.

16. $x^3 + 6x^2 + 9x$.

17. $2x^3 - 2x^2 - 18x + 9$.

18. $x^3 + x^2 - 12x + 12$.

19. $2 - 3x + \frac{3}{2}x^2 - \frac{1}{4}x^3$.

20. $\frac{1}{4}x^2 (16 - x^2)$.

21. $16x^2 - 3x^4$.

22. $x^3 - x^2 - 13x - 3$.

23. $\dfrac{x}{100}(2x + 5)^2(x - 4)$.

24. $x^5 - 4x^3$.

25. $\dfrac{8x - 16}{x^2 + 4}$.

26. $\dfrac{9 - x^2}{x^2 + 3}$.

27. $\dfrac{x}{x - 3}$.

28. $\dfrac{x^2}{2x - 6}$.

29. $\dfrac{4x^2 - 9}{x^2 - 9}$.

30. $\dfrac{x^3 - 3x^2}{x^2 + 4}$.

31. Make graphs showing how the circumference and area of a circle vary with its radius.

32. A rectangular box with a square base, and without a top, is to be made from lumber. The volume is to be 256 cu. ft. Express its surface area as a function of the length x of an edge of its base. Make the graph and estimate the value of x for which the amount of lumber required will be a minimum.

33. A piece of tin is 21 in. square. A box is to be made from it by cutting a small square from each corner and turning up the sides as indicated in Fig. 16. Express the volume V of the box as a function of the length x of the side of the square cut out. Make the graph and study the way in which V varies with x.

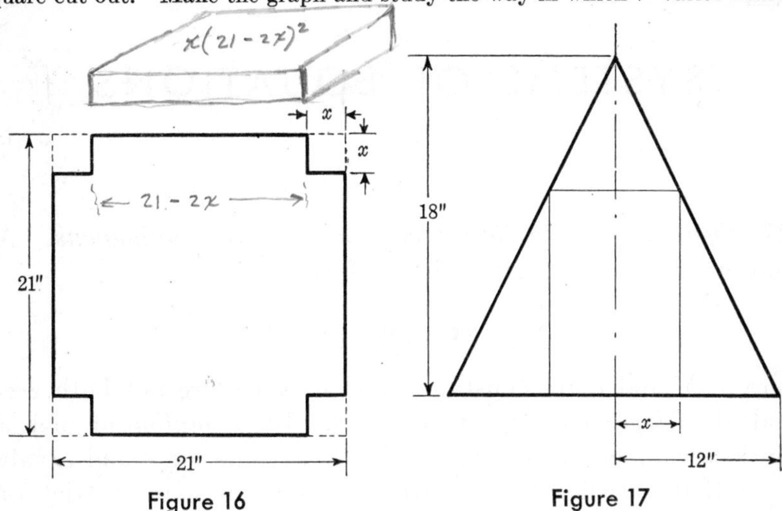

Figure 16 Figure 17

34. A right circular cylinder of radius x is inscribed in a right circular cone of radius 12 in. and height 18 in., as shown in Fig. 17. Express the volume of the cylinder as a function of x and make a graph of the function. What value of x appears to give the largest cylinder?

35. A right circular cone of height x in. is inscribed in a sphere of radius 6 in. Express the volume V of the cone as a function of x. Draw the graph of this function and estimate the value of x for which V is largest.

36. A cylindrical can is to have a volume of 800 cu. in. The material used for the top costs 6 cents per 100 sq. in. and that used for the rest of the can costs 2 cents per 100 sq. in. Express the cost C of material as a function of the radius x of the can. Draw the graph of the function. For about what value of x does the cost appear to be smallest?

CHAPTER V

SYSTEMS OF EQUATIONS

37. *Systems of two linear equations in two unknowns.* An equation that can be written in the form

$$ax + by = c$$

where a, b, and c are constants, and a and b are not both zero, is called a *linear* equation in x and y. By a *solution* of such an equation, we mean a pair of numbers—a value for x and a value for y—that satisfies the equation. We shall here restrict our attention to the case in which a, b, and c are real numbers and shall consider only solutions consisting of pairs of real numbers.

An equation of the above type has infinitely many solutions. For example, the equation $x + 2y = 8$ has the solution $x = 0$ and $y = 4$, or $x = 2$ and $y = 3$, or $x = -8$ and $y = 8$, etc. We may, in fact, choose any value that we please for x, and this number together with $y = \frac{1}{2}(8 - x)$ will constitute a solution. It thus appears that the problem of "solving" a single linear equation in two unknowns is rather trivial. Suppose, however, that we have *two* such equations, for example:

$$x + 2y = 8, \tag{1}$$
$$11x + 4y = -11. \tag{2}$$

We may ask whether or not there are pairs of values of x and y that satisfy *both* of these equations, and how such pairs may be found, if they exist. The answer is obtained by eliminating one of the unknowns and reducing the system to a single equation in the other. There are two methods:

First method. *Elimination by substitution.* Solve one of the equations for y in terms of x or for x in terms of y, whichever is easier, and substitute this result into the other equation. Solve this resulting equation, which involves only one unknown. Substitute the value found for this unknown back into one of the given equations and then solve it for the other unknown. Finally, check the result by substituting the values of both unknowns back into both of the given equations.

In the above example, we solve (1) for x in terms of y:

$$x = 8 - 2y. \tag{3}$$

We substitute this in (2) for x, and then solve the resulting equation:

$$\begin{aligned} 11(8 - 2y) + 4y &= -11, \\ -18y &= -99, \\ y &= 5\tfrac{1}{2}. \end{aligned} \tag{4}$$

Next, we substitute this result for y in (1) or (3) and solve for x:

$$x = 8 - 2(5\tfrac{1}{2}) = -3.$$

Finally, we substitute the pair $x = -3$ and $y = 5\frac{1}{2}$ into (1) and (2) and find that they actually satisfy both equations:

$$\begin{aligned} -3 + 2(5\tfrac{1}{2}) &= 8, \\ 11(-3) + 4(5\tfrac{1}{2}) &= -11. \end{aligned}$$

The reasoning is as follows: If (1) is true, then (3) must be true; if (2) and (3) are true, then (4) must be true. Hence, if there is a number pair that satisfies both (1) and (2), the pair must be $y = 5\frac{1}{2}$ and $x = -3$. It does not follow automatically that if $y = 5\frac{1}{2}$ and $x = -3$, then (1) and (2) must be true. We have shown the truth of this converse theorem by direct substitution.

SECOND METHOD. *Elimination by use of multipliers followed by addition or subtraction.* If we multiply both sides of equation (1) above by -2, we can replace the given equations (1) and (2) with the equivalent pair

$$\begin{aligned} -2x - 4y &= -16, \\ 11x + 4y &= -11, \end{aligned}$$

in which the coefficients of y are numerically equal but opposite in sign. If these relations are true, the sum of the left members will be equal to the sum of the right members. This gives the equation

$$9x = -27$$

from which we get immediately the result $x = -3$. We may substitute this back into (1) or (2) and solve the resulting equation for y, or we may start over and eliminate x between (1) and (2), and thus solve for y in the way that we just found the value of x.

38. ***Graphical interpretation.*** It can be proved that the graph of any equation of the form $ax + by = c$ in which a, b, and c are real, and a and b are not both zero, is a straight line. It is for this reason that the equation is called linear. The proof belongs

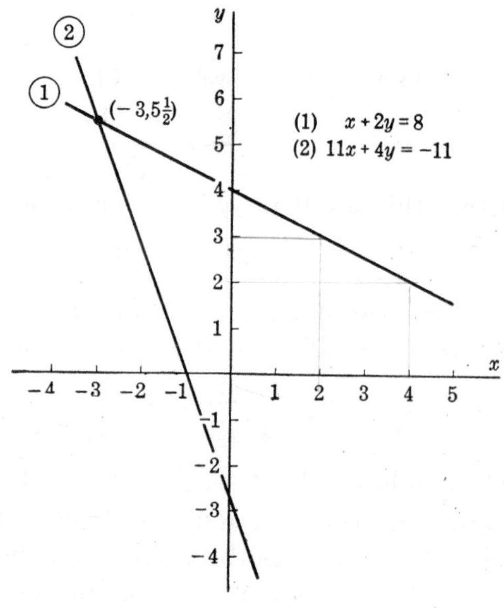

Figure 18

to the subject of analytic geometry and will not be given here, but we shall assume this fact.

The graphs of equations (1) and (2) are the straight lines ① and ②, respectively, in Fig. 18. From our definition of the graph of an equation, we know that equation (1) is satisfied by the coordinates of all points on line ① and by no other pair of real numbers. Similarly, equation (2) is satisfied by the coordinates of all points on line ② and by no other pair of real numbers. It follows that both equations are satisfied by the coordinates of the point of intersection of these lines and by no other pair of real numbers. We may therefore interpret the problem of solving the system of equations (1) and (2) as that of finding the coordinates of the point of intersection of the corresponding graphs.

39. ***The general case.*** Either of the two procedures outlined above may be employed to solve the general system of equations:

$$a_1x + b_1y = c_1, \tag{1}$$
$$a_2x + b_2y = c_2. \tag{2}$$

We can eliminate y by multiplying (1) by b_2 and (2) by b_1, and subtracting. We can similarly eliminate x by multiplying (1) by a_2 and (2) by a_1, and subtracting. The resulting equations are:

$$(a_1b_2 - a_2b_1)x = c_1b_2 - c_2b_1, \tag{3}$$
$$(a_1b_2 - a_2b_1)y = a_1c_2 - a_2c_1. \tag{4}$$

If (1) and (2) are to be true, then (3) and (4) must be true. Now if $a_1b_2 - a_2b_1 \neq 0$, we can divide by this coefficient to obtain the following results:

$$x = \frac{c_1b_2 - c_2b_1}{a_1b_2 - a_2b_1}; \qquad y = \frac{a_1c_2 - a_2c_1}{a_1b_2 - a_2b_1}. \tag{5}$$

It can be shown by direct substitution that the values of x and y given by (5) satisfy both (1) and (2), and we can conclude that the system has this solution and no other if $a_1b_2 - a_2b_1 \neq 0$. In this case the graphs of (1) and (2) are two nonparallel lines intersecting at the point whose coordinates are given by (5).

If $a_1b_2 - a_2b_1 = 0$, then equations (3) and (4) become

$$0 \cdot x = c_1b_2 - c_2b_1,$$
$$0 \cdot y = a_1c_2 - a_2c_1.$$

These equations are not satisfied by any pair of values of x and y if either of the right members is different from zero. Hence the given system has no solution if

$$a_1b_2 - a_2b_1 = 0 \qquad \text{and} \qquad c_1b_2 - c_2b_1 \neq 0;$$

i.e., if

$$\frac{a_1}{a_2} = \frac{b_1}{b_2} \neq \frac{c_1}{c_2}.$$

In this case the equations are said to be *inconsistent*. The graphs are parallel lines that have no point in common.

Finally, if

$$a_1b_2 - a_2b_1 = 0 \quad \text{and} \quad c_1b_2 - c_2b_1 = 0;$$

i.e., if

$$\frac{a_1}{a_2} = \frac{b_1}{b_2} = \frac{c_1}{c_2},$$

then the graphs coincide. In this case any pair of values of x and y that satisfy (1) will satisfy (2). Under this condition, the equations are said to be *dependent.*

Examples

The equations

$$x + 2y = 6,$$
$$2x + 4y = 7,$$

are inconsistent. The equations

$$x + 2y = 6,$$
$$2x + 4y = 12,$$

are dependent. In the first case the graphs are two parallel lines having no point in common. In the second case the graphs are coincident lines having all their points in common.

PROBLEMS

1. Describe the graph of the equation $ax + by = c$ for the case in which $a = 0$ but b and c are different from zero; for the case in which $c = 0$ but a and b are different from zero; for the case in which $a = 0$ and $c = 0$ but $b \neq 0$. In each case sketch the graph of a typical example.

2. By direct substitution we find that $x = 2$ and $y = 3$ satisfies the pair of equations (1) $4x - y = 5$ and (2) $x + 4y = 14$. Which of the following propositions is thereby proved?

(*a*) If $x = 2$ and $y = 3$, then (1) and (2) are true.
(*b*) If (1) and (2) are true, then $x = 2$ and $y = 3$.
Prove the other proposition.

3. By direct substitution, we find that $x = 1$ and $y = 5$ satisfies the pair of equations (1) $3x + y = 8$ and (2) $6x + 2y = 16$. Which of the following propositions is thereby proved?

(*a*) If $x = 1$ and $y = 5$, then (1) and (2) are true.
(*b*) If (1) and (2) are true, then $x = 1$ and $y = 5$.
Is the other proposition true?

4. Prove the following propositions concerning the pair of linear equations

$$a_1x + b_1y = 0,$$
$$a_2x + b_2y = 0.$$

(*a*) The equations have the solution $x = 0, y = 0$.
(*b*) The equations have no solution other than $x = 0, y = 0$, if $a_1b_2 - a_2b_1 \neq 0$.

Solve each of the following systems if it has a unique solution; otherwise characterize it as inconsistent or dependent. In each case make the corresponding graph:

5. $7x - 3y = 36,$
$2x + 5y = 22.$

6. $3x - y = 1,$
$5x + 2y = -13.$

7. $4x + 2y = -7,$
$5x - y = -14.$

8. $3x - y = 6,$
$9x - \frac{1}{2}y = 13.$

9. $4x - 3y = 20,$
$2x - \frac{3}{2}y = 10.$

10. $9x - 3y = 14,$
$x - y = 6.$

11. $\dfrac{3x}{7} + \dfrac{9y}{4} = 10,$
$\dfrac{5x}{2} - \dfrac{4y}{3} = \dfrac{1}{2}.$

12. $\dfrac{4u}{3} + \dfrac{3v}{2} = \dfrac{1}{6},$
$\dfrac{-u}{2} + \dfrac{v}{4} = \dfrac{11}{6}.$

Solve each of the following systems:

13. $2(x - 1) + 2y = 4x,$
$3(2x - 3) - y = 15.$

14. $2(3 - 2x) + 3(y - 1) = 22,$
$4x - 3(2 - y) = \dfrac{14x}{5}.$

15. $\frac{2}{3}(4x - \frac{3}{2}y) = \frac{5}{2}(2x + y + 7),$
$9x + 2(y - \frac{7}{4}) = y + 2(2x - 1).$

16. $\frac{3}{2}(x + 4) - 4 + \frac{2}{3}y = 5x - 16,$
$\frac{5}{4}x + 6(\frac{2}{9}y + 3) = \frac{2}{3}(x + y).$

17. $\dfrac{7x + 6y}{2x + y + 5} = 6,$
$\dfrac{2y - x + 4}{x + y - 1} = 8.$

18. $\dfrac{7x - 5y}{x + y} = 1.8,$
$\dfrac{8x - y}{13x - 12y - 7} = 4.1.$

Solve the following systems by making use of the fact that the given equations are linear in $1/x$ and $1/y$:

19. $\dfrac{1}{x} + \dfrac{1}{y} = \dfrac{1}{4},$
$\dfrac{1}{x} - \dfrac{1}{y} = \dfrac{3}{8}.$

20. $\dfrac{3}{2x} + \dfrac{6}{y} = 12,$
$\dfrac{1}{x} + \dfrac{4}{y} = 8.$

21. $\dfrac{8}{3x} + \dfrac{2}{5y} = \dfrac{5}{3},$
$\dfrac{24}{x} + \dfrac{5}{4y} = \dfrac{391}{48}.$

22. $\dfrac{7}{2x} = 5\left(\dfrac{1}{y} - \dfrac{1}{2}\right),$
$\dfrac{4}{x} + \dfrac{5}{y} = \dfrac{25}{7}.$

23. It is desired to make 100 lb. of a 50 per cent acid solution by diluting a 60 per cent solution with a 20 per cent solution. How many pounds of each should be used?

24. The sum of the digits in a two-place number is 11. If the digits are reversed, the number is decreased by 45. Find the number. HINT: Let x denote the digit in the tens place and y the digit in the units place. Then the number is $10x + y$. Thus $27 = 10(2) + 7$.

25. The quotient of two numbers is equal to $\frac{13}{3}$. If the numerator and the denominator are each increased by 18, the quotient becomes equal to $\frac{7}{3}$. Find the numbers.

26. A boat can go 36 miles downstream in 1 hr. 48 min., but requires 4 hr. for the return trip upstream. Assuming that the boat travels at a constant speed and that the stream also has a constant speed, find the speed of the stream and that of the boat in still water. HINT: Assume that if the speed of the boat in still water is x m.p.h. and that of the stream is y m.p.h., then the boat can travel at $x + y$ m.p.h. downstream and at $x - y$ m.p.h. upstream.

27. A swimming pool can be filled by either or both of two pipes which feed into it. When both pipes are used, it can be filled in $4\frac{4}{7}$ hr. If both pipes run for 3 hr. and one is then shut off, the other finishes filling the pool in another $3\frac{2}{3}$ hr. How long would it take each pipe alone to fill the pool? HINT: Let x = number of hours required for pipe A to fill the pool; then $1/x$ is the fractional part of the pool that is filled by this pipe in 1 hr. Use y in the same way for pipe B. First solve for $1/x$ and $1/y$ as in Probs. 19 to 22 above.

28. Two given points (x_1,y_1) and (x_2,y_2) determine a line lying in the xy-plane. It can be proved that if the line is not parallel to the y-axis, it has an equation of the form $y = mx + b$. Find the values of m and b, and thus find the equation of the line, determined by the points (2,1) and (11,4). HINT: In the equation $y = mx + b$, we must determine m and b so that when $x = 2$, $y = 1$ and when $x = 11$, $y = 4$. Hence we must have

$$1 = 2m + b,$$
$$4 = 11m + b.$$

These equations may be solved for m and b.

29. Using the method indicated in Prob. 28, find the equation of the line determined by the given pair of points:

(*a*) $(-1,1)$; $(8,4)$. (*b*) $(-2,4)$; $(6,-4)$.
(*c*) $(-2,-6)$; $(8,-4)$. (*d*) $(6,4)$; $(-4,-\frac{8}{3})$.
(*e*) $(-5,0)$; $(0,2)$. (*f*) $(2,-9)$; $(-6,-1)$.

30. Show that the system of three equations

$$4x - 3y = 10, \qquad 5y - x = 6, \qquad 3x + 2y = 16,$$

is consistent by showing that the solution of two of them satisfies the other one. What is the graphical interpretation? Discuss the situation that results when the number 16 in the third equation is replaced by 8.

31. Show that the system of three equations

$$3y - 2x = 12, \qquad y - x = 5, \qquad 3y + 10x = -24,$$

is consistent, by showing that the solution of two of them satisfies the other one. Show the graphical interpretation. Discuss the situation that results if -24 is replaced by -30 in the third equation.

32. Show that the system of three equations

$$5y = 4(x - 6), \qquad 4x + y = 0, \qquad 7y + 4x = -24,$$

is consistent, by showing that the solution of two of them satisfies the other one. Show the graphical interpretation. Discuss the situation that results if -24 is replaced by -28 in the third equation.

40. ***Systems of three linear equations in three unknowns.*** An equation that can be written in the form

$$ax + by + cz = d$$

where a, b, c, and d are constants, and a, b, and c are not all zero, is called a *linear* equation in x, y, and z. By a *solution* of such an equation, we mean a set of three numbers—a value for each of the variables x, y, and z—that satisfies the equation. As in the corresponding case of an equation in two unknowns, we restrict our attention to the case in which a, b, c, and d are real, and consider only solutions consisting of real numbers.

A single equation of the above type has infinitely many solutions. For example, in the equation

$$4x + 8y + z = 2,$$

we may choose any values we please for x and y, and these numbers together with $z = 2 - 8y - 4x$ will constitute a solution. If we have two such equations, there is still, except in special cases, an infinite number of solutions. Thus in the equations

$$\begin{aligned} 4x + 8y + z &= 2, \\ x + 7y - 3z &= -14, \end{aligned}$$

we may assign any value we please to one of the unknowns, say z, and the two equations may then be solved to find the corresponding values of x and y.

If we have three such equations, there is, except in special cases, just one solution. Thus the three equations

$$\begin{aligned} &(1) \qquad & 4x + 8y + z &= 2, \\ &(2) \qquad & x + 7y - 3z &= -14, \\ &(3) \qquad & 2x - 3y + 2z &= 3, \end{aligned}$$

are satisfied by $x = -3$, $y = 1$, and $z = 6$, and by no other set of numbers. We may arrive at this result as follows:

We may eliminate z between (1) and (2) by multiplying (1) by 3, and adding. Similarly, we may eliminate z between (1) and (3) by multiplying (1) by 2, and subtracting. This gives the two equations

$$13x + 31y = -8, \tag{4}$$
$$6x + 19y = 1. \tag{5}$$

This system can of course be solved by the usual methods, the solution being

$$x = -3, \qquad y = 1. \tag{6}$$

If we substitute these results into (1), we find that

$$z = 6. \tag{7}$$

We may reason as follows: If (1), (2), and (3) are true, then (4) and (5) must be true. If (4) and (5) are true, then (6) must be true. If (6) and (1) are true, then (7) must be true. We thus know that (1), (2), and (3) cannot be true for any values of x, y, and z other than $x = -3$, $y = 1$, and $z = 6$. That the given equations are true if x, y, and z have these values can be shown by direct substitution.

The general case of three linear equations in three unknowns will be discussed in Chap. XV (page 243). We shall find that under certain special conditions the system may have no solution, and that under other special conditions it may have many solutions. In the first of these cases, the equations are said to be *inconsistent*, and in the second case they are said to be *dependent*. For example, the equations

$$\begin{aligned} x + 2y + z &= 3, \\ 3x - y + 2z &= 8, \\ 2x + 4y + 2z &= 4, \end{aligned}$$

are obviously inconsistent, for any set of values of x, y, and z that satisfies the first equation cannot possibly satisfy the third. On the other hand, the system

$$\begin{aligned} x + 2y + z &= 3, \\ 2x + 4y + 2z &= 6, \\ 3x + 6y + 3z &= 9, \end{aligned}$$

is dependent because any set of values of x, y, and z that satisfies the first equation will certainly satisfy both of the others.

41. ***Systems involving nonlinear equations.*** A pair of nonlinear equations in two unknowns can be attacked by the method used on linear equations; *i.e.*, we may try to eliminate one of the unknowns between the two equations and then try to solve the resulting equation. If one of the equations is linear, we may accomplish the elimination by solving this equation for y in terms of x, or for x in terms of y, and substituting into the other equation.

Example 1

Solve the system

$$x^2 + y^2 = 10x, \tag{1}$$
$$4y = 3x - 8. \tag{2}$$

Solution

We shall solve (2) for y in terms of x, substitute this for y in (1) and then proceed to solve the resulting equation for x. Thus, from (2) we have

$$y = \frac{3x - 8}{4}. \tag{3}$$

Substituting this into (1) and simplifying, we get

$$x^2 + \left(\frac{3x - 8}{4}\right)^2 = 10x,$$
$$x^2 + \frac{9x^2 - 48x + 64}{16} = 10x,$$
$$16x^2 + 9x^2 - 48x + 64 = 160x,$$

or, finally,

$$25x^2 - 208x + 64 = 0. \tag{4}$$

This quadratic equation can be solved by using the quadratic formula or by factoring:

$$(25x - 8)(x - 8) = 0,$$
$$x = \tfrac{8}{25} = 0.32 \quad \text{or} \quad x = 8. \tag{5}$$

If we substitute 0.32 for x in (2), we get $y = -1.76$; similarly, corresponding to $x = 8$, we find $y = 4$.

We may reason that if (2) is true, then (3) must be true. If (1) and (3) are true, then (4) must be true, etc. We are thus assured that if (1) and (2) are to be true then we must have either

(6) $x = 0.32$ and $y = -1.76$ or $x = 8$ and $y = 4$.

As we have noted in previous cases, it does not follow automatically that the number pairs given in (6) will satisfy (1) and (2). We have shown that if (1) and (2) are to be true, we must have the conditions given by (6), but we have not shown that if (6) is

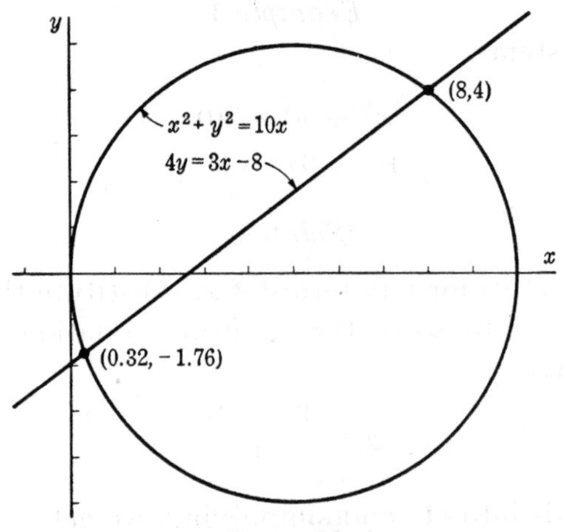

Figure 19

true, then (1) and (2) must be true. This is most easily done by direct substitution of the values given by (6) into (1) and (2). It turns out that both pairs given by (6) satisfy both (1) and (2).

A geometrical interpretation is given by Fig. 19. Equation (1) is satisfied by the coordinates of all points on the circle which is its graph, and by no other pair of real numbers. Equation (2) is satisfied by the coordinates of every point on the straight line which is its graph, and by no other pair of real numbers. *Both* equations are satisfied by the coordinates of those points that lie on both the circle and the line, and by no other pair of real numbers. We may therefore interpret the problem of finding the pairs of real numbers that satisfy both equations as that of finding the points of intersection of the corresponding graphs.

If neither equation is linear, we may still employ the same general procedure. It may of course lead us to an equation that we cannot solve by the methods that we now have.

Example 2

Solve the system

$$y = \frac{x^2}{4}, \qquad y = \frac{9x}{x^2 + 3}.$$

Solution

We may eliminate y by equating the two right-hand members. We then have the equation

$$\frac{x^2}{4} = \frac{9x}{x^2 + 3}$$

which can be reduced to

$$x^4 + 3x^2 - 36x = 0.$$

It is obvious that x is a factor of the left member, and it turns out that $x - 3$ is also a factor. The equation can be written in the form

$$x(x - 3)(x^2 + 3x + 12) = 0.$$

The only real roots are $x = 0$ and $x = 3$, because the roots of the remaining quadratic are imaginary. Corresponding to $x = 0$, we get $y = 0$; corresponding to $x = 3$, we find $y = 2\frac{1}{4}$. Both of these pairs satisfy both equations, so we have the two real solutions:

$$x = 0 \text{ and } y = 0 \qquad \text{or} \qquad x = 3 \text{ and } y = 2\tfrac{1}{4}.$$

An examination of the steps used assures us that there can be no other pair of real numbers satisfying both equations.

The corresponding graphs are shown in Fig. 20. The two curves intersect at (0,0) and at $(3,2\frac{1}{4})$.

It has probably occurred to the student by now that a way of finding (or at least approximating) the real solutions would be to draw the graphs and pick off the coordinates of the points of intersection. This method is frequently used. In example 2 we have shown that if the two given equations are to be true, then x must be a number such that

$$x^4 + 3x^2 - 36x = 0.$$

If we had not been able to factor the left member of this equation, we might have tried to find its real roots, at least approximately, by making the graph of this polynomial. Other methods are discussed in Chap. XIV.

The graphs of course exhibit only the pairs of *real* numbers that satisfy the equations and give no interpretation of solutions involving imaginary numbers. Thus if we should replace the line $4y = 3x - 8$ in Fig. 19 by some line that has no point in

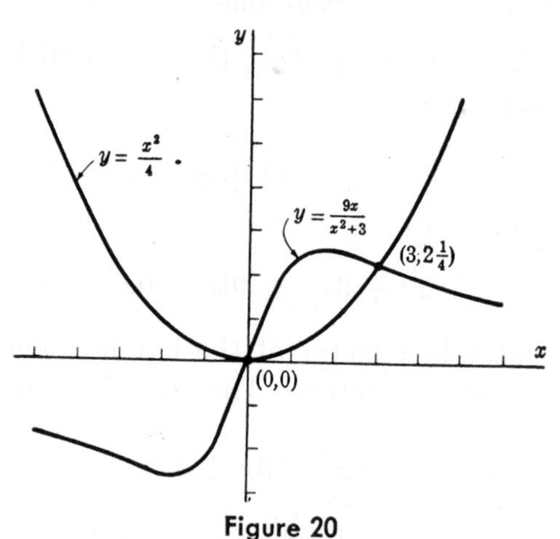

Figure 20

common with the circle, we would find that the corresponding equations have no common solution in the field of real numbers.

PROBLEMS

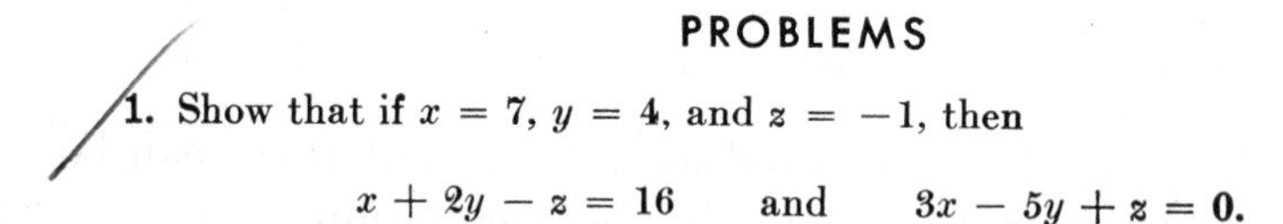

1. Show that if $x = 7$, $y = 4$, and $z = -1$, then

$$x + 2y - z = 16 \quad \text{and} \quad 3x - 5y + z = 0.$$

Write down the converse and prove that it is not true.

2. Show that if $x = -2$, $y = 5$, and $z = 4$, then

$$3x + 2y - z = 0,\ x + y + z = 7, \text{ and } 5x + 8y - z = 26.$$

State and prove the converse.

3. Show that if $x = 1$, $y = 1$, and $z = 1$, then

$$3x - 2y - z = 0,\ x - 2y + z = 0, \text{ and } 2x - y - z = 0.$$

State the converse and show that it is not true by finding other solutions of the system. What condition is both necessary and sufficient in order that the three equations may be satisfied?

Solve each of the following systems if it has a unique solution; otherwise, classify the system as inconsistent or dependent:

4. $x + 6y - 2z = 7,$
$5x + y + 7z = 25,$
$x - 3y + z = 13.$

5. $2x - 8y - z = 11,$
$3x + 7y - z = 0,$
$x - 3y - 5z = -18.$

6. $4x - 3y + 2z = -10,$
$2x + 6y - 3z = 1,$
$12x + y + 7z = 16.$

7. $3x - 6y + z = -9,$
$-6x + 12y - 5z = 21,$
$x + 2y + 6z = -6.$

8. $2x + 5y - 2z = 10,$
$-x - y + 4z = -14,$
$7x + 12y - 10z = 32.$

9. $5u + 11v - 2w = -12,$
$-3u + v + 4w = 10,$
$4u - 3v - w = -9.$

10. $x - 3y + z = 4,$
$3x - 9y + 3z = 6,$
$x + 5y - z = 2.$

11. $x + y + 2z = 4,$
$3x - y + z = 5,$
$2x + 2y + 4z = 8.$

12. $3x + 5y = 1,$
$x + 7z = -2,$
$y - 3z = -1.$

13. $2u - w = 0,$
$w + 6v = -2,$
$4u + 3v = 8.$

14. Show that the following system of four equations in three unknowns has a unique solution:

$$\begin{aligned} 2x + 5y + 2z &= 4, \\ x - y + z &= 9, \\ 3x + 7y + 3z &= 7, \\ 6x - y - 5z &= 11. \end{aligned}$$

15. Show that the following system has no solution:

$$\begin{aligned} 2x + 2y + 5z &= 5, \\ 3x - 5y - 12z &= 12, \\ 8x + 6y - 7z &= -1, \\ 5x + 4y + 4z &= 5. \end{aligned}$$

Solve each of the following systems:

16. $2w + 3x - y + 4z = -7,$
$w - 3x - 4y - 8z = -2,$
$w + x + y + z = 6,$
$3w + 5x + 2y + 4z = 8.$

17. $4s + 5t - 2u - 2v = 6,$
$s - 3t + 2u + v = 4,$
$s + t - u + v = 10,$
$2s + 3t + u - 3v = -11.$

18. Show that the equations

$$\begin{aligned} x + y + z &= 4, \\ x + 2y + 2z &= 8, \\ x + 3y + 3z &= 12, \end{aligned}$$

have the solution $x = 0$, $y + z = 4$.

19. Show that the two equations

$$\begin{aligned} 2x + 2y + z &= 12, \\ x + y + 2z &= 12, \end{aligned}$$

have the solution $z = 4$, $x + y = 4$.

20. Find three numbers such that the sum of the first and second is 88, the sum of the second and third is 83, and the sum of the first and third is 44.

21. A reservoir has one supply pipe A through which it is filled, and two drain pipes, B and C. With pipes A and B open the reservoir can be filled in 12 days. With A and C open it can be filled in 8 days. When A is closed and B and C are open, it can be emptied in 6 days. How many days would be required to fill the reservoir through A if B and C were closed? (Assume a constant rate of flow for all the pipes.)

In each of the following cases find all solutions of the given system, including those that involve imaginary numbers. Draw the corresponding graphs:

22. $y = \frac{1}{4}x^2 - 1,$
$2y = x + 2.$

23. $y = 6x - x^2,$
$3x + 2y + 8 = 0.$

24. $4y + x^2 = 8x + 9,$
$x + 4y = 9.$

25. $y = \frac{1}{4}(x + 2)^2,$
$x + 2y = 0.$

26. $2y = 9 - x^2,$
$x + y = 5.$

27. $2y = x^2 + 8,$
$y = x - 1.$

28. $x^2 + 2y = 3x + 10,$
$y = 2x + 6.$

29. $y + 8 = x^2 - 2x,$
$y + 2x + 8 = 0.$

30. $8y = x^2,$
$y^2 = x.$

31. $y = 10 + 3x - x^2,$
$y + 5x = x^2.$

32. $y^2 = x + 4,$
$y = (x + 4)^2.$

33. $x^2 + y^2 = 100,$
$2y = x + 10.$

34. $x^2 + y^2 = 8y,$
$y = x + 4.$

35. $x^2 + y^2 = 40,$
$x^2 = 6y + 24.$

36. $2x^2 + y^2 = 36,$
$x^2 + y^2 = 20.$

37. $9x^2 + 25y^2 = 225,$
$3x + 9 = y^2.$

38. $y = \dfrac{12}{x^2 + 2},$
$y = 2x^2 - 6.$

39. $8x^2 + 3y^2 = 120,$
$2x^2 = 3y + 6.$

40. The area of a rectangle is 144 sq. in., and its perimeter is 50 in. What are its dimensions?

41. The perimeter of a rectangle is 136 in., and the length of a diagonal is 52 in. What are its dimensions?

42. The product of two positive numbers is equal to $2\frac{1}{2}$ times their sum. The difference of their squares is equal to 12 times their sum. What are the numbers?

43. The sum of the squares of two positive numbers is 208. The sum of the cubes of the numbers is equal to 112 times the sum of the numbers. Find the numbers.

44. The area of a rectangular sheet of tin is 384 sq. in. If we cut a 2-in. square from each corner and fold up the sides, an open box is formed having a volume of 480 cu. in. What are the dimensions of the sheet?

45. If a train were to increase its average speed between two towns by 12 m.p.h., the time required would be 1 hr. less. On the other hand if it were to decrease its average speed by 3 m.p.h., the time required would be $\frac{1}{3}$ hr. more. What is its average speed and how far apart are the towns?

Solve each of the following systems:

46. $4xy + x^2 - y^2 = 36,$
$3xy - y^2 = 18.$

47. $x^2 + 4xy - 3y^2 = 12y,$
$2x + 5y = 2.$

48. $x^2 + 4y^2 = 52,$
$x^2 - xy + 2y^2 = 22.$

49. $2x^2 + xy + y^2 = 28,$
$xy + 3y^2 = 4.$

50. $x^2 + y^2 + z^2 = 14,$
$x + y + z = 4,$
$2x + y = 3.$

51. $3x^2 + 4y^2 - 4x + z = 21,$
$x - 6y - 8z = 10,$
$x + 2y + 2z = 0.$

CHAPTER VI

INEQUALITIES

42. ***Introduction.*** In our study of real numbers we agreed to say, regarding two real numbers a and b, that a *is greater than* $b (a > b)$ if $a - b$ is a positive number. Thus $8 > 5$, $0 > -3$, and $-4 > -5$. The statement that b is less than $a (b < a)$ is equivalent to $a > b$.

If the real numbers are associated with the points of a directed line on which the positive direction is to the right (*e.g.*, the x-axis), then $a > b$ means that a lies to the right of b on the line.

We may state that x lies between 2 and 4 by writing $2 < x < 4$ which says that x is greater than 2 but less than 4. We use the symbols $\leqq$ and $\geqq$ to mean "less than or equal to" and "greater than or equal to," respectively. We thus indicate the points in the interval from 2 to 4, including the left end point of this interval, by writing $2 \leqq x < 4$.

Since the ideas of *greater than* and *less than* apply only to real numbers, it will be understood that all the letters used in this chapter to represent numbers stand for real numbers.

43. ***Inequalities.*** We defined an equation as a statement that two numbers, or two expressions representing numbers, are equal. Correspondingly we define an inequality as a statement that one real number, or an expression representing a real number, is greater than or less than another such number or expression.

If the inequality involves only constants, or if it is true for all admissible values of the variables involved, it is called an *absolute inequality*. Thus $8 > 3$ is an absolute inequality; so also is $x^2 + 4 > 0$, it being understood that x is a real number, for x^2 is either 0 or a positive number, and consequently $x^2 + 4$ is positive, for all real values of x. Absolute inequalities correspond to identities in the case of equations.

If the inequality fails to be true for one or more admissible values of the variable or variables involved, it is called a *conditional inequality.* Thus $3x > 6$ is a conditional inequality; it is true for all values of x greater than 2 but is not true for x equal to or less than 2. To *solve* a conditional inequality means to find those values of the variable or variables for which it is true.

Before we can solve inequalities successfully, we must know something about their fundamental properties. We must know, for example, whether or not we may add the same number or expression to both sides, and whether or not we may multiply or divide both sides by the same number. It seems plausible that we may solve the inequality $3x > 6$ simply by dividing both sides by 3 to get $x > 2$; but if we divide both sides of the inequality $-3x < 9$ by -3, and write $x < -3$, our result is obviously wrong because for $x < -3$, $-3x > 9$. We suspect then that the inequality symbol should be reversed when we divide both sides by a negative number.

In the next section we shall state several theorems covering the fundamental properties of inequalities. We shall have use for the following terminology: Two inequalities are said to have the *same sense* if their inequality symbols point in the same direction; they are said to have *opposite senses* if their inequality symbols point in opposite directions. Thus the two inequalities $2x < 6$ and $x < 3$ have the same sense, while $3 < 5$ and $7 > 2$ have opposite senses.

44. *Properties of inequalities.* We shall give proofs for only the first two of the following theorems:

***Theorem* I.** *The sense of an inequality is not changed if the same real number, or expression representing a real number, is added to both sides; i.e., if* $\mathbf{a} > \mathbf{b}$*, then* $\mathbf{a} + \mathbf{c} > \mathbf{b} + \mathbf{c}$.

Proof: Assume that $a > b$. Then $a - b = p$ where p denotes a positive number. Now from the fact that

$$(a + c) - (b + c) = a - b$$

it follows that

$$(a + c) - (b + c) = p.$$

But this means that $(a + c) > (b + c)$, which was to be proved.

Since adding a negative number is equivalent to subtracting a positive number, this theorem also covers the case of *subtracting* the same number or expression from both sides. For example, if

$$5x - 3 > 2x + 15,$$

we may add 3 to both sides and subtract $2x$ from both sides, and thus conclude that

$$3x > 18.$$

***Theorem* II.** *The sense of an inequality is not changed if both sides are multiplied by the same positive number, or expression representing a positive number; i.e., if* $a > b$, *then* $ac > bc$ *if* c *is positive.*

Proof: Assume that $a > b$. Then $a - b = p$ where p denotes a positive number. Now from the fact that

$$ac - bc = (a - b)c$$

it follows that

$$ac - bc = pc.$$

If c is any positive number, pc is positive, and this means that $ac > bc$ which was to be proved.

Since dividing by a positive number c is equivalent to multiplying by $1/c$, this theorem also covers the case of *dividing* both sides by the same positive number. Thus if we have $3x > 18$, we may divide both sides by 3, or multiply both sides by $\frac{1}{3}$, to conclude that $x > 6$.

***Theorem* III.** *The sense of an inequality is reversed if both sides are multiplied by the same negative number, or expression representing a negative number; i.e., if* $a > b$, *then* $ac < bc$ *if* c *is negative.*

The proof is left as an exercise for the student. The theorem of course also covers the case of dividing both sides by the same negative number.

***Theorem* IV.** *The sense of an inequality whose members are positive numbers is not changed by taking the same positive power or the same positive root of both sides, it being understood in the latter case that the principal root is taken; i.e., if* a *and* b *are positive and* $a > b$, *then* $a^n > b^n$ *and* $\sqrt[n]{a} > \sqrt[n]{b}$, *if* n *is positive.*

45. *Solving conditional inequalities. Graphical interpretation.* The following examples illustrate methods that may be used in solving conditional inequalities.

Example 1

Find all values of x for which

$$\frac{7}{3}x - 4 > \frac{1}{2}(3x + 7). \tag{1}$$

Solution

First multiply both sides by 6 in order to remove the denominators; we then have

$$14x - 24 > 9x + 21. \tag{2}$$

If we now subtract $9x$ from both sides and add 24 to both sides, we get

$$5x > 45. \tag{3}$$

Finally, dividing both members of (3) by 5, we have

$$x > 9. \tag{4}$$

As in the case of similar operations on equations, we know that if (1) is to be true, then (4) must be true. If we start with the hypothesis that $x > 9$ and reverse the steps, we see that (1) is a consequence of (4); *i.e.*, (1) is true if x is any number greater than 9.

Example 2

For what values of x is $3x^2 < 4x + 7$?

Solution

If the given inequality is to be true, then we must have

$$3x^2 - 4x - 7 < 0$$

or

$$(3x - 7)(x + 1) < 0.$$

Now this product will be negative for those values of x for which one factor is positive and the other negative. Furthermore, the factor $x + 1$ is positive for $x > -1$ and negative for $x < -1$, while the factor $3x - 7$ is positive for $x > \frac{7}{3}$ and negative for $x < \frac{7}{3}$. We may construct the following table:

Value of x	Sign of $(x + 1)$	Sign of $(3x - 7)$	Sign of $(x + 1)(3x - 7)$
$x < -1$	$-$	$-$	$(-)\cdot(-) = +$
$-1 < x < \frac{7}{3}$	$+$	$-$	$(+)\cdot(-) = -$
$x > \frac{7}{3}$	$+$	$+$	$(+)\cdot(+) = +$

We thus arrive at the conclusion that, if the given inequality is to be true, then x must be between -1 and $\frac{7}{3}$. From the reversibility of the steps employed, it is clear that if $-1 < x < \frac{7}{3}$, then $3x^2 < 4x + 7$.

We could have solved example 2 graphically in the following way: If

$$3x^2 < 4x + 7$$

then

$$3x^2 - 4x - 7 < 0.$$

We then make the graph of the equation

$$y = 3x^2 - 4x - 7$$

and pick off from the graph the range of values of x for which y is negative. It is clear from the graph (Fig. 21) that the value of

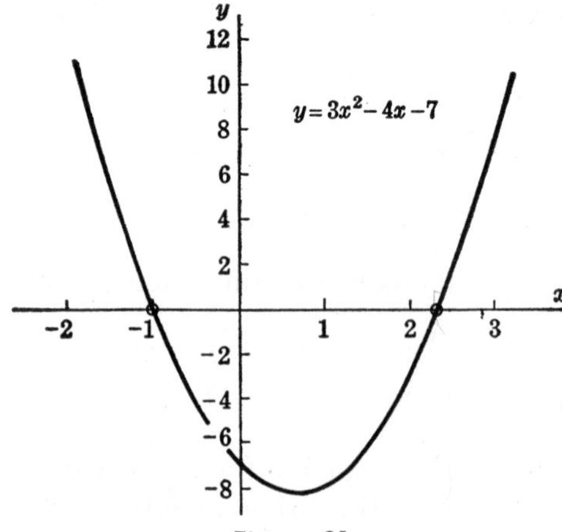

Figure 21

the function $3x^2 - 4x - 7$ is negative for values of x between -1 and $\frac{7}{3}$. For $x = -1$ and for $x = \frac{7}{3}$, it is zero. For $x < -1$ and for $x > \frac{7}{3}$, it is positive; *i.e.*, for these values of x, $3x^2 > 4x + 7$.

It is obvious that any conditional inequality involving one variable can be reduced to the form $f(x) > 0$ or $f(x) < 0$. The values of x that satisfy it can then be found by making the graph of the equation $y = f(x)$ and picking out the range of values of x for which y is positive or negative as the case may be.

Example 3

Find all values of x for which

$$\frac{9}{x-3} > \frac{16}{3x+2}. \tag{1}$$

Solution

We may first remove the denominators by multiplying both members by $(x-3)^2(3x+2)^2$, *which is positive for all values of* x *except* 3 and $-\frac{2}{3}$. The result is

$$9(x-3)(3x+2)^2 > 16(3x+2)(x-3)^2 \tag{2}$$

or

$$9(x-3)(3x+2)^2 - 16(3x+2)(x-3)^2 > 0. \tag{3}$$

We may now factor the left member of (3) and thus obtain the inequality

$$(x-3)(3x+2)[9(3x+2) - 16(x-3)] > 0, \tag{4}$$

or, after dividing by 11,

$$(x-3)(3x+2)(x+6) > 0. \tag{5}$$

We may reason that if (1) is to be true, then (5) must be true; conversely, (1) is true for every value of x that satisfies (5). These values can be found by considering the signs of the three factors of (5), as follows:

Value of x	Sign of $(x-3)$	Sign of $(3x+2)$	Sign of $(x+6)$	Sign of $(x-3)(3x+2)(x+6)$
$x > 3$	+	+	+	+
$-\frac{2}{3} < x < 3$	−	+	+	−
$-6 < x < -\frac{2}{3}$	−	−	+	+
$x < -6$	−	−	−	−

We can conclude that (1) is true for all values of x greater than 3 and also for all values of x between -6 and $-\frac{2}{3}$. If we wished to use the graphical method, we would sketch the graph of the equation

$$y = (x - 3)(3x + 2)(x + 6)$$

and observe that y is positive for $x > 3$ and for $-6 < x < -\frac{2}{3}$. This is left to the student.

Another way of picturing the situation is indicated in Fig. 22. Here each factor is represented by a line parallel to the x-axis,

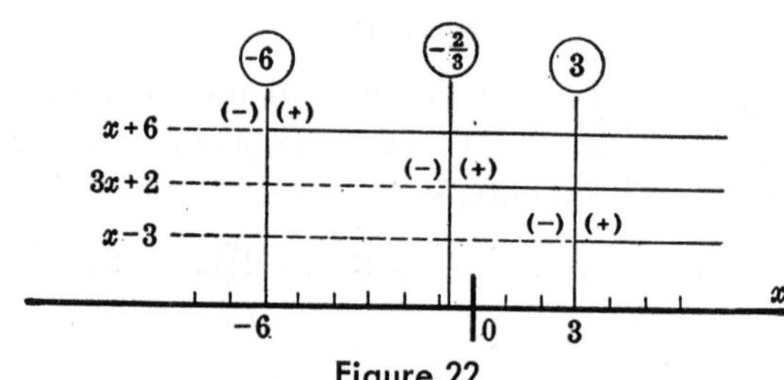

Figure 22

the line being solid over the interval in which the factor is positive and dotted where it is negative. It shows that for $x > 3$ all three factors are positive (three solid lines), and that for $-6 < x < -\frac{2}{3}$ one factor is positive and two are negative (one solid and two dotted lines). In these intervals only, the product is positive.

PROBLEMS

1. Prove that if $a > b$ and c is a negative number, then $ac < bc$.

2. If a is any positive number different from 1, then $(a - 1)^2 > 0$. Deduce from this inequality that $a + (1/a) > 2$, thus proving that the sum of any positive number ($\neq 1$) and its reciprocal is greater than 2.

3. Let a and b be unequal positive numbers. Then of course $(\sqrt{a} - \sqrt{b})^2 > 0$. Deduce from this that $a + b > 2\sqrt{ab}$, thus proving that the sum of any two unequal positive numbers is greater than twice the square root of their product.

4. Prove that if a is any real number then

$$a^2 + 2a + 3 > 0.$$

HINT: Write the left side as $(a + 1)^2 + 2$.

5. Using the method suggested in the hint in Prob. 4, show that $4x^2 > 12x - 13$ is an absolute inequality, x being a real number.

Solve each of the following conditional inequalities:

6. $7x + 5 < 3x + 43$.

7. $12x - 17 > 6(3x + 8)$.

8. $2(3x - 4) > 7x - 19$.

9. $\frac{2}{3}x + 5 < \frac{1}{4}(x + 10)$.

10. $16 - \frac{2}{15}x < \frac{1}{5}x - 6$.

11. $\frac{8}{3}x + 26 > 18 - \frac{2}{9}x$.

12. $-3 < x - 5 < +3$.

13. $-5 < x + 2 < +5$.

14. $2x^2 - 9 > 0$.

15. $9x^2 < 25$.

16. $x(x - 4) > 0$.

17. $x(9 - 2x) < 0$.

18. $x^2(x - 4) > 0$.

19. $x(3 - x)^2 > 0$.

20. $(2x + 5)(x - 1) < 0$.

21. $(x + 4)(2x + 5) > 0$.

22. $x(x - 1)(x + 2) > 0$.

23. $x(x - 1)^2(x + 2) < 0$.

24. $\dfrac{x(x + 1)}{x - 2} > 0$.

25. $\dfrac{x(x - 4)}{x - 5} < 0$.

26. $\dfrac{(2x - 3)(x - 2)^2}{x + 1} > 0$.

27. $\dfrac{(2x + 5)(x - 5)}{x - 3} < 0$.

28. $3x^2 + 11x < 20$.

29. $2x^2 > 3(9 - x)$.

30. $25 > 4x(5 - x)$.

31. $x^2 < \frac{1}{2}(11x + 21)$.

32. $x^3 < 6x^2 - 8x$.

33. $x^3 > x^2 + 2x$.

It will be recalled that if x is real, the absolute value of x, denoted by $|x|$, is simply x if x is positive, and is the positive number $-x$ if x is negative. Thus $|3| = 3$ and $|-3| = 3$. Also $|0| = 0$. Solve the following inequalities:

34. $|x - 3| < 5$. Hint: This means that $x - 3$ is between -5 and $+5$ so that $x - 3 < 5$ and $x - 3 > -5$.

35. $|2x - 5| < 7$.

36. $\left|\dfrac{2x - 1}{2}\right| < 1$.

37. $\left|\dfrac{x - 6}{3}\right| < 1$.

38. $\left|\dfrac{2x + 3}{2}\right| < 3$.

39. $\left|\dfrac{2x - 1}{5}\right| > 3$.

40. $\left|\dfrac{2x - 6}{3}\right| > 4$.

Solve each of the following inequalities:

41. $\dfrac{x}{x - 2} > 5$.

42. $\dfrac{3x}{x - 6} > 8$.

43. $\dfrac{x + 4}{x - 3} < 2$.

44. $\dfrac{2x + 5}{3x - 1} > 4$.

45. $\dfrac{2x + 9}{x + 2} < x$.

46. $\dfrac{4.5 - 6x}{x - 3} < 2x$.

47. $\dfrac{13}{x + 4} < \dfrac{15}{2x - 3}$.

48. $\dfrac{6}{2x - 1} > \dfrac{5}{x - 2}$.

49. $\dfrac{7}{3x - 1} < \dfrac{12}{5x + 6}$.

50. $\dfrac{8}{2x + 5} > \dfrac{7}{3x - 2}$.

Prove that each of the following inequalities is true if a and b represent unequal positive numbers:

51. $\frac{1}{2}(a^2 + b^2) > ab.$

52. $\frac{a}{b} + \frac{b}{a} > 2.$

53. $a + b > \frac{4ab}{a + b}.$

54. $a^3 + b^3 > a^2b + ab^2.$

55. $\frac{a^2}{b} + \frac{b^2}{a} > a + b.$

56. $\frac{a}{b^2} + \frac{b}{a^2} > \frac{1}{a} + \frac{1}{b}.$

CHAPTER VII

NEGATIVE AND FRACTIONAL EXPONENTS. THE EXPONENTIAL FUNCTION

46. ***Introduction.*** In Chap. I we defined the symbol a^n, for n a positive integer, as follows:

$$a^n = a \cdot a \cdot a \cdot \cdot \cdot a \qquad (n \text{ factors}).$$

We called this the *nth power* of a. We called a the *base* and n the *exponent* of the power.

We found that operations with these positive integral exponents are governed by the following laws:

(I) $$a^m \cdot a^n = a^{m+n}.$$

(II) $$\frac{a^m}{a^n} = a^{m-n} \quad \text{if } m > n \text{ and } a \neq 0.$$
$$= \frac{1}{a^{n-m}} \quad \text{if } m < n \text{ and } a \neq 0.$$

(III) $$(a^m)^n = a^{mn}.$$

(IV) $$(a \cdot b)^n = a^n \cdot b^n.$$

(V) $$\left(\frac{a}{b}\right)^n = \frac{a^n}{b^n} \quad \text{if } b \neq 0.$$

It is now desirable to extend our definition of the symbol a^n so as to give meanings to symbols such as $4^{\frac{3}{2}}$, 2^{-3}, and 5^0. In our attempt to arrive at suitable definitions, we shall be guided by a desire that the above laws, which apply to positive integral exponents, shall apply also to negative and fractional exponents, for it would obviously be awkward if we should arrive at a situation in which $2^x \cdot 2^y$ were equal to 2^{x+y} for positive integral values of x and y, but were equal to something else if x and y were fractions.

First we shall consider negative integers and zero as exponents and then extend our definitions so as to include all rational numbers.

47. ***Negative integers and zero as exponents.*** If we wish that a^m/a^n should be equal to a^{m-n} in all cases, we shall have, for example,

$$\frac{a^2}{a^5} = a^{2-5} = a^{-3}.$$

But we know that

$$\frac{a^2}{a^5} = \frac{a \cdot a}{a \cdot a \cdot a \cdot a \cdot a} = \frac{1}{a^3}.$$

If we are to give to the symbol a^{-3} a meaning consistent with these results, we must have $a^{-3} = 1/a^3$. We are led, therefore, to set up the following *definition:*

$$\mathbf{a^{-n} = \frac{1}{a^n} \qquad (a \neq 0).}$$

For the present n is restricted to the field of positive integers, but that restriction will be removed shortly.

Examples

$$4^{-2} = \frac{1}{4^2} = \frac{1}{16}; \qquad (-2)^{-5} = \frac{1}{(-2)^5} = \frac{1}{-32}.$$

In order to decide upon a suitable definition of the symbol a^0, we proceed in a similar fashion. If law II is to hold for $m = n$, we should have

$$\frac{a^n}{a^n} = a^{n-n} = a^0.$$

But we know that, if $a \neq 0$, then

$$\frac{a^n}{a^n} = 1.$$

This suggests that we should have $a^0 = 1$ if $a \neq 0$. Similarly, if law I is to apply, then

$$a^n \cdot a^0 = a^{n+0} = a^n,$$

and if $a \neq 0$, this is true if and only if $a^0 = 1$. We therefore set up the *definition:*

$$\mathbf{a^0 = 1} \qquad \text{if } a \neq 0.$$

We shall assign no meaning at all to a^0 for the case in which $a = 0$; *i.e.*, the symbol 0^0 will remain undefined—it will not stand for any number.

Examples

$$(4\pi)^0 = 1; \qquad (-3)^0 = 1; \qquad (a - b)^0 = 1, \text{ if } a \neq b.$$

In the various branches of science, such as physics, chemistry, and astronomy, one frequently deals with numbers that are extremely large and also with numbers that are numerically very small. Such a number can be written in a compact form by using powers of 10 as indicated in the following examples:

$$6{,}410{,}000{,}000 = 641 \times 10^7 \quad \text{or} \quad 6.41 \times 10^9.$$
$$0.000{,}000{,}037 = 37 \times 10^{-9} \quad \text{or} \quad 3.7 \times 10^{-8}.$$

Any positive number can thus be written as the product of a number that lies in the interval from 1 to 10 and a suitable power of 10, and a number is sometimes said to be in *standard form* when it is expressed in this fashion. In problems involving multiplication and division of large and small numbers, the task of locating the decimal point in the answer may be simplified by writing the numbers in this standard form.

Example

$$\begin{aligned}\frac{(39{,}000)\cdot(0.000{,}05)}{(0.025)\cdot(1{,}200)} &= \frac{(3.9 \times 10^4)\cdot(5 \times 10^{-5})}{(2.5 \times 10^{-2})\cdot(1.2 \times 10^3)} \\ &= \frac{(3.9)\cdot(5)}{(2.5)\cdot(1.2)} \times 10^{4-5+2-3} \\ &= 6.5 \times 10^{-2} = 0.065.\end{aligned}$$

48. ***Rational fractions as exponents.*** We have said that if $x^2 = a$, then x is a square root of a; if $x^3 = a$, then x is a cube root of a, etc. We have observed that if a is positive, there is precisely one positive number that is an nth root of a, and we have called this the *principal nth root* and denoted it by the symbol $\sqrt[n]{a}$. If a is negative, there is no positive nth root, but if n is odd, there is a negative one; for this case we called this negative one the principal nth root. Thus $\sqrt[3]{-8} = -2$.

Let us now seek a suitable definition for the symbol $a^{1/n}$ where n is a positive integer. If the usual rules are to hold, we should have

$$(a^{1/n})^n = a^1 = a.$$

This suggests that we should define $a^{1/n}$ to mean an nth root of a, and in order that there may be no ambiguity, we define it to stand for the *principal* nth root in the cases mentioned above. No other cases need be considered at the present time. We have then, by definition,

$$a^{1/n} = \sqrt[n]{a},$$

this being the positive nth root if a is positive and the negative one if a is negative and n is odd.

Examples

$$(-32)^{\frac{1}{5}} = -2; \qquad (225)^{\frac{1}{2}} = 15; \qquad (\tfrac{1}{64})^{\frac{1}{3}} = \tfrac{1}{4}.$$

Consider now the symbol $a^{m/n}$ where a is positive and m and n are natural numbers. If law III is to hold, we should have

$$(a^{1/n})^m = (a^m)^{1/n} = a^{m/n}.$$

Consequently we define $a^{m/n}$ to denote the mth power of the principal nth root of a, or the principal nth root of a^m, these two numbers being the same for $a > 0$. Thus

$$9^{\frac{3}{2}} = (\sqrt{9})^3 = 3^3 = 27 \qquad \text{or} \qquad 9^{\frac{3}{2}} = \sqrt{9^3} = \sqrt{729} = 27.$$

We shall not define $a^{m/n}$ for the case in which a is negative and both m and n are > 1. We do not need to consider it at this time, and there are several difficulties involved.*

We extend the definition already given for negative integral exponents to negative fractional exponents in an obvious way. Thus

$$(64)^{-\frac{2}{3}} = \frac{1}{(64)^{\frac{2}{3}}} = \frac{1}{16}.$$

It can be proved that, with the above definitions regarding negative, zero, and fractional exponents, the following rules apply

* For example, we should naturally set up the definition so that $(-1)^{\frac{5}{3}} = -1$. Then we should require that $(-1)^{\frac{10}{6}} = -1$ also, because we would insist upon the principle of substituting equals for equals. Now it is true that $(\sqrt[6]{-1})^{10} = -1$ for a proper choice of $\sqrt[6]{-1}$, but $\sqrt[6]{(-1)^{10}} = +1$ by our previous definition. The rule $(a^{1/n})^m = (a^m)^{1/n}$ would then not hold.

to operations with exponents. We assume for the present that a and b are positive and that x and y are rational numbers.

(I) $$a^x \cdot a^y = a^{x+y}.$$

(II) $$\frac{a^x}{a^y} = a^{x-y} \qquad \text{if } a \neq 0.$$

(III) $$(a^x)^y = (a^y)^x = a^{xy}.$$

(IV) $$(a \cdot b)^x = a^x \cdot b^x.$$

(V) $$\left(\frac{a}{b}\right)^x = \frac{a^x}{b^x} \qquad \text{if } b \neq 0.$$

PROBLEMS

Perform the indicated operations and express the result in a simple form. (In each problem assume appropriate restrictions in regard to the field of numbers that the letters may represent.)

1. $(-\frac{1}{2}x^{-\frac{1}{2}})^2$.
2. $(\frac{1}{4}x^{-\frac{1}{3}})^3$.
3. $\frac{2}{5}(8)^{-\frac{2}{3}}x^2$.
4. $\left(\frac{9x^4}{y^2}\right)^{-\frac{1}{2}}$.
5. $\left(\frac{1}{4a^2b^2}\right)^{-\frac{1}{2}}$.
6. $\frac{5x^2y^3}{(x^2y^2)^{-\frac{1}{2}}}$.
7. $\left(\frac{27x^{-3}}{y^{-6}}\right)^{-\frac{2}{3}}$.
8. $(4x + 4y)(x^2 - y^2)^{-1}$.
9. $\frac{x^{-1} + y^{-1}}{(x + y)^{-1}}$.
10. $\frac{4x^2 + 4}{1 + x^{-2}}$.
11. $(1 + x^{-\frac{1}{2}})(1 - x^{-\frac{1}{2}})$.
12. $\frac{9a^3 - a^{-1}}{3 - a^{-2}}$.
13. $(x^{-\frac{1}{2}} + y^{-\frac{1}{2}})^2$.
14. $(x^{\frac{1}{2}} + y^{\frac{1}{2}})(x^{\frac{1}{2}} - y^{\frac{1}{2}})$.
15. $\frac{1 - x^{-\frac{1}{2}}}{1 + x^{-\frac{1}{2}}} + \frac{1 + x^{-\frac{1}{2}}}{1 - x^{-\frac{1}{2}}}$.
16. $\frac{z^{\frac{1}{1-n}-1} - z^{\frac{1}{1-n}}}{z^{\frac{n}{1-n}}}$.
17. $\frac{(27)^{-\frac{2}{3}} + (27)^{\frac{2}{3}}}{(7)4^0 + \frac{2}{3}(3)^{-1}}$.
18. $(a^{-1} - b^{-1})(a^{-2} + a^{-1}b^{-1} + b^{-2})$.

Express each of the following numbers as the product of a number between 1 and 10 and a positive or negative power of 10:

19. 6,400,000.
20. 400,000,000.
21. 52,600,000,000.
22. 0.000,002,4.
23. 0.000,000,07.
24. 0.000,012,3.

Solve each of the following equations for x:

25. $5x^{-\frac{1}{2}} = 2$.
26. $x^{-\frac{3}{2}} = 27$.
27. $x^{-\frac{5}{2}} = 32$.
28. $3x^{\frac{1}{3}} = 2$.
29. $(4x - 7)^{\frac{3}{2}} = 125$.

30. $(2x + 5)^{-\frac{5}{2}} = 2^{-5}$. **31.** $(\sqrt{2x + 5} + 7)^{\frac{1}{2}} = 4$.
32. $(3 - \sqrt{x - 2})^{\frac{1}{2}} = 2$.

In each of the following cases, transform the given expression into an equivalent one that has no radicals or fractional exponents in the denominator:

33. $\dfrac{1}{\sqrt{3}}$. HINT: $\dfrac{1}{\sqrt{3}} = \dfrac{1}{\sqrt{3}} \cdot \dfrac{\sqrt{3}}{\sqrt{3}} = \dfrac{\sqrt{3}}{3}$.

34. $\dfrac{4}{\sqrt{6}}$. **35.** $\sqrt{\dfrac{9}{32}}$. **36.** $\dfrac{8}{\sqrt{50}}$. **37.** $\dfrac{4}{\sqrt{6} + \sqrt{2}}$.

38. $\dfrac{\sqrt{80}}{\sqrt{80} - \sqrt{10}}$. **39.** $\dfrac{x}{\sqrt{x} - \sqrt{y}}$.

40. Using the laws of exponents, show that if x and y are positive numbers, then $\sqrt{9x^2y^3} = 3xy\sqrt{y}$. Is this same result true if x is negative and y is positive? Evaluate both sides of the equation for $x = -3$, $y = 2$.

41. Simplify the expression

$$\frac{\sqrt{1 - 4y^2} - y \cdot \frac{1}{2}(1 - 4y^2)^{-\frac{1}{2}}(-8y)}{1 - 4y^2}$$

and show in particular that it is equivalent to $(1 - 4y^2)^{-\frac{3}{2}}$. What restriction should be made on the value of y?

42. Simplify the expression

$$\frac{1}{2}(a^2 - x^2)^{-\frac{1}{2}}(-2x) + a\frac{1/a}{\sqrt{1 - (x/a)^2}}$$

and show in particular that it is equivalent to $\sqrt{(a - x)/(a + x)}$. Assume that $a > 0$ and $|x| < a$.

43. Simplify the expression

$$\frac{1}{2}(x^2 - 4)^{-\frac{1}{2}}(2x) - 2\frac{\frac{1}{4}(x^2 - 4)^{-\frac{1}{2}}(2x)}{1 + \frac{1}{4}(x^2 - 4)}.$$

What restriction on x is necessary?

44. Simplify the expression

$$\frac{\frac{1}{2}}{\sqrt{1 - (x/2)^2}} + \frac{x \cdot \frac{1}{2}(4 - x^2)^{-\frac{1}{2}}(-2x) - \sqrt{4 - x^2}}{x^2}$$

and show in particular that it can be reduced to $-\sqrt{4 - x^2}/x^2$. What restriction on x is necessary?

45. Discuss the following sequence of "equalities":

$$-1 = (-1)^1 = (-1)^{\frac{2}{2}} = \sqrt{(-1)^2} = \sqrt{+1} = +1.$$

49. *Irrational exponents. The exponential function.* We have assigned a meaning to the symbol $a^x (a > 0)$ for all rational values of x. At this stage the student does not have sufficient mathematical training to profit from a detailed discussion of the case in which x is irrational. He is probably willing to assume that there is a unique number that is denoted by the symbol $5^{\sqrt{2}}$, for example, and since $\sqrt{2}$ lies between 1.41 and 1.42 he would probably feel that $5^{\sqrt{2}}$ should be between $5^{1.41}$ and $5^{1.42}$. Now

$$5^{1.41} = 5^{\frac{141}{100}} = 9.673^* \qquad \text{and} \qquad 5^{1.42} = 5^{\frac{142}{100}} = 9.830.$$

Hence $5^{\sqrt{2}}$ should be between 9.673 and 9.830. A closer approximation results from the observation that $\sqrt{2}$ lies between 1.414 and 1.415. Since

$$5^{1.414} = 5^{\frac{1,414}{1,000}} = 9.735 \qquad \text{and} \qquad 5^{1.415} = 5^{\frac{1,415}{1,000}} = 9.751,$$

we should conclude that $5^{\sqrt{2}}$ lies between 9.735 and 9.751. If we continue in this way, we obtain an increasing sequence of numbers, $5^{1.4}$, $5^{1.41}$, $\cdots$, and a decreasing sequence, $5^{1.5}$, $5^{1.42}$, $\cdots$. It is perhaps reasonable to expect (and it can be proved) that these sequences both have the *same* number as their limit. It is this number that we desire to represent by the symbol $5^{\sqrt{2}}$.

Let us assume without further discussion that if a is any positive number, the symbol a^x denotes a definite number for every real value of x—irrational as well as rational—and let us consider the problem of making the graph of the equation $y = a^x$. Points may be obtained in the usual way by substituting values for x and computing the corresponding values of y. Thus for the equation

$$y = (1.5)^x$$

we may construct the following table of values:

x	0	1	2	3	4	5	−1	−2	−3	−4
y	1	1.50	2.25	3.38	5.06	7.59	0.67	0.44	0.30	0.20

* This computation is carried out by means of logarithms. The procedure will be explained in the next chapter.

If we plot the corresponding points and draw a smooth curve through them, we have an approximation to the graph (Fig. 23). We could of course obtain points between those actually plotted

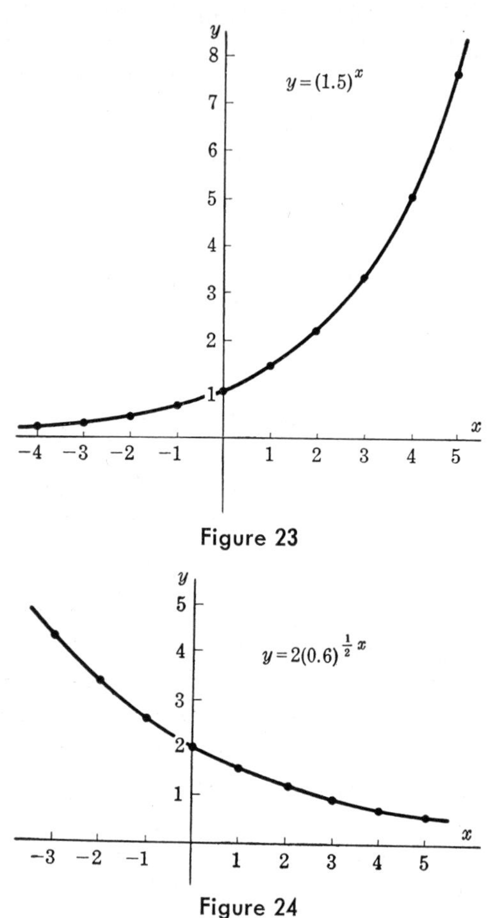

Figure 23

Figure 24

by using nonintegral values of x. Thus, corresponding to $x = 2.5$, we would find that

$$y = (1.5)^{2.5} = (1.5)^2 \sqrt{1.5} = 2.76.$$

The function a^x where a is positive is called an *exponential function*. The value of the function is positive for all real values of x and increases as x increases if $a > 1$, the graph having the general shape shown in Fig. 23. If a is between 0 and 1, the value of a^x decreases as x increases, and the graph has the general form shown in Fig. 24.

50. ***Graphical solution of exponential equations.*** An equation in which the unknown occurs in an exponent is usually called an *exponential equation.* Examples are

$$5^x + 2x^2 = 16.$$
$$(14)^{\frac{1}{3}x} - 3x = x^2 + 7.$$

The real roots of such an equation can be found, in general, only by some method of successive approximations. The usual first step is that of locating each real root between two consecutive integers. In the following example we show how this may be accomplished with the aid of a graph.

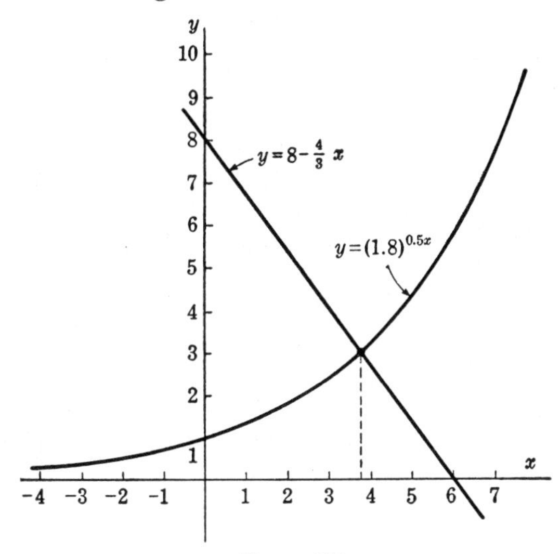

Figure 25

Example

Determine the number of real roots of the equation

$$(1) \qquad (1.8)^{0.5x} + \tfrac{4}{3}x = 8,$$

and locate each such root between two consecutive integers.

Solution

Equation (1) is equivalent to

$$(2) \qquad (1.8)^{0.5x} = 8 - \tfrac{4}{3}x.$$

We may then interpret our problem as that of seeking all real values of x for which the two sides of (2) are equal. In Fig. 25

we have plotted on the same axes the graphs of the two equations

$$y = (1.8)^{0.5x} \qquad \text{and} \qquad y = 8 - \tfrac{4}{3}x.$$

Any real value of x that satisfies (2) is the abscissa (or x-coordinate) of a point that is common to the two graphs. In this case, the graphs have only one point of intersection, and its abscissa is between 3 and 4. Hence the given equation (1) has one and only one real root, a number between 3 and 4.

PROBLEMS

In each of the following cases draw the graph of the given equation over the specified interval:

1. $y = (1.6)^x \quad -3 \leqq x \leqq 5.$

2. $y = \frac{1}{4}(2)^x \quad -2 \leqq x \leqq 6.$

3. $Q = 1.5(1.2)^t \quad -3 \leqq t \leqq 8.$

4. $S = 3(2)^{-\frac{1}{2}x} \quad -4 \leqq x \leqq 4.$

5. $y = \frac{1}{8}(3.6)^{\frac{1}{2}x} \quad -2 \leqq x \leqq 6.$

6. $y = 4(1.6)^{-2x} \quad -2 \leqq x \leqq 3.$

7. $y = 2.8(0.8)^{\frac{1}{2}x} \quad -3 \leqq x \leqq 5.$

8. $y = 4 + \frac{1}{2}(2^{-x}) \quad -4 \leqq x \leqq 4.$

9. $y = 2[(0.6)^x - 1] \quad -3 \leqq x \leqq 5.$

10. $y = 3 - (1.5)^x \quad -4 \leqq x \leqq 5.$

11. $y = \frac{1}{4}x + (\frac{1}{2})^x \quad -4 \leqq x \leqq 8.$

12. $y = \frac{1}{2}(2^x + 2^{-x}) \quad -4 \leqq x \leqq 4.$

13. $y = \frac{1}{2}(2^x - 2^{-x}) \quad -4 \leqq x \leqq 4.$

14. Using the laws of exponents, show that the functions $(\frac{1}{2})^x$ and 2^{-x} are identical.

15. Using the laws of exponents, show that the functions $(2)^{x-3}$ and $\frac{1}{8}(2^x)$ are identical.

16. Using the laws of exponents, show that the function 2^{3x} can be written in the form 8^x.

17. Simplify the expression

$$\sqrt{\left(\frac{3^x - 3^{-x}}{2}\right)^2 + 1}$$

and show in particular that it is equivalent to $\frac{1}{2}(3^x + 3^{-x})$.

18. For a certain value of k, the value of 4^k is equal to 5. What is the value of 2^{3k}?

19. Locate between two consecutive integers the real root of the equation

$$(1.5)^x = 5 - x$$

by making the graphs of the equations $y = (1.5)^x$ and $y = 5 - x$ on the same axes and finding approximately the point of intersection. As a better approximation, determine whether the root is slightly less or slightly more than 2.5.

20. Determine the number of real roots of the equation $x^2 + 2^x - 4 = 0$, and locate each root between two consecutive integers. HINT: If $x^2 + 2^x - 4 = 0$

then $2^x = 4 - x^2$. Make the graphs of the equations $y = 2^x$ and $y = 4 - x^2$ on the same axes and locate the intersections approximately.

21. Using a method similar to that suggested in Prob. 20, show that the equation

$$2(0.6)^{\frac{1}{2}x} - \tfrac{2}{3}x - 4 = 0$$

has a real root in the neighborhood of -1.5. Make the necessary computations to determine whether the root is a little less than or a little greater than -1.5.

22. Locate each real root of the equation

$$2(1.3)^x + \tfrac{1}{4}x^2 = 6$$

between two consecutive integers.

23. Show that the equation

$$4(0.5)^x + (0.5)x + 1 = 0$$

has no real root.

24. Locate each real root of the equation

$$(1.8)^{0.5x} - 0.75x = 3$$

between two consecutive integers.

CHAPTER VIII

LOGARITHMS

51. *Introduction.* In this chapter we shall be concerned primarily with the use of logarithms as an aid in certain types of computations. For example, it was mentioned on page 121 that logarithms had been used in finding that

$$5^{1.41} \text{ or } 5^{\frac{141}{100}} = 9.673.$$

The student has perhaps concluded from this example that logarithms must have afforded a considerable simplification in what would otherwise have been a very difficult and tedious calculation. This is indeed the case.

Before taking up the definition of the logarithm of a number, we shall discuss a few preliminary matters in regard to the subject of computations.

52. *Significant digits. Approximate numbers.* The number of *significant figures* or *significant digits* in a number expressed in decimal form is, with an exception to be noted below, simply the number of digits used in writing it. Thus 65.36 or 6.536 has four significant digits and 50,768.2 has six.

The exception is that zeros are not significant digits when used just for the purpose of placing the decimal point. Thus 0.0067 has only two significant digits—just as if it were 6.7 or 67. Final zeros may or may not be significant digits. If one says that the population of a certain town is 24,000, the zeros are not significant digits if it is the intention that this number represents the population only to the nearest thousand. If it is the intention to state that the population is precisely 24,000, then the zeros are significant figures.

Suppose now that a measurement of a distance d is taken with an instrument that gives the result to the nearest hundredth of an inch, and that the reading is 6.47 in. We do not know, as a result of this measurement, that d is exactly 6.47 in., but only that it is nearer to 6.47 than to 6.46 or 6.48. We are thus using 6.47 to represent a number d which may actually be anywhere between 6.465 and 6.475. When a number is used in this sense, it is often called an *approximate number.*

If a length, measured to the nearest hundredth of an inch, is 4.7 in., we would usually write it as 4.70 in. If, measured to the nearest thousandth, it were 3.9 in., we would write it as 3.900 in., etc. The zeros used here are of course significant digits.

53. ***Multiplication and division of approximate numbers.*** Let it be assumed that we have measured the base b and altitude h of a triangle with an instrument that gives the results to the nearest hundredth of an inch, and that the readings are

$$b = 6.47 \text{ in.}, \qquad h = 5.93 \text{ in.}$$

If we use the formula $A = \frac{1}{2}bh$ to compute the area and retain all the digits, we have

$$A = (0.5)(6.47)(5.93) = 19.18355 \text{ sq. in.}$$

It would obviously be improper to say that the area is 19.18355 sq. in., for this would be claiming an accuracy of knowledge about the area that we cannot possibly have. All that we really know about it is that it lies between

$$(0.5)(6.465)(5.925) = 19.1525625 \text{ sq. in.}$$

and

$$(0.5)(6.475)(5.935) = 19.2145625 \text{ sq. in.}$$

We should then "round off" our above result and write

$$A = (0.5)(6.47)(5.93) = 19.2 \text{ sq. in.}$$

Observe now that 6.47 and 5.93 are approximate numbers, given to three significant digits. The other factor is not an approximate number in the sense defined above—it represents exactly $\frac{1}{2}$. In the above result we have retained the same number of significant figures as there are in the approximate numbers. Stated somewhat more generally, the rule that we use is as follows:

In expressing the result of a multiplication or division of approximate numbers, retain only as many significant digits as there are in the number having the smallest number of significant digits.

This rule usually results in a degree of accuracy that is consistent with that of the measured data. Note that the position of the decimal point is not involved in this matter and that "exact" numbers, such as the factor 0.5 in the above example, have no effect.

54. *The logarithm of a number.* Let a be a positive number that is not equal to 1. Then *if* $a^x = N$, *we shall agree to call* x *the logarithm of* N *to the base* a. If we use the abbreviation $\log_a N$ to mean the logarithm of N to the base a, we can write our definition as follows:

$$\text{If} \qquad a^x = N, \qquad \text{then} \qquad x = \log_a N.$$

Examples

$$2^5 = 32, \qquad \text{therefore, } \log_2 32 = 5;$$
$$10^2 = 100, \qquad \text{therefore, } \log_{10} 100 = 2;$$
$$4^{-2} = \tfrac{1}{16}, \qquad \text{therefore, } \log_4 \tfrac{1}{16} = -2;$$
$$10^{0.716} = 5.20, \qquad \text{therefore, } \log_{10} 5.20 = 0.716;^*$$
$$10^{2.055} = 113.5, \qquad \text{therefore, } \log_{10} 113.5 = 2.055.$$

The student should study the above definition and examples until he has the situation clearly in mind.

It can be proved that any positive number N can be expressed in the form a^x if a is any positive number not equal to 1. For example, we can express 78 in the form 6^x by properly choosing the value of x. The value of x must of course be between 2 and 3; in fact, it must be between 2.4 and 2.5 because

$$6^{2.4} = 73.7 \qquad \text{and} \qquad 6^{2.5} = 88.2.$$

It turns out that to three significant digits the value of x is 2.43 and to five it is 2.4315. Thus $6^{2.43} = 77.8$, and $6^{2.4315} = 78.00$.

* This is an approximate number in the sense of the definition on p. 127. It means that $\log_{10} 5.20$ is between 0.7155 and 0.7165. Most of the logarithms with which we shall be concerned are irrational numbers, and we shall use their decimal approximations to the degree of accuracy needed for our purposes.

We can then say that, correct to five significant digits, the logarithm of 78 to the base 6 is 2.4315.

55. *Properties of logarithms.* The basic properties that make logarithms useful in certain kinds of computations are contained in the following three theorems:

***Theorem* I.** *The logarithm of the product of two numbers is equal to the sum of the logarithms of the factors; i.e.,*

$$\log_a (M \cdot N) = \log_a M + \log_a N.$$

***Theorem* II.** *The logarithm of the quotient of two numbers is equal to the logarithm of the numerator minus the logarithm of the denominator; i.e.,*

$$\log_a \frac{M}{N} = \log_a M - \log_a N.$$

***Theorem* III.** *The logarithm of the pth power of a number is equal to p times the logarithm of the number; i.e.,*

$$\log_a N^p = p \cdot \log_a N.$$

The proof of Theorem I is as follows: Let $\log_a M = x$ and $\log_a N = y$; then

$$M = a^x \quad \text{and} \quad N = a^y.$$

It follows that

$$M \cdot N = a^x \cdot a^y = a^{x+y}.$$

This last result means that

$$\log_a (M \cdot N) = x + y,$$

and since $x = \log_a M$ and $y = \log_a N$, we have

$$\log_a (M \cdot N) = \log_a M + \log_a N.$$

The proof of Theorem II is similar and the details will be left to the student. The proof of Theorem III is as follows: If $\log_a N = y$, then $N = a^y$. It follows that

$$N^p = (a^y)^p = a^{py}.$$

This result means that

$$\log_a (N^p) = py,$$

and since $y = \log_a N$, we have

$$\log_a (N^p) = p \cdot \log_a N.$$

The applications of these theorems to problems of computation will be taken up in the next few sections. We may note immediately, however, the following consequence of the theorems: *If we know the logarithms of the prime numbers, we can find the logarithm of any positive number.* Thus, for example, since $35 = 5 \cdot 7$,

$$\log_a 35 = \log_a 5 + \log_a 7.$$

Similarly,

$$\begin{aligned}\log_a \frac{8}{25} &= \log_a 8 - \log_a 25 \\ &= \log_a (2^3) - \log_a (5^2) \\ &= 3 \log_a 2 - 2 \log_a 5.\end{aligned}$$

PROBLEMS

1. Write out the proof of Theorem II of Sec. 55.

2. In defining the logarithm of a number to the base a, why do we require that a should be different from 1?

In each of the following cases express the given statement in a form involving a logarithm:

3. $4^2 = 16$; $5^{-2} = \frac{1}{25}$; $3^0 = 1$.
4. $8^{\frac{1}{3}} = 2$; $16^{-\frac{1}{4}} = \frac{1}{2}$; $2^{-3} = \frac{1}{8}$.
5. $10^2 = 100$; $10^{-2} = 0.01$; $10^3 = 1{,}000$.
6. $7^0 = 1$; $100^{\frac{1}{2}} = 10$; $1{,}000^{-\frac{1}{3}} = 0.1$.
7. $10^{0.9085} = 8.1$; $10^{1.273} = 18.75$; $10^{2.736} = 544.5$.
8. $10^{0.902} = 7.98$; $10^{1.902} = 79.8$; $10^{2.902} = 798$.

In each of the following cases express the given statement in exponential form:

9. $\log_2 64 = 6$; $\log_5 25 = 2$; $\log_4 1 = 0$.
10. $\log_2 (\frac{1}{8}) = -3$; $\log_{10} (0.1) = -1$; $\log_8 4 = \frac{2}{3}$.
11. $\log_4 32 = 2.5$; $\log_{100} 10 = 0.5$; $\log_{10} (0.01) = -2$.
12. $\log_{10} 861 = 2.935$; $\log_{10} 8.77 = 0.943$; $\log_{10} 35.4 = 1.549$.

13. Prove that if $10^{0.702} = 5.035$, then $10^{1.702} = 50.35$ and $10^{2.702} = 503.5$. Write the corresponding statements concerning logarithms to the base 10.

Find the value of each of the following:

14. $\log_7 7$; $\log_3 81$; $\log_{10} 0.001$.
15. $\log_{25} 125$; $\log_8 (0.5)$; $\log_{10} 1{,}000$.

16. $\log_{16} 64$; $\log_{32} 4$; $\log_{10} 1$.
17. $\log_{10} 100{,}000$; $\log_{625} 25$; $\log_9 \frac{1}{3}$.

Each of the following equations is of the form $\log_a N = x$. In each case find the missing value of x, N, or a:

18. $\log_9 27 = x$; $\log_4 N = -2.5$; $\log_a 100 = 0.5$.
19. $\log_{81} 9 = x$; $\log_{16} N = \frac{1}{4}$; $\log_a 32 = 5$.
20. $\log_{10} 0.001 = x$; $\log_9 N = -\frac{3}{2}$; $\log_a 8 = 1.5$.

Given that $\log_{10} 2 = 0.3010$, $\log_{10} 3 = 0.4771$, $\log_{10} 5 = 0.6990$, $\log_{10} 7 = 0.8451$ find the logarithm to the base 10 of each of the following numbers:

21. 35; 0.03; 98.
22. 105; 1.4; $\frac{2}{7}$.
23. 375; 6,300; 5,000.
24. $17\frac{1}{2}$; 34.3; $\frac{1}{3}$.
25. 2,800; 1.75; 0.005.
26. $\frac{35}{12}$; $\frac{21}{40}$; 6.3.
27. 8,000; 0.015; $\frac{6}{70}$.
28. $\frac{50}{42}$; $\sqrt{35}$; $\sqrt[3]{6}$.
29. $(75)^{1.4}$; $\sqrt[5]{80}$; $(84)^{\frac{2}{3}}$.
30. $\sqrt{10{,}500}$; $\frac{1}{\sqrt{6}}$; $\sqrt[3]{100}$.

31. Sketch the graph of the equation $N = 2^x$. In what way does the graph exhibit the fact that there is no real number which is the logarithm to the base 2 of a negative number?

32. Using Theorem I of Sec. 55, prove that if x, y, and z are positive numbers, then

$$\log_a (xyz) = \log_a x + \log_a y + \log_a z.$$

33. Using Theorems I and II of Sec. 55, show that if x, y, and z are positive numbers, then

$$\log_a \frac{xy}{z} = \log_a x + \log_a y - \log_a z.$$

56. ***Common logarithms. Characteristic and mantissa.*** For purposes of computation, it is convenient to use a system of logarithms to the base 10. Logarithms to this base are called *common logarithms* or *Briggsian logarithms.**

We shall henceforth use the symbol **log** N *with no base written down* to mean $\log_{10} N$. Thus we shall write $\log 10 = 1$ and $\log 100 = 2$, since $10^1 = 10$ and $10^2 = 100$. Similarly,

$$\log 0.1 = -1,$$

$\log 0.01 = -2$, $\log 0.001 = -3$, etc.

* After Henry Briggs, 1556–1631, who first used them.

Let us now write the numbers 56.2, 562, 0.00562, in what we have called the *standard form* (page 117). Thus

$$\begin{aligned} 56.2 &= 5.62 \times 10; \\ 562 &= 5.62 \times 10^2; \\ 0.00562 &= 5.62 \times 10^{-3}. \end{aligned}$$

It follows immediately that

$$\begin{aligned} \log 56.2 &= \log 5.62 + \log 10 = \log 5.62 + 1; \\ \log 562 &= \log 5.62 + \log (10^2) = \log 5.62 + 2; \\ \log 0.00562 &= \log 5.62 + \log (10^{-3}) = \log 5.62 - 3. \end{aligned}$$

It thus appears that, if we knew the value of log 5.62, we could find that of 56.2, 562, 0.00562, etc., simply by adding a positive or negative integer. As we shall see later, $\log 5.62 = 0.7497$ so that

$$\begin{aligned} \log 56.2 &= 0.7497 + 1 = 1.7497; \\ \log 562 &= 0.7497 + 2 = 2.7497; \\ \log 0.00562 &= 0.7497 - 3 = -2.2503. \end{aligned}$$

This last result of course means that

$$10^{0.7497-3} = \frac{10^{0.7497}}{10^3} = 10^{-2.2503} = 0.00562.$$

For most purposes we would prefer to leave the logarithm in the form $0.7497 - 3$, or write it as $7.7497 - 10$. Either of these is equivalent to -2.2503, and there is some advantage in writing log 0.00562 in a form in which the decimal part is equal to log 5.62. We now set up the following:

Definition. *When log N is written in a form in which the decimal part is zero or a positive number less than* 1, *the integral part is called the* **characteristic** *and the decimal part is called the* ***mantissa*** *of log N.*

Examples

$\log 56.2 = 1.7497$. The characteristic is 1 and the mantissa is 0.7497.

$\log 0.00562 = 0.7497 - 3$ or $7.7497 - 10$. The characteristic is -3(or $7 - 10$) and the mantissa is 0.7497.

$\log 10{,}000 = 4.0000$. The characteristic is 4 and the mantissa is zero.

The characteristic of log N is determined by the position of the decimal point in the number N. If the decimal point follows the first significant digit, as in the numbers 2.48, 7.635, and 9.826, the characteristic is zero because the common logarithm of such a number is between 0 and 1. The student should show that the following rule takes care of all other cases.

Write down the given number and draw a vertical line just to the right of the first nonzero digit as indicated below. The characteristic is numerically equal to the number of digits between this line and the decimal point. It is positive if the decimal point is to the right of the line and negative if to the left:

Number:	2|659.4	0.7|82	0.0006|5
Characteristic:	3	-1	-4

It may also be observed that, if a given number is written in what we have called standard form, the characteristic of its common logarithm is simply the exponent of 10. Thus,

$$2659.4 = 2.6594 \times 10^3;$$
$$0.782 = 7.82 \times 10^{-1};$$
$$0.00065 = 6.5 \times 10^{-4}.$$

The exponents 3, -1, and -4 are the characteristics, and the logarithms of 2.6594, 7.82, and 6.5 are the corresponding mantissas.

57. ***Use of tables.*** We have seen that we can write down the logarithms of the numbers 56.2, 0.00562, 5620, etc., if we know the value of log 5.62, for this is the mantissa of the logarithm for any of these numbers, and we know a rule for determining the characteristic.

Tables of mantissas have been computed by methods that are beyond the scope of this text. On page 135 is shown a part of a table that gives mantissas to four decimal places for numbers containing three digits. In order to find log 5.62, or the mantissa of log 56.2 or log 562, etc., from this table, one runs down the first column until he comes to the number 56. He then moves to the right along this row until he reaches the column that has 2 at the top. At this place he finds the number 7497. This means that the mantissa is .7497. The decimal point is omitted in printing the table. Then

$$\begin{aligned} \log 5.62 &= 0.7497, & \log 5620 &= 3.7497, \\ \log 0.562 &= 0.7497 - 1 & \text{or} \quad & 9.7497 - 10, \\ \log 0.000562 &= 0.7497 - 4 & \text{or} \quad & 6.7497 - 10. \end{aligned}$$

58. ***Interpolation.*** In order to find from this table the mantissa corresponding to a number containing four digits, we may use an approximation process called *interpolation.* This process is based on the assumption that for a small change in the number N, the corresponding change in $\log N$ is proportional to the change in N. Thus in order to find log 56.24, we proceed as follows: Disregarding decimal points for the moment, we note from our table that the mantissa corresponding to 5620 is 7497 and that corresponding to 5630 is 7505. Now since our number, 5624, is four-tenths of the way from 5620 to 5630, we take as its mantissa the number that is four-tenths of the way from 7497 to 7505. We find this number by taking four-tenths of $(7505 - 7497)$, which is four-tenths of 8, and adding it to 7497. Now four-tenths of 8 is 3 to the nearest integer. We therefore take $7497 + 3$ or 7500 as our mantissa. The characteristic is 1, and hence we have

$$\log 56.24 = 1.7500.$$

The assumption on which this procedure is based is not strictly true; for example, log 6.5 is not exactly halfway between log 6 and log 7, but the process leads to results that are sufficiently accurate for our present purposes.

There are tables of mantissas to five, six, and even more places. They are all used in essentially the same way. Consequently one who understands the use of a four-place table can readily use a five- or six-place table when he needs the greater accuracy. For example, one finds from a five-place table, without interpolation, that $\log 56.24 = 1.75005$. If he wanted to find log 56.248 from a five-place table, he would have to use interpolation, taking as his mantissa the number that is eight-tenths of the way from that given for 56240 to that given for 56250.

59. ***Finding a number when its logarithm is given.*** If $\log N$ is given and we wish to find N, we must reverse the above proce-

Common Logarithms.

N	0	1	2	3	4	5	6	7	8	9
55	7404	7412	7419	7427	7435	7443	7451	7459	7466	7474
56	7482	7490	7497	7505	7513	7520	7528	7536	7543	7551
57	7559	7566	7574	7582	7589	7597	7604	7612	7619	7627
58	7634	7642	7649	7657	7664	7672	7679	7686	7694	7701
59	7709	7716	7723	7731	7738	7745	7752	7760	7767	7774
60	7782	7789	7796	7803	7810	7818	7825	7832	7839	7846
61	7853	7860	7868	7875	7882	7889	7896	7903	7910	7917
62	7924	7931	7938	7945	7952	7959	7966	7973	7980	7987
63	7993	8000	8007	8014	8021	8028	8035	8041	8048	8055
64	8062	8069	8075	8082	8089	8096	8102	8109	8116	8122
65	8129	8136	8142	8149	8156	8162	8169	8176	8182	8189
66	8195	8202	8209	8215	8222	8228	8235	8241	8248	8254
67	8261	8267	8274	8280	8287	8293	8299	8306	8312	8319
68	8325	8331	8338	8344	8351	8357	8363	8370	8376	8382
69	8388	8395	8401	8407	8414	8420	8426	8432	8439	8445
70	8451	8457	8463	8470	8476	8482	8488	8494	8500	8506
71	8513	8519	8525	8531	8537	8543	8549	8555	8561	8567
72	8573	8579	8585	8591	8597	8603	8609	8615	8621	8627
73	8633	8639	8645	8651	8657	8663	8669	8675	8681	8686
74	8692	8698	8704	8710	8716	8722	8727	8733	8739	8745
75	8751	8756	8762	8768	8774	8779	8785	8791	8797	8802
76	8808	8814	8820	8825	8831	8837	8842	8848	8854	8859
77	8865	8871	8876	8882	8887	8893	8899	8904	8910	8915
78	8921	8927	8932	8938	8943	8949	8954	8960	8965	8971
79	8976	8982	8987	8993	8998	9004	9009	9015	9020	9025
80	9031	9036	9042	9047	9053	9058	9063	9069	9074	9079
81	9085	9090	9096	9101	9106	9112	9117	9122	9128	9133
82	9138	9143	9149	9154	9159	9165	9170	9175	9180	9186
83	9191	9196	9201	9206	9212	9217	9222	9227	9232	9238
84	9243	9248	9253	9258	9263	9269	9274	9279	9284	9289
85	9294	9299	9304	9309	9315	9320	9325	9330	9335	9340
86	9345	9350	9355	9360	9365	9370	9375	9380	9385	9390
87	9395	9400	9405	9410	9415	9420	9425	9430	9435	9440
88	9445	9450	9455	9460	9465	9469	9474	9479	9484	9489
89	9494	9499	9504	9509	9513	9518	9523	9528	9533	9538
90	9542	9547	9552	9557	9562	9566	9571	9576	9581	9586
91	9590	9595	9600	9605	9609	9614	9619	9624	9628	9633
92	9638	9643	9647	9652	9657	9661	9666	9671	9675	9680
93	9685	9689	9694	9699	9703	9708	9713	9717	9722	9727
94	9731	9736	9741	9745	9750	9754	9759	9763	9768	9773
95	9777	9782	9786	9791	9795	9800	9805	9809	9814	9818
96	9823	9827	9832	9836	9841	9845	9850	9854	9859	9863
97	9868	9872	9877	9881	9886	9890	9894	9899	9903	9908
98	9912	9917	9921	9926	9930	9934	9939	9943	9948	9952
99	9956	9961	9965	9969	9974	9978	9983	9987	9991	9996
N	0	1	2	3	4	5	6	7	8	9

dure. Suppose, for example, that we wish to find the number N given that

$$\log N = 7.8138 - 10.$$

At first we disregard the characteristic, which is to be used only in locating the decimal point in N, and hunt for the number that has the mantissa 8138. We do not find this mantissa in the table, but we do find the mantissas 8136 and 8142. These correspond to the numbers 651 and 652, respectively. Now our given mantissa, 8138, is two-sixths or one-third of the way from 8136 to 8142. We therefore take as our number the one that is one-third of the way from 6510 to 6520. This is 6513. In this number we place the decimal point so that the characteristic is -3. Finally then, if $\log N = 7.8138 - 10$, $N = 0.006513$.

PROBLEMS

1. Using Theorems I, II, and III of Sec. 55, show that if x, y, and z are positive numbers, then

$$\log \frac{xy^2}{z\sqrt{x^2 + y^2}} = \log x + 2 \log y - \log z - \tfrac{1}{2} \log (x^2 + y^2).$$

2. Using Theorems I, II, and III of Sec. 55, show that if x, y, and z are positive numbers, then

$$\log \frac{x\sqrt[3]{yz}}{\sqrt{y^2 + z^2}} = \log x + \tfrac{1}{3} \log y + \tfrac{1}{3} \log z - \tfrac{1}{2} \log (y^2 + z^2).$$

3. Classify the following statement as true or false: If x and y are positive numbers, then $\log \sqrt{x} = \frac{1}{2} \log x$ and $\log \sqrt{y} = \frac{1}{2} \log y$, and hence

$$\log (\sqrt{x} + \sqrt{y}) = \tfrac{1}{2} \log x + \tfrac{1}{2} \log y.$$

4. Classify the following statement as true or false: If x and y are positive numbers, then $\log x^3 = 3 \log x$ and $\log y^3 = 3 \log y$, and hence

$$\log (x^3 + y^3) = 3 \log x + 3 \log y.$$

5. Write down the common logarithm of each of the following numbers: 1,000,000; 100,000,000; 0.0001; 0.0000001; 10^{12}; 10^{-9}; 10^0.

In each of the following cases find the common logarithms of the given numbers using a four-place table (page 302):

6. 23.8; 0.0721; 16,400.
7. 9.64; 0.0352; 1,500.
8. 473; 0.882; 0.00039.
9. 75.6; 40,000; 0.00181.
10. 3.87; 0.00094; 47,500.
11. 978; 0.423; 290,000.
12. 10.4; 3,000; 0.00725.
13. 14.63; 0.7237; 3,942.
14. 9.835; 0.004624; 7,007.
15. 21.78; 0.5982; 683,500.
16. 1.634; 0.09042; 47,200.
17. 3.142; 0.002184; 8,888.
18. 2.718; 0.0375; 81,170.

In each of the following cases find the common logarithms of the given numbers using a five-place table:

19. 7.242; 0.003821; 71,650.
20. 14.56; 0.8254; 54,760.
21. 9.872; 0.0324; 6,185.
22. 35.55; 0.0007285; 247,600.
23. 16.254; 0.38542; 52,754.
24. 9.8916; 0.072347; 355,550.
25. 3.1416; 0.0082571; 46,425.

In each of the following cases use a four-place table to find the numbers whose common logarithms are given:

26. $\log A = 1.7642$; $\log B = 8.1724 - 10$.
27. $\log A = 3.7605$; $\log B = 6.4185 - 10$.
28. $\log A = 0.4621$; $\log B = 9.2746 - 10$.
29. $\log x = 4.5342$; $\log y = 7.9261 - 10$.
30. $\log x = 3.2324$; $\log y = 0.7652 - 2$.
31. $\log r = 0.3675$; $\log s = 8.6494 - 10$.
32. $\log M = 2.0465$; $\log N = 0.9146 - 1$.
33. $\log M = 3.1800$; $\log N = 2.8010 - 5$.
34. $\log M = 1.3476$; $\log N = 6.5457 - 9$.

In each of the following cases use a five-place table to find the numbers whose common logarithms are given:

35. $\log A = 3.71628$; $\log B = 8.41265 - 10$.
36. $\log A = 1.47124$; $\log B = 9.23571 - 10$.
37. $\log A = 5.64722$; $\log B = 0.72841 - 2$.
38. $\log x = 0.87885$; $\log y = 6.18284 - 10$.
39. $\log x = 2.23714$; $\log y = 0.05261 - 3$.
40. $\log x = 4.44642$; $\log y = 17.90105 - 20$.

41. If $\log N = -2.4374$, find the characteristic and mantissa of $\log N$ and then find N.

42. If $\log N = -3.8132$, find the characteristic and mantissa of $\log N$ and then find N.

60. ***Computation with logarithms.*** We give below several examples illustrating the use of logarithms in problems of multiplication and division, and in the operations of finding powers and roots of numbers.

Example 1

Calculate $\frac{(7.34)(87.6)}{0.529}$.

Solution

If we denote the result by N so that

$$N = \frac{(7.34)(87.6)}{0.529},$$

then, from Theorems I and II (page 129),

$$\begin{aligned} \log N &= \log [(7.34)(87.6)] - \log 0.529 \\ &= \log 7.34 + \log 87.6 - \log 0.529. \end{aligned}$$

$$\begin{aligned} \log 7.34 &= 0.8657 \\ \log 87.6 &= \underline{1.9425} \\ \log 7.34 + \log 87.6 &= 2.8082 \\ \log 0.529 &= \underline{9.7235 - 10} \\ \log N &= 3.0847^{*} \end{aligned}$$

$$N = 1{,}215.$$

Example 2

Compute $\sqrt[3]{74{,}650}$.

Solution

If we let x denote the required number so that $x = \sqrt[3]{74{,}650}$, then from Theorem III (page 129),

$$\log x = \tfrac{1}{3} \log 74{,}650.$$

$$\begin{aligned} \log 74{,}650 &= 4.87303; \\ \log x = \tfrac{1}{3}(4.87303) &= 1.62434. \\ x &= 42.105. \end{aligned}$$

(In this case we have used a five-place table.)

Example 3

Compute $(17.6)^{1.4}$.

* In order to perform the subtraction, we may write 2.8082 as 12.8082 − 10.

Solution

If we let N denote the required result so that $N = (17.6)^{1.4}$, then by Theorem III (page 129),

$$\log N = 1.4 \log 17.6.$$

$$\begin{aligned} \log 17.6 &= 1.2455 \\ \log N = 1.4(\log 17.6) &= (1.4)(1.2455) \\ &= 1.7437. \\ N &= 55.4. \end{aligned}$$

Example 4

Compute $(0.834)^{\frac{2}{7}}$.

Solution

If we denote the required number by x, then

$$\log x = \tfrac{2}{7} \log 0.834.$$

$$\begin{aligned} \log 0.834 &= 0.9212 - 1 \\ &= 6.9212 - 7.^{*} \\ \log x = \tfrac{2}{7}(6.9212 - 7) &= 1.9775 - 2 \\ &= 0.9775 - 1. \\ x &= 0.949. \end{aligned}$$

Example 5

Compute $\dfrac{(379.5)(6.089)(0.3214)}{(0.08935)\sqrt{17.65}}$.

Solution

It is convenient to let x denote the required result and arrange the work as follows:

Numerator

$$\begin{aligned} \log 379.5 &= 2.57921 \\ \log 6.089 &= 0.78455 \\ \log 0.3214 &= 9.50705 - 10 \\ \hline \log \text{num.} &= 12.87081 - 10 \\ \log \text{denom.} &= 9.57446 - 10 \\ \hline \log x &= 3.29635. \\ x &= 1{,}978.5. \end{aligned}$$

Denominator

$$\begin{aligned} \log 0.08935 &= 8.95109 - 10 \\ \tfrac{1}{2} \log 17.65 &= 0.62337 \\ \hline \log \text{denom.} &= 9.57446 - 10 \end{aligned}$$

* We replace the characteristic -1 by $6 - 7$ in order to make it easier to multiply by $\frac{2}{7}$.

PROBLEMS

In each of the following problems perform the indicated computation using a four-place table:

1. $(37.8)(0.216)(0.00943)$.
2. $(0.644)(98.2)(1.46)$.
3. $(26{,}100)(0.0816)(0.334)$.
4. $(43.2)(9.15)(0.00744)$.
5. $(54.63)(165.4)(0.03562)$.
6. $(2.574)(0.8125)(783.7)$.
7. $\dfrac{(37.46)(0.06510)}{0.08793}$.
8. $\dfrac{(751.4)(0.09251)}{1.353}$.
9. $\dfrac{(0.03582)(0.8768)}{0.004144}$.
10. $\dfrac{654.6}{(2.147)(0.03771)}$.
11. $\dfrac{(56.2)(1.65)(0.0941)}{(0.00246)(0.0835)}$.
12. $\dfrac{(21.2)(426)(.00415)}{(6.95)(0.0127)}$.
13. $\dfrac{(2.67)(528)(0.255)}{(0.0365)(23.9)}$.
14. $\dfrac{(894)(0.215)(0.0771)}{(0.832)(0.00616)}$.
15. $\sqrt[3]{714.2}$.
16. $\sqrt[5]{85.6}$.
17. $\sqrt[3]{0.218}$.
18. $\sqrt{825.7}$.
19. $(16.62)^{0.4}$.
20. $(5.74)^{2.5}$.
21. $(2.53)^{3.6}$.
22. $(0.834)^{2.4}$.
23. $(0.0944)^{0.3}$.
24. $(1.725)^{6}$.
25. $\dfrac{(26.43)(1.892)^3}{(0.842)\sqrt{14.34}}$.
26. $\dfrac{(8.92)^3\sqrt{0.0722}}{365.8}$.

In each of the following problems perform the indicated computation using a five-place table:

27. $\sqrt[3]{\dfrac{(82{,}506)(12.624)}{7.1252}}$.
28. $(9.8026)(15.435)^{0.4}$.
29. $\dfrac{(456.28)\sqrt{22.465}}{(7.2146)(0.038524)}$.
30. $\dfrac{(9.4275)(763.48)}{(88.534)\sqrt[3]{17.622}}$.

31. If P dollars are invested at 4 per cent compounded annually, the amount at the end of n years is given by the formula $A = P(1.04)^n$. If $P = \$5{,}000$ and $n = 12$ years, find A.

32. Using the formula given in Prob. 31, find A if $P = \$750$ and $n = 20$ years.

33. Compute the circumference of a circle whose area is 1,624 sq. in.

34. The area of the surface of a sphere of radius r is $4\pi r^2$. Assuming that the earth is a sphere of radius 3,960 miles, compute the area of its surface in square miles.

35. If a, b, and c are the lengths of the sides of a triangle, the radius of the inscribed circle is given by the formula

$$r = \sqrt{\frac{(s-a)(s-b)(s-c)}{s}}, \qquad \text{where } s = \tfrac{1}{2}(a+b+c).$$

Compute r, given that $a = 63.42$ ft., $b = 73.84$ ft., $c = 43.62$ ft.

36. If a, b, and c are the lengths of the sides of a triangle, the area k of the triangle is given by the formula

$$k = \sqrt{s(s-a)(s-b)(s-c)} \qquad \text{where } s = \tfrac{1}{2}(a+b+c).$$

Compute the area of a triangle whose sides are 16.84 ft., 12.26 ft., and 21.82 ft.

37. The period T of a simple pendulum of length L ft. is the time in seconds required for the pendulum to make one complete oscillation. This period is given by the formula

$$T = 2\pi \sqrt{L/g} \qquad \text{where } g = 32.2.$$

Assuming that $\pi = 3.14$, compute T for the case in which $L = 4.36$ ft.

38. Using the formula of Prob. 37, compute the length L of a simple pendulum for which the period is 2.75 sec.

61. *Change of base.* We have mentioned that any positive number other than 1 can be used as a base for a system of logarithms, and we have seen that for computational purposes it is convenient to use 10 as a base. A system in which the irrational number $2.718 \cdots$ is the base will be discussed briefly in Sec. 62. This system is simpler in certain respects than the system of common logarithms.

In the present section we wish to derive a relation between $\log_a N$ and $\log_b N$, where a and b are any two bases—a relation by means of which we can, for example, compute $\log_{17} 385$ from the known values of $\log_{10} 385$ and $\log_{10} 17$. We have the following:

Theorem. *The logarithm of any positive number* **N** *to any base* **b** *is equal to the logarithm of* **N** *to any other base* **a** *divided by the logarithm of* **b** *to the base* **a**; *i.e.,*

$$\log_b N = \frac{\log_a N}{\log_a b}.$$

Proof: Let $\log_b N = x$; then $N = b^x$ and, consequently,

$$\begin{aligned} \log_a N &= \log_a (b^x) \\ &= x \log_a b. \end{aligned}$$

Solving for x we get

$$x = \frac{\log_a N}{\log_a b}.$$

But we had let x denote $\log_b N$, so we have

$$\log_b N = \frac{\log_a N}{\log_a b}.$$

Example

$$\log_{17} 385 = \frac{\log_{10} 385}{\log_{10} 17} = \frac{2.58546}{1.23045} = 2.10123.$$

The student should be careful not to confuse this operation with that in which we subtract log 17 from log 385 to obtain log $\frac{385}{17}$. In the present operation we *divide* log 385 by log 17 to obtain $\log_{17} 385$. One can of course use logarithms in this division if he wishes; he is in fact free to divide 2.58546 by 1.23045 by any method that suits his fancy.

By letting $N = a$, we obtain the following corollary to the above theorem:

$$\log_b a = \frac{1}{\log_a b}.$$

Thus, for example, $\log_{10} 100$ is the reciprocal of $\log_{100} 10$.

62. ***Natural logarithms.*** Logarithms to the base e, where e is an irrational number whose value to five decimal places is 2.71828, are called *natural* or *Napierian** logarithms. This number is defined by the infinite series

$$e = 1 + \frac{1}{1} + \frac{1}{1 \cdot 2} + \frac{1}{1 \cdot 2 \cdot 3} + \frac{1}{1 \cdot 2 \cdot 3 \cdot 4} + \cdots,$$

and its value can be obtained to any degree of accuracy by taking a sufficient number of terms. It can be shown by methods of calculus that, if x is any number between -1 and $+1$, then

$$\log_e \frac{1 + x}{1 - x} = 2\left(x + \frac{x^3}{3} + \frac{x^5}{5} + \frac{x^7}{7} + \cdots\right).$$

From this series one can compute the natural logarithms of numbers, and then by using the formula for change of base that we developed in Sec. 61, he can find the corresponding common logarithms. Thus

$$\log_{10} N = \frac{\log_e N}{\log_e 10}.$$

* After J. Napier, 1550–1617, to whom the invention of logarithms is usually credited.

LOG

We are more frequently concerned with the problem of computing $\log_e N$ from our table of common logarithms. For this purpose we of course have the relation

$$\log_e N = \frac{\log_{10} N}{\log_{10} e}.$$

Now $\log_{10} e = \log_{10} 2.71828 = 0.43429$, so that

$$\boldsymbol{\log_e N = \frac{\log_{10} N}{0.43429} = 2.3026 \log_{10} N.}$$

Thus *one multiplies the common logarithm of any number by 2.3026 to obtain the corresponding natural logarithm.*

Examples

$$\log_e 65 = 2.3026(\log 65) = 2.3026(1.81291) = 4.1744.$$
$$\log_e 650 = 2.3026(\log 650) = 2.3026(2.81291) = 6.4770.$$

It is to be noted that the natural logarithms of 65 and 650 do not have the same decimal part. In fact, since $650 = (65)(10)$,

$$\begin{aligned}\log_e 650 &= \log_e 65 + \log_e 10 \\ &= \log_e 65 + 2.3026.\end{aligned}$$

Multiplying a number by 10 thus increases its natural logarithm by 2.3026 instead of by 1 as was the case for common logarithms.

Table IV (page 310) in this text gives the natural logarithms of three-digit numbers from 1 to 10, to four decimal places. One would find log 650 from this table thus: Since $650 = 6.50 \times 10^2$,

$$\begin{aligned}\log_e 650 &= \log_e 6.50 + 2(\log_e 10) \\ &= 1.8718 + 2(2.3026) \\ &= 6.4770.\end{aligned}$$

63. *The logarithmic function.* In the field of real numbers the logarithmic function $\boldsymbol{\log_a x}$ $(a > 1)$ is defined for all positive values of x. Its value increases as x increases—from a large negative value for x near zero up to zero at $x = 1$, and on up through positive values for $x > 1$. The graph has the form shown in Fig. 26.

64. ***Exponential and logarithmic equations.*** An equation in which the unknown appears in an exponent is usually called an

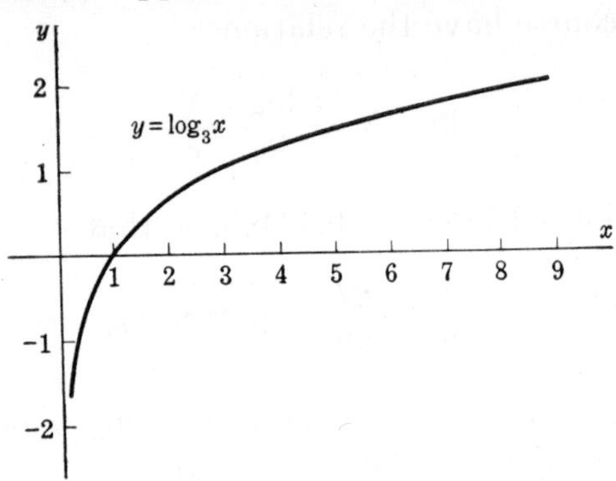

Figure 26

exponential equation. Such an equation can sometimes be solved by means of logarithms.

Example 1

Find any solutions that may exist for the equation

$$12^{2x-1} = 7(2.6)^x. \tag{1}$$

Solution

If x is a number such that the two members of (1) are equal, then their common logarithms must be equal; *i.e.*,

$$\log [12^{2x-1}] = \log [7 \cdot (2.6)^x] \tag{2}$$

or

$$(2x - 1) \log 12 = \log 7 + x \log (2.6). \tag{3}$$

This is a linear equation in x which we may solve in the usual way:

$$(2 \log 12 - \log 2.6)x = \log 7 + \log 12; \tag{4}$$

$$x = \frac{\log 7 + \log 12}{2 \log 12 - \log 2.6} = \frac{1.9243}{1.7434} = 1.1038. \tag{5}$$

If (1) is to be true, then (5) must be true. This means that no number other than $x = 1.1038$ can satisfy (1). If we start with (5) as our hypothesis and reverse the steps, we see that (1) is a consequence of (5). This means that 1.1038 is actually a root.

The term *logarithmic equation* is usually applied to an equation in which the logarithm of an expression involving the unknown occurs.

Example 2

Find any solutions that may exist for the equation

$$\log (6x + 40) = 2 + \log \tfrac{1}{3}x. \tag{1}$$

Solution

If x is a number such that (1) is true, then

$$\log (6x + 40) - \log \tfrac{1}{3}x = 2; \tag{2}$$

$$\log \frac{6x + 40}{\frac{1}{3}x} = 2; \tag{3}$$

$$\log \frac{18x + 120}{x} = 2. \tag{4}$$

Now if (4) is to be true, we must have

$$\frac{18x + 120}{x} = 10^2 = 100. \tag{5}$$

From (5) we find immediately that

$$82x = 120 \quad \text{or} \quad x = \tfrac{60}{41}. \tag{6}$$

Thus if (1) is to be true, then we must have $x = \frac{60}{41}$. The student should show that this number is actually a root by direct substitution.

Example 3

Find all real values of x, if there are any, for which

$$4^x = 18 + 3(2^x). \tag{1}$$

Solution

Since $4 = 2^2$, $4^x = (2^2)^x = 2^{2x} = (2^x)^2$. We can therefore write (1) as a quadratic equation in 2^x and then solve it for 2^x:

$$(2^x)^2 - 3(2^x) - 18 = 0; \tag{2}$$

$$(2^x - 6)(2^x + 3) = 0; \tag{3}$$

$$2^x = 6 \quad \text{or} \quad 2^x = -3. \tag{4}$$

There is no value of x for which $2^x = -3$, but we can have $2^x = 6$ if

$$x \log 2 = \log 6, \tag{5}$$

or

$$x = \frac{\log 6}{\log 2} = \frac{0.77815}{0.30103} = 2.5850. \tag{6}$$

That (1) is true if x has the value given by (6) follows immediately from the reversibility of the steps.

PROBLEMS

Compute the following logarithms to four significant digits:

1. $\log_{14} 725$. **2.** $\log_{34} 2{,}670$. **3.** $\log_{85} 58$.
4. $\log_{25} 10{,}000$. **5.** $\log_6 774$. **6.** $\log_{20} 4.82$.
7. $\log_e 2$. **8.** $\log_e 1{,}000$. **9.** $\log_e 1.84$.
10. $\log_e 384$. **11.** $\log_e 49.8$. **12.** $\log_e 1{,}225$.

13. Draw the graph of the equation $y = \log_2 x$ over the interval from $x = \frac{1}{8}$ to $x = 8$.

14. Draw the graph of the equation $y = \log_4 x$ over the interval from $x = \frac{1}{16}$ to $x = 8$.

15. Sketch the graph of the equation $y = \log_{10} x$. Explain the graphical interpretation of the interpolation process given in this chapter. If one should find log 6.5 by interpolation from log 6 and log 7, would his result be a little too small or a little too large?

Solve each of the following equations for x:

16. $6^{2x-1} = 216$. **17.** $4^{x^2} = 8 \cdot 2^{-x}$. **18.** $5^x = 32$.
19. $(15.2)^x = 7.64$. **20.** $(1.05)^x = 4.00$. **21.** $3^{2x-1} = 78$.
22. $9^{x-2} = 7^x$. **23.** $5^{2x-4} = 3^x$. **24.** $124^x = 12(5^x)$.
25. $(1.04)^{-x} = 0.75$. **26.** $16^{3x-2} = 8(4.2)^x$. **27.** $10^x = 50^{x+2}$.
28. $(8.4)^x = (5.9)(3.4)^x$. **29.** $67^x = 3(5.8)^{x+1}$.

Solve each of the following equations for x:

30. $\log (x + 1) - \log 2 = \log 8$. **31.** $\log x^2 = 6 \log 2$.
32. $\log (2x + 7) = \log 4 + \log (2 - x)$.
33. $\log_5 (3x + 2) = 3$. **34.** $\log (50x + 10) = 2 + \log 2x$.
35. $\log (30x + 100) = 1 + 2 \log x$.

36. If $\frac{1}{2}x + \log x = 3$, then $\log x = 3 - \frac{1}{2}x$. Sketch the graphs of the equations $y = \log x$ and $y = 3 - \frac{1}{2}x$ on the same axes, and infer from these graphs that the equation $\frac{1}{2}x + \log x = 3$ has one and only one real root. Between what two consecutive integers does it lie?

37. If $x = 3 + \log x$, then $x - 3 = \log x$. Draw the graphs of the equations $y = x - 3$ and $y = \log x$ on the same axes, and from these graphs infer that the equation $x = 3 + \log x$ has exactly two real roots. Locate each root between two consecutive integers.

38. Find all real roots of the equation $3^x + \dfrac{6}{3^x} = 5$.

39. Find all real roots of the equation $\frac{1}{2}(e^x - e^{-x}) = 4$.

40. Solve the equation $A = P(1 + r)^n$ for n in terms of A, P, and r.

41. Solve the equation $y = \frac{1}{2}(e^x + e^{-x})$ for x in terms of y.

42. Solve the equation $y = \dfrac{e^x - e^{-x}}{e^x + e^{-x}}$ for x in terms of y.

CHAPTER IX

VARIATION

65. ***The terminology of variation.*** The topic of this chapter is of little importance from the viewpoint of the mathematician. It is discussed briefly here because the student will encounter the terminology frequently in his courses in physics, chemistry, and other sciences. He may read, for example, that "the electrical resistance of a wire varies directly as its length and inversely as its cross-sectional area," or that 'the square of the time it takes a planet to make a circuit about the sun varies as the cube of its mean distance from the sun," or that "the force of attraction between two spheres varies jointly as their masses and inversely as the square of the distance between their centers." Our present task is that of learning how to interpret and use such statements.

The statement *y varies directly as x*, or *y is proportional to x*, means that the relation between x and y is of the form

$$y = kx$$

where k is a constant. Similarly, the statement *y varies directly as the square of x*, or *y is proportional to the square of x*, means that

$$y = kx^2,$$

etc.

Examples

The circumference of a circle varies directly as its radius; the area of a circle varies directly as the square of its radius; the volume of a sphere varies directly as the cube of its radius.

The statement *y varies inversely as x* means that the relation between x and y is of the form

$$y = \frac{k}{x}$$

where k is a constant.

A typical problem involving these ideas is the following:

Example

A cylinder contains 400 cu. in. of air at a pressure of 15 lb. per sq. in. It is known that, if the air is compressed while its temperature is held fixed, the pressure varies inversely as the volume. Find the pressure when the air has been compressed into a volume of 250 cu. in.

Solution

Since the pressure varies inversely as the volume, we have the relation

$$p = \frac{k}{v}.$$

We find the value of k from the fact that when $v = 400$ cu. in., $p = 15$ lb. per sq. in.

$$15 = \frac{k}{400} \qquad \text{or} \qquad k = 6{,}000.$$

The relation between p and v is then

$$p = \frac{6{,}000}{v}.$$

From this we can find the pressure corresponding to any specified volume; in particular, for $v = 250$ cu. in.,

$$p = \frac{6{,}000}{250} = 24 \text{ lb. per sq. in.}$$

The statement w *varies jointly as* x *and* y, or w *is proportional to* x *and* y, means that

$$w = kxy$$

where k is a constant. For example, the volume of a right circular cylinder varies jointly as its height and the square of its radius, the relation being $v = \pi r^2 h$.

Finally, the statement w *varies jointly as* x *and* y *and inversely as* z means that

$$w = k\frac{xy}{z}$$

where k is a constant.

Example 1

The load that can safely be applied to a beam of rectangular cross section varies jointly as the width and square of the depth of the cross section, and inversely as the length of the beam. If a 2- by 4-in. beam 6 ft. long safely supports a load of 400 lb., what is the corresponding safe load on a 2- by 6-inch beam 10 ft. long? Assume in each case that the shorter dimension of the cross section is the width.

Solution

If we let b and d be the width and depth of the cross section in inches, L the length in feet, and w the safe load, then we have

$$w = k\,\frac{bd^2}{L}.$$

We determine k from the fact that $w = 400$ lb. when $b = 2$ in., $d = 4$ in., and $L = 6$ ft.:

$$400 = k\,\frac{2(4^2)}{6} \qquad \text{or} \qquad k = \frac{6(400)}{2(16)} = 75.$$

We then have

$$w = 75\,\frac{bd^2}{L}.$$

Now if $b = 2$ in., $d = 6$ in., and $L = 10$ ft.,

$$w = 75\,\frac{2(6^2)}{10} = 540 \text{ lb.}$$

Example 2

Other conditions being equal, the thrust T exerted by a propeller varies jointly as the fourth power of its diameter and the square of the number of revolutions per minute. What would be the effect of increasing the number of revolutions per minute by 50 per cent and decreasing the diameter by $33\frac{1}{3}$ per cent?

Solution

For a propeller of diameter D_1 turning at n_1 r.p.m., the thrust is

$$T_1 = kD_1^4n_1^2.$$

If for another propeller we have $D_2 = \frac{2}{3}D_1$ and $n_2 = \frac{3}{2}n_1$, then its thrust is

$$T_2 = kD_2{}^4n_2{}^2 = k(\tfrac{2}{3}D_1)^4(\tfrac{3}{2}n_1)^2 = \tfrac{4}{9}kD_1{}^4n_1{}^2 = \tfrac{4}{9}T_1.$$

Thus the thrust of the second propeller would be four-ninths that of the first one, so that the effect would be to decrease the thrust by five-ninths or about 56 per cent.

PROBLEMS

1. If y is proportional to $\sqrt{x}$ and $y = 14$ when $x = 9$, find the value of y when $x = 25$.

2. If w varies directly as x^2 and inversely as h and if $w = 62$ when $x = 4$ and $h = 3$, find the value of w when $x = 3$ and $h = 31$.

3. If Q varies jointly as x and y^2 and inversely as z and if $Q = 128$ when $x = 4$, $y = 3$, and $z = 6$, find the value of Q when $x = 18$, $y = 2$, and $z = 12$.

4. If y is proportional to $\log x$ and if $y = 30$ when $x = 6$, find the value of y when $x = 216$.

5. If the horsepower required to propel a certain ship is proportional to the cube of its speed and if the ship requires 2,400 hp. for a speed of 12 knots, what horsepower would be needed for a speed of 15 knots?

6. Suppose that y varies directly as x and that $y = y_1$ when $x = x_1$, and $y = y_2$ when $x = x_2$. Show that

$$\frac{y_1}{y_2} = \frac{x_1}{x_2}.$$

7. Suppose that y varies inversely as x and that $y = y_1$ when $x = x_1$ and $y = y_2$ when $x = x_2$. Show that

$$\frac{y_1}{y_2} = \frac{x_2}{x_1}.$$

8. Show that if w varies directly as y^2 and if y varies directly as x^3, then w varies directly as x^6.

9. Show that if w varies inversely as y and if y varies inversely as x, then w varies directly as x.

10. If the volume of a gas remains constant, its pressure varies directly as its absolute temperature. The pressure of the gas in a 600 cu. ft. container is 28 lb. per sq. in. when the absolute temperature is 520°F. What will be the pressure if the temperature is raised to 600°F.?

11. The kinetic energy of a moving body (which is not rotating) is proportional to the square of its speed. Compare the kinetic energy of a body moving at 20 m.p.h. with that of the same body moving at 60 m.p.h.

12. The horsepower that a rotating shaft can safely transmit varies jointly as the cube of its diameter and the number of revolutions it makes per minute. If a 2-in. shaft rotating at 800 r.p.m. can transmit 180 hp., what horsepower can a $1\frac{1}{2}$-in. shaft transmit at 1,200 r.p.m.?

13. The pressure of a given quantity of a gas varies directly as its absolute temperature and inversely as its volume. A cylinder contains 80 cu. in. of gas at an absolute temperature of 600°F. and the pressure is 48 lb. per sq. in. Find the pressure if the gas is compressed to 60 cu. in. and the temperature raised to 660°F.

14. The period of a simple pendulum, which is the time required for one complete swing, is proportional to the square root of its length. If the period is 2.72 sec. for a pendulum 6 ft. long, what should be the period if the length is 8 ft.?

15. The weight of an object above the surface of the earth varies inversely as the square of its distance from the center of the earth. If a man weighs 200 lb. at the earth's surface, how much would he weigh if he were 240 miles above the surface? Take the radius of the earth as 3,960 miles.

16. Other conditions being equal, the thrust of a propeller varies jointly as the fourth power of its diameter and the square of the number of revolutions per minute. Show that the effect of decreasing the diameter by 50 per cent and doubling the number of revolutions per minute would be to decrease the thrust by 75 per cent.

17. The electrical resistance of a copper wire varies directly as its length and inversely as the square of its diameter. What will be the effect of replacing a certain wire with another whose length is 50 per cent greater and whose diameter is 25 per cent greater?

18. The square of the time required for a planet to make one circuit about the sun varies as the cube of its mean distance from the sun. The mean distance of the earth from the sun 92.9 million miles and that of Saturn is 886 million miles. Find the time required for Saturn to make one circuit about the sun.

CHAPTER X

PROGRESSIONS

66. ***Definition of a sequence.*** Consider the function

$$f(x) = 3x - 1$$

and let x be restricted to the set of natural numbers; *i.e.*, let x take only the values 1, 2, 3, 4, 5, and so on. The corresponding values of the function are

$$f(1) = 2, f(2) = 5, f(3) = 8, f(4) = 11, f(5) = 14, \text{ etc.}$$

These numbers constitute a *sequence* in which the first term is 2, the second term is 5, the third term is 8, and so on, the nth term being $3n - 1$. The definitions are as follows:

Let $f(x)$ be a function whose range of definition is the set of all natural numbers. The corresponding set of function values is called an *infinite sequence.* The individual function values are called *terms* of the sequence, $f(1)$ being the first term, $f(2)$ the second term, and so on.

An infinite sequence is often designated by writing down the first few terms and the nth term in the form illustrated below. A shorter notation consists simply of writing the expression for the nth term and enclosing it in braces.

Example

The above sequence may be designated by writing

$$2, 5, 8, \cdots, 3n - 1, \cdots$$

or, more briefly, by writing

$$\{3n - 1\}.$$

If the range of definition of the function is the set of all natural numbers less than or equal to some specified natural number k, the corresponding set of function values is called a *finite sequence.* Consider, for example, the function

$$g(x) = \frac{2^x}{4}$$

and let x be restricted to the natural numbers less than or equal to 8. The corresponding function values are $g(1) = \frac{1}{2}$, $g(2) = 1$, $g(3) = 2$, $g(4) = 4$, and so on, up to and including $g(8) = 64$. We might designate this sequence by writing down the first few terms, the nth term, and the last term or last two terms, as follows:

$$\frac{1}{2}, 1, 2, \cdots, \frac{2^n}{4}, \cdots, 32, 64.$$

A shorter notation would be

$$\left\{\frac{2^n}{4}\right\}_{n \leqq 8} \qquad \text{or even} \qquad \left\{\frac{2^n}{4}\right\}_8$$

it being understood in this last notation that the sequence consists of the function values corresponding to $n = 1$ to 8, inclusive.

Any finite set of numbers $a_1, a_2, a_3, \cdots, a_n$, taken in order, can of course be regarded as a sequence. In this case the values of the function $f(x)$ for $x = 1, 2, \cdots, n$ have been specified to be $a_1, a_2, \cdots, a_n$, respectively, without specifying the form of the function.

Example

The numbers

$$3, 4, 5, 8, 17, 40$$

taken in this order, constitute a sequence. To ask for *the* function $f(x)$ whose values for $x = 1, 2, \cdots, 6$ are these numbers is nonsense for there are infinitely many such functions. One of them is

$$2x - x^2 + 2^x;$$

another is

$$2^x + (2 - x)[x + (x - 1)(x - 3)(x - 4)(x - 5)(x - 6)].$$

67. ***Arithmetic progressions.*** An *arithmetic progression* is a sequence in which every term after the first can be obtained from the term immediately preceding by adding to it a fixed number d. This number d is called the *common difference.*

Examples

The sequence

$$7, 10, 13, 16, 19, 22, 25, 28$$

is an arithmetic progression having eight terms; the first term is 7 and the common difference is 3.

The sequence $\{11 - 5n\}$ is an infinite arithmetic progression whose first term is 6 and whose common difference is -5. The first seven terms are $6, 1, -4, -9, -14, -19, -24$.

68. ***Formula for the nth term of an arithmetic progression.*** Let a_1 denote the first term, d the common difference, and n the number of terms of an arithmetic progression. We wish to derive a formula for the nth term in terms of this data. We observe that

$$\begin{aligned} \text{Second term} &= a_1 + d, \\ \text{Third term} &= (a_1 + d) + d = a_1 + 2d, \\ \text{Fourth term} &= (a_1 + 2d) + d = a_1 + 3d, \\ \text{Fifth term} &= (a_1 + 3d) + d = a_1 + 4d, \end{aligned}$$

and so on. If we denote the nth term by a_n, we see immediately that

$$a_n = a_1 + (n - 1)d.$$

Example

Find the 225th positive odd integer.

Solution

If we were to write down the first 225 terms of the arithmetic progression

$$1, 3, 5, 7, 9, \cdots, 2n - 1, \cdots,$$

we would have $a_1 = 1$, $d = 2$, and $n = 225$. The 225th term would be

$$a_{225} = 1 + (224) \cdot 2 = 449.$$

69. ***Formula for the sum of the first n terms of an arithmetic progression.*** We wish next to derive a formula by which we can compute rather easily the sum of the first n terms of a given finite arithmetic progression—a formula that gives the sum in terms of the first and last terms and the number of terms.

If we denote this sum by S_n, then of course

$$S_n = a_1 + (a_1 + d) + (a_1 + 2d) + \cdots + [a_1 + (n - 1)d].$$

An equally valid formula for S_n is obtained if we add up all the terms in the reverse order; *i.e.*, starting with the last term, which we denote by a_n. The next to last term is then $a_n - d$, the next is $a_n - 2d$, etc., and, finally, the first term is $a_n - (n - 1)d$. The sum is

$$S_n = a_n + (a_n - d) + (a_n - 2d) + \cdots + [a_n - (n - 1)d].$$

If we add these two expressions for S_n, term by term, we have

$$\begin{aligned} 2S_n &= (a_1 + a_n) + (a_1 + a_n) + (a_1 + a_n) + \cdots + (a_1 + a_n) \\ &= n(a_1 + a_n). \end{aligned}$$

Therefore,

$$\mathbf{S_n = \frac{n}{2}(a_1 + a_n).}$$

Example

Find the sum of the first 225 positive odd integers.

Solution

We have $a_1 = 1$, $n = 225$, and in the above example we found that $a_n = 449$. Then

$$S = \frac{225}{2}(1 + 449) = 50{,}625.$$

If we wish, we may replace a_n in the above formula for S_n by $a_1 + (n - 1)d$, and thus obtain a formula for S_n in terms of a_1, n, and d, instead of in terms of a_1, n, and a_n. The result is

$$S_n = \frac{n}{2}[2a_1 + (n - 1)d].$$

PROBLEMS

Write down the first five terms of each of the following sequences:

1. $\{2n - 5\}$.
2. $\{3n + 1\}$.
3. $\left\{\frac{n}{2} + 3\right\}$.
4. $\left\{\frac{4 - 2n}{3}\right\}$.
5. $\{2^n - 2n^2\}$.
6. $\{3n - 2^{-n}\}$.
7. $\{6n - n^2\}$.
8. $\{n^3 - 8n\}$.
9. $\left\{\frac{\log n}{2n}\right\}$.

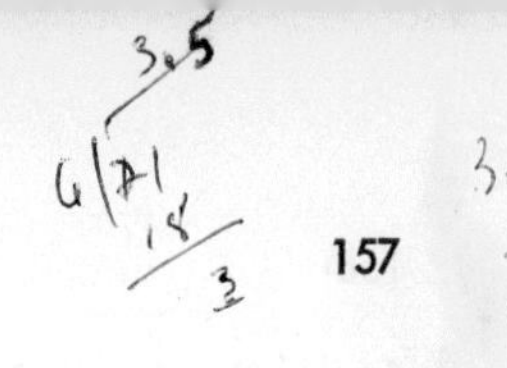

10. $\{n \log n - n\}$.

11. Which of the sequences of Probs. 1 to 10 are arithmetic progressions?

Find the formula for the nth term of each of the following arithmetic progressions:

12. $13, 8, 3, \cdots$.

13. $\frac{5}{32}, \frac{1}{16}, -\frac{1}{32}, \cdots$.

14. $12, 13\frac{1}{2}, 15, \cdots$.

15. $1{,}001, 1{,}002, 1{,}003, \cdots$.

16. $2z, 2z - 2, 2z - 4, \cdots$.

17. $2z - 3, 2z - 1, 2z + 1, \cdots$

Find the indicated term in each of the following arithmetic progressions:

18. $-8, -3, 2, \cdots$, 18th.

19. $2, 4, 6, \cdots$, 50th.

20. $12, 13.5, 15, \cdots$, 20th.

21. $19, 17, 15, \cdots$, 36th.

22. $-2\frac{2}{3}, -1\frac{1}{3}, 0, \cdots$, 14th.

23. $2.35, 2.40, 2.45, \cdots$, 46th.

24. $1.20, 1.08, 0.96, \cdots$, 42nd.

25. $-4, -2\frac{1}{2}, -1, \cdots$, 60th.

26. $a + 2b, 3a + 3b, 5a + 4b, \cdots$, 12th.

27. $3a + 4b, 2b, -3a, \cdots$, 8th.

28. Find the sum of all positive integers less than 100.

29. Find the sum of all positive odd integers less than 200.

Find the sum of each of the following arithmetic progressions:

30. $\{2n - 5\}_8$.

31. $\{3n + 1\}_{20}$.

32. $a_1 = 4, d = \frac{3}{2}, n = 18$.

33. $a_1 = 12, d = -\frac{1}{2}, n = 30$.

34. $a_1 = 14, a_{20} = 72, n = 20$.

35. $a_1 = 8, a_{12} = -20, n = 12$.

36. The 10th term of an arithmetic progression is 32, and the 18th term is 48. Find the sum of the first 12 terms.

37. The 14th term of an arithmetic progression is 72.5, and the 20th term is 93.5. Find the sum of the first 4 terms.

38. Show that the sum of the first n positive odd integers is equal to n^2.

39. Show that the sum of the first n positive even integers is equal to $n(n + 1)$.

40. Show that the sum of the first n positive integers is equal to $\frac{1}{2}n(n + 1)$.

41. For what value or values of k is the sequence $k - 3, k + 5, 2k - 1$, an arithmetic progression?

42. For what value or values of k is the sequence $2k + 4, 3k - 7, k + 12$, an arithmetic progression?

43. Show that the numbers $3k + 5, 2k + 7, k - 8$, do not form an arithmetic progression for any value of k.

44. If the numbers a, b, c, form an arithmetic progression, then b is called the arithmetic mean of a and c. For what value or values of k is $4k - 1$ the arithmetic mean of $2(3k + 1)$ and $k + 12$?

45. For what value or values of k is $k^2 - 1$ the arithmetic mean of $2k + 3$ and $2k - 5$? (See Prob. 44.)

70. *Geometric progressions.*

A *geometric progression* is a sequence in which every term, after the first, can be obtained from the term immediately preceding by multiplying it by a fixed number r. This number is called the *common ratio.*

Examples

The sequence $\{2^n\}$ or

$$2, 4, 8, \cdots, 2^n, \cdots$$

is a geometric progression with common ratio 2.

The numbers

$$6, -2, \tfrac{2}{3}, -\tfrac{2}{9}, \tfrac{2}{27}$$

form a geometric progression having five terms. The common ratio is $-\frac{1}{3}$.

71. ***Formula for the nth term of a geometric progression.*** Let a_1 denote the first term, r the common ratio, and n the number of terms of a geometric progression. We wish to derive a formula for the nth term in terms of this data. We observe that

$$\begin{aligned} \text{Second term} &= a_1r, \\ \text{Third term} &= (a_1r)r = a_1r^2, \\ \text{Fourth term} &= (a_1r^2)r = a_1r^3, \\ \text{Fifth term} &= (a_1r^3)r = a_1r^4, \end{aligned}$$

and so on. If we denote the nth term by a_n we see immediately that

$$\boldsymbol{a_n = a_1r^{n-1}}.$$

Example

Find the 10th term of the geometric progression $8, 4, 2, 1, \frac{1}{2}, \cdots$.

Solution

In this case $a_1 = 8$ and $r = \frac{1}{2}$; the 10th term is

$$a_{10} = 8 \cdot (\tfrac{1}{2})^9 = \tfrac{1}{64}.$$

72. ***Formula for the sum of the first n terms of a geometric progression.*** We wish now to derive a formula for the sum of the first n terms of a geometric progression. We have of course

$$S_n = a_1 + a_1r + a_1r^2 + \cdots + a_1r^{n-2} + a_1r^{n-1}. \tag{1}$$

If we multiply both members of this equation by r, we get

$$rS_n = a_1r + a_1r^2 + a_1r^3 + \cdots + a_1r^{n-1} + a_1r^n. \tag{2}$$

By subtracting (2) from (1) we obtain the relation

$$S_n - rS_n = a_1 - a_1r^n$$

or

$$S_n(1 - r) = a_1 - a_1r^n. \tag{3}$$

From (3) we have immediately

$$S_n = \frac{a_1 - a_1r^n}{1 - r}, \qquad \text{if } r \neq 1.$$

This formula gives the sum in terms of a_1, r, and n. We may obtain from it a formula for the sum in terms of a_1, r, and a_n by observing that since $a_1r^{n-1} = a_n$, $a_1r^n = ra_n$. We then have

$$S_n = \frac{a_1 - ra_n}{1 - r}, \qquad \text{if } r \neq 1.$$

Example

In the geometric progression

$$8, 4, 2, 1, \tfrac{1}{2}, \cdots$$

we have $a_1 = 8$, and $r = \frac{1}{2}$. The sum of the first 10 terms is

$$S_{10} = \frac{8 - 8(\frac{1}{2})^{10}}{1 - \frac{1}{2}} = 16[1 - (\tfrac{1}{2})^{10}]$$
$$= \frac{1{,}023}{64} = 15\tfrac{63}{64}.$$

73. ***Sum of an infinite geometric progression.*** Consider now the infinite geometric progression

$$1, \frac{1}{2}, \frac{1}{4}, \frac{1}{8}, \frac{1}{16}, \cdots, \frac{1}{2^{n-1}}, \cdots.$$

Let S_1 denote the first term, S_2 the sum of the first two terms, S_3 the sum of the first three terms, etc. Then

$$S_1 = 1,$$
$$S_2 = 1 + \tfrac{1}{2} = 1\tfrac{1}{2},$$
$$S_3 = 1 + \tfrac{1}{2} + \tfrac{1}{4} = 1\tfrac{3}{4},$$
$$S_4 = 1 + \tfrac{1}{2} + \tfrac{1}{4} + \tfrac{1}{8} = 1\tfrac{7}{8},$$
$$S_5 = 1 + \tfrac{1}{2} + \tfrac{1}{4} + \tfrac{1}{8} + \tfrac{1}{16} = 1\tfrac{15}{16},$$

and so on. It is intuitively evident that in this particular case the sum of n terms does not become indefinitely large as n increases. In fact, the sum is less than 2 no matter how many terms we take. It is furthermore intuitively clear that S_n is arbitrarily near 2 if n is sufficiently large; *i.e.*, if we take enough terms, we can make the sum as near 2 as we please.

We express the above situation by saying that S_n *approaches 2 as a limit as* **n** *becomes indefinitely large*, or by saying that "*the limit of* S_n *as* **n** *approaches infinity* is 2.*" We abbreviate this statement by writing

$$\lim_{n \to \infty} S_n = 2.$$

This limit, when it exists, is called the *sum* of the infinite geometric progression. The student will recognize that this is a new definition—a new use of the word sum. We of course cannot add up all the terms in the progression; in fact it is meaningless even to speak of "all the terms." But when the sum of n terms comes closer and closer to a fixed number K in the sense described above, as more and more terms are taken, we agree to call this number K the sum of the infinite geometric progression.

That not all such progressions have a sum in this sense is evident if we consider the progression

$$2, 4, 8, 16, 32, 64, \cdots, 2^n, \cdots.$$

In this case $S_1 = 2$, $S_2 = 6$, $S_3 = 14$, $S_4 = 30$, etc. It is obvious that S_n becomes arbitrarily large as n increases indefinitely. We sometimes express this by saying that S_n *approaches infinity as* **n** *approaches infinity*, and writing

$$\lim_{n \to \infty} S_n = \infty.$$

Our next task is that of determining the conditions under which an infinite geometric progression has a sum in the above sense,

* The student should observe that we do not here give any meaning to the word "infinity" standing alone. We have merely coined the phrase "approaches infinity" to describe a rather simple state of affairs—that in which the value of a variable continues to increase indefinitely. There is no justification whatever for the notion that infinity means a very large number. It doesn't denote a number at all.

and finding a general formula for that sum. For this purpose, we study the expression for the sum of n terms:

$$S_n = \frac{a_1 - a_1 r^n}{1 - r}.$$

It can be proved, although we shall not give the proof here, that as n becomes larger and larger r^n becomes arbitrarily near to zero in numerical value *if r is any number between* -1 *and* $+1$. Thus, for example, if $r = \frac{1}{2}$, then $r^2 = \frac{1}{4}$, $r^3 = \frac{1}{8}$, $r^4 = \frac{1}{16}$, etc.; if $r = -\frac{2}{3}$, then $r^2 = \frac{4}{9}$, $r^3 = -\frac{8}{27}$, $r^4 = \frac{16}{81}$, etc.; if $r = 0.9$, then $r^2 = 0.81$, $r^3 = 0.729$, $r^4 = 0.6561$, etc. In each case r^n is arbitrarily near zero in numerical value if n is sufficiently large.

Returning to the above expression for S_n, we see that if r is any number between -1 and $+1$, then $a_1 r^n$ approaches zero as n increases indefinitely. This means that value of S_n approaches nearer and nearer to $\dfrac{a_1}{1 - r}$. If we let S_∞ denote the sum of the infinite geometric progression, we have the formula

$$S_\infty = \frac{a_1}{1 - r}, \qquad \text{if } |r| < 1.$$

Examples

The sum of the infinite geometric progression

$$1, \tfrac{1}{2}, \tfrac{1}{4}, \tfrac{1}{8}, \tfrac{1}{16}, \cdots,$$

is

$$\frac{a_1}{1 - r} = \frac{1}{1 - \frac{1}{2}} = 2.$$

The sum of the infinite geometric progression

$$9, -6, 4, -\tfrac{8}{3}, \tfrac{16}{9}, \cdots,$$

is

$$\frac{a_1}{1 - r} = \frac{9}{1 - (-\frac{2}{3})} = \frac{27}{5} = 5.4.$$

In this latter case $S_1 = 9$, $S_2 = 3$, $S_3 = 7$, $S_4 = 4.33$, $S_5 = 6.11$, $S_6 = 4.93$, $S_7 = 5.72$, $S_8 = 5.19$, etc. Thus S_n is alternately larger and smaller than 5.4.

If $|r| > 1$, the sum of n terms increases indefinitely in absolute value as n increases. In this case, the infinite geometric progression has no sum. If $|r| = 1$, the terms are all equal to $\pm a_1$, where a_1 is the first term. Again, the infinite progression has no sum.

74. ***Periodic decimals.*** It was stated in Chap. I (page 26) that every periodic decimal represents a rational number which is the quotient of two integers. We shall see now how to find this number.

Example

The periodic decimal $0.69\dot{3}$, or $0.69333 \cdots$, is equivalent to

$$\tfrac{69}{100} + 0.003 + 0.0003 + 0.00003 + \cdots$$

wherein the part starting with 0.003 is an infinite geometric progression with $a_1 = 0.003$ and $r = 0.1$. Therefore

$$0.69\dot{3} = \frac{69}{100} + \frac{0.003}{1 - 0.1}$$
$$= \frac{69}{100} + \frac{3}{900} = \frac{52}{75}.$$

Conversely, if one should take the rational fraction $\frac{52}{75}$ and reduce it to decimal form by means of division, the result would be $0.69\dot{3}$.

PROBLEMS

Write down the next three terms and the nth term of each of the following geometric progressions:

1. 3, 9, 27, $\cdots$. **2.** 4, 8, 16, $\cdots$. **3.** 6, -2, $\frac{2}{3}$, $\cdots$.
4. 8, -4, 2, $\cdots$. **5.** $\sqrt{2}$, 2, $2\sqrt{2}$, $\cdots$. **6.** 100, -10, 1, $\cdots$.

7. If log 2 is the first term and log 4 the second term of a geometric progression, what are the third and fourth terms?

8. If log 8 is the first term and log 2 the second term of a geometric progression, what are the third and fourth terms?

9. If the sequence a, b, c, is an arithmetic progression, then b is called the *arithmetic mean* of the numbers a and c; if the sequence is a geometric progression, b is called the *geometric mean* of a and c. Show that the arithmetic mean is $\frac{1}{2}(a + c)$ and the geometric mean is $\pm\sqrt{ac}$.

10. Using the definitions of Prob. 9, find both the arithmetic mean and the geometric mean of the numbers 3 and 27.

11. For what value or values of k is the sequence $k - 2$, $k - 6$, $2k + 3$, a geometric progression?

12. For what value or values of k is the sequence $3k + 4$, $k - 2$, $5k + 1$, a geometric progression?

13. For what value or values of k is $k + 1$ the geometric mean of $4k + 2$ and $(k + 5)$? See the definition in Prob. 9.

14. Show that the sum of the first n terms of the geometric progression $1, \frac{1}{2}, \cdots$ is given by the formula

$$S_n = 2[1 - (\tfrac{1}{2})^n].$$

Discuss the way in which S_n varies with n as n increases indefinitely.

15. Show that the sum of the first n terms of the geometric progression $1, \frac{2}{3}, \frac{4}{9}, \ldots$ is given by the formula

$$S_n = 3[1 - (\tfrac{2}{3})^n].$$

Discuss the way in which S_n varies with n as n becomes larger and larger.

16. Find the sum of the first six terms of the geometric progression $9, 3, 1, \ldots$. What is the sum of the infinite progression?

In each of the following cases compute the sum of the specified number of terms of the given geometric progression:

17. $50, 10, 2, \cdots$, 5 terms.
18. $2, -6, 18, \cdots$, 8 terms.
19. $16, -4, 1, \cdots$, 6 terms.
20. $27, 9, 3, \cdots$, 5 terms.
21. $\frac{3}{4}, 3, 12, \cdots$, 5 terms.

Find the sum of each of the following infinite geometric progressions:

22. $1, -\frac{1}{5}, \frac{1}{25}, \cdots$.
23. $1, -\frac{1}{2}, \frac{1}{4}, \cdots$.
24. $\frac{3}{2}, 1, \frac{2}{3}, \cdots$.
25. $27, -18, 12, \cdots$.
26. $12, 9, \frac{27}{4}, \cdots$.
27. $1, 0.8, 0.64, \cdots$.

In each of the following cases find the rational fraction which, when reduced to decimal form, gives the specified periodic decimal:

28. $1.\dot{4}$.
29. $3.\dot{6}$.
30. $8.\dot{7}$.
31. $14.\dot{5}$.
32. $2.\dot{4}\dot{5}$.
33. $8.\dot{5}\dot{1}$.
34. $0.7\dot{4}\dot{3}$.
35. $0.1\dot{4}\dot{2}$.
36. $1.23\dot{1}\dot{2}\dot{3}$.
37. $6.\dot{4}0\dot{5}$.

38. A rubber ball is dropped from a height of 6 ft. Each time it strikes the floor, it rebounds to approximately three-fourths of the height from which it fell. What, approximately, is the total distance traveled by the ball?

39. Show that the formula for S_n can be written in the form

$$S_n = \frac{a_1 - ra_n}{1 - r}, \qquad \text{if } r \neq 1.$$

Discuss the case in which $r = 1$.

40. The sequence a, b, c is an arithmetic progression whose sum is 18. If a and b are each increased by 4 and if c is increased by 36, the new numbers form a geometric progression. Find a, b, and c.

41. A tank holds 64 gal. of a certain liquid. Suppose that 16 gal. are removed and the tank is refilled by replacing this liquid with water. Then 16 gal. of this resulting mixture are removed, and the tank is again filled with water. This operation is repeated until 6 batches have been removed. How much of the original liquid remains in the tank?

42. Air is removed from a tank by means of a pump which removes one-tenth of the remaining air at each stroke. What fractional part of the original amount remains after 5 strokes? After how many strokes will less than half of the original amount be present?

CHAPTER XI

MATHEMATICAL INDUCTION

75. ***Introduction.*** In this chapter we introduce an important method of reasoning known as *mathematical induction.* Its field of usefulness lies in the study of theorems concerning functions of a variable n whose range is the field of natural numbers.

Let us suppose, for example, that one had noticed the following relations:

$$\begin{aligned} 1 &= 1^2, \\ 1 + 3 &= 2^2, \\ 1 + 3 + 5 &= 3^2, \\ 1 + 3 + 5 + 7 &= 4^2, \\ 1 + 3 + 5 + 7 + 9 &= 5^2. \end{aligned}$$

He would begin to suspect that the sum of the first n positive odd integers may be equal to n^2 for all values of n. How could he go about trying to prove this? In this particular illustration we are dealing with the sum of the terms of an arithmetic progression, and it would be possible to use the formula derived in Chap. X (page 156) to effect a proof. We introduce in this chapter a method of attack that would be applicable even if this were not the case.

By the method of mathematical induction, we shall be able to prove not only the above relation but many others of a similar nature. For example,

$$1^3 + 2^3 + 3^3 + \cdots + n^3 = \frac{n^2(n+1)^2}{4}.$$

$$\frac{1}{1 \cdot 2} + \frac{1}{2 \cdot 3} + \frac{1}{3 \cdot 4} + \cdots + \frac{1}{n(n+1)} = \frac{n}{n+1}.$$

76. ***Proof by mathematical induction.*** A proof by mathematical induction of a theorem of the kind described above requires the following two parts:

PART I. *Prove that if the relation were true for any positive integral value of* $\mathbf{n}$, *say for* $\mathbf{n} = \mathbf{k}$, *then it would be true for the next larger value of* $\mathbf{n}$; *i.e., for* $\mathbf{n} = \mathbf{k} + 1$.

PART II. *Show by direct substitution that the theorem is true for* $n = 1$.

It is one of the basic axioms of algebra that the given relation is true for all positive integral values of n if the above two steps can be effected.*

Part I is of course the more difficult part of the proof, and it should be emphasized that in this part we are not at all concerned with whether or not the given relation is true for *any* value of n. We are concerned only with the proposition: ***if it were true for*** $\mathbf{n} = \mathbf{k}$, ***then it would have to be true for*** $\mathbf{n} = \mathbf{k} + \mathbf{1}$.

Example

Carry out Part I of the proof of the relation

$$1^3 + 2^3 + 3^3 + \cdots + n^3 = \frac{n^2(n+1)^2}{4}.$$

Solution

If the relation were true for $n = k$, then we would have

$$1^3 + 2^3 + 3^3 + \cdots + k^3 = \frac{k^2(k+1)^2}{4}. \tag{1}$$

This is our *hypothesis*. We are not concerned with whether or not there is any integer k for which it is actually true. We wish to prove only that if (1) *were* true then it would also be true that

$$1^3 + 2^3 + 3^3 + \cdots + k^3 + (k+1)^3 = \frac{(k+1)^2(k+2)^2}{4}. \tag{2}$$

In order to prove this, we take (1) and *add the next term to both sides*. Thus if (1) were true, then it would have to be true that

* One might have a theorem to be proved true for all $n > 1$, or all $n > 3$, etc. The corresponding alteration in our statement is obvious.

$$1^3 + 2^3 + 3^3 + \cdots + k^3 + (\boldsymbol{k} + \mathbf{1})^3 = \frac{k^2(k+1)^2}{4} + (\boldsymbol{k} + \mathbf{1})^3$$
$$= \frac{k^2(k+1)^2 + 4(k+1)^3}{4}$$
$$= \frac{(k+1)^2[k^2 + 4(k+1)]}{4}$$
$$= \frac{(k+1)^2(k^2 + 4k + 4)}{4}$$
$$= \frac{(k+1)^2(k+2)^2}{4}.$$

This last result is identical with (2). We have therefore proved that if (1) were true, then (2) would be true.

The above proof does not, in itself, assure us of the truth of our theorem. It merely assures us that *if it is true for some positive integer* **k**, *then it must be true for the next larger integer.* Thus if the theorem were found to be true for $n = 3$ it must be true for $n = 4$, and then since it is true for $n = 4$, it must be true for $n = 5$, etc. The proof of Part I therefore assures us of the truth of the theorem for all values of n larger than the smallest one for which it is actually verified. In carrying out Part II we verify the theorem for the smallest positive integer for which we are to establish its truth; ordinarily this is $n = 1$. In the above case we have, for $n = 1$,

$$1^3 = \frac{1^2 \cdot 2^2}{4} = 1.$$

The truth of the theorem for $n = 1$, and hence its truth for all positive integral values of n, is thus established.

The work can be conveniently arranged as in the following:

Example

Prove that the sum of the squares of the first n positive odd integers is equal to $\frac{1}{3}n(4n^2 - 1)$; *i.e.*, prove that

$$1^2 + 3^2 + 5^2 + \cdots + (2n-1)^2 = \frac{n(4n^2-1)}{3}.$$

Solution

PART I. *Hypothesis:*

$$1^2 + 3^2 + 5^2 + \cdots + (2k-1)^2 = \frac{k(4k^2-1)}{3}.$$

Conclusion:

$$1^2 + 3^2 + 5^2 + \cdots + (2k - 1)^2 + (2k + 1)^2 = \frac{(k + 1)[4(k + 1)^2 - 1]}{3} = \frac{(k + 1)(4k^2 + 8k + 3)}{3}.*$$

Proof: If

$$1^2 + 3^2 + 5^2 + \cdots + (2k - 1)^2 = \frac{k(4k^2 - 1)}{3}$$

then, adding the next term to both sides,

$$\begin{aligned} 1^2 + 3^2 + 5^2 + \cdots + (2k - 1)^2 + (\mathbf{2k + 1})^2 &= \frac{k(4k^2 - 1)}{3} + (\mathbf{2k + 1})^2 \\ &= \frac{k(4k^2 - 1) + 3(2k + 1)^2}{3} \\ &= \frac{(2k + 1)[k(2k - 1) + 3(2k + 1)]}{3} \\ &= \frac{(2k + 1)(2k^2 + 5k + 3)}{3} \\ &= \frac{(2k + 1)(2k + 3)(k + 1)}{3} \\ &= \frac{(k + 1)(4k^2 + 8k + 3)}{3}. \end{aligned}$$

This last result coincides with our conclusion, and hence we know that if the given theorem is true for $n = k$ it is true for $n = k + 1$.

PART II. For $n = 1$ we have

$$1^2 = \frac{1(4 - 1)}{3} = 1$$

which is true. The theorem is thus established.

It should be remarked that if one attempts to prove a false theorem by this method he will either fail to prove Part I, or fail in the verification of Part II. The student can readily see,

* The hypothesis is the statement of the given theorem for the case $n = k$. The conclusion is the corresponding statement for the case $n = k + 1$.

or example, that if the right member in the theorem of the above xample were

$$\frac{n(4n^2 - 1)}{3} + 1 \qquad \text{instead of} \qquad \frac{n(4n^2 - 1)}{3},$$

he theorem would be false. The proof of Part I would go through recisely as for the true theorem, but we would fail in Part II; .e., we could prove that *if* the theorem were true for $n = k$, it *vould be* true for $n = k + 1$, but we would be unable to find any umber for which it is true. On the other hand if the right member vere

$$\frac{n(4n^2 - 1)}{3} + n - 1$$

he verification in Part II for $n = 1$ would check. In this case ve would fail to establish Part I; for after adding the next term o both sides, we would find that the right side would not reduce o the required result. Incidentally, if the right side were

$$\frac{n(4n^2 - 1)}{3} + (n - 1)(n - 2) \cdots (n - 1{,}000)$$

t would give correct results in Part II for all values of n from 1 o 1,000, and it still would of course be a false theorem.

We conclude by proving a theorem of a somewhat different kind.

Example

Prove that $a^n - b^n$ is divisible by $a - b$ for all positive integral values of n $(a \neq b)$.

Solution

PART I. *Hypothesis:* $a^k - b^k$ is divisible by $a - b$.

Conclusion: $a^{k+1} - b^{k+1}$ is divisible by $a - b$.

Proof:
$$a^{k+1} - b^{k+1} = a^{k+1} - ab^k + ab^k - b^{k+1}.$$
$$= a(a^k - b^k) + b^k(a - b).$$

Now if $a^k - b^k$ is divisible by $a - b$, then the entire right side, and consequently the left side, is divisible by $a - b$; *i.e.*, if $a^k - b^k$ is divisible by $a - b$, then $a^{k+1} - b^{k+1}$ is divisible by $a - b$.

PART II. For $n = 1$, $a^n - b^n$ is simply $a - b$ which is of course divisible by $a - b$. We used this fact in Part I. The proof is thus complete.

PROBLEMS

Prove the following relations by means of mathematical induction:

1. $1 + 2 + 3 + \cdots + n = \dfrac{n(n+1)}{2}$.

2. $1 + 3 + 5 + \cdots + (2n - 1) = n^2$.

3. $2 + 4 + 6 + \cdots + 2n = n(n + 1)$.

4. $1 + 3 + 6 + \cdots + \frac{1}{2}n(n + 1) = \frac{1}{6}n(n + 1)(n + 2)$.

5. $1 \cdot 2 + 2 \cdot 3 + 3 \cdot 4 + \cdots + n(n + 1) = \frac{1}{3}n(n + 1)(n + 2)$.

6. $1 \cdot 2 + 3 \cdot 4 + 5 \cdot 6 + \cdots + (2n - 1)(2n) = \frac{1}{3}n(4n^2 + 3n - 1)$.

7. $1 \cdot 3 + 2 \cdot 4 + 3 \cdot 5 + \cdots + n(n + 2) = \frac{1}{6}n(n + 1)(2n + 7)$.

8. $1 \cdot 6 + 4 \cdot 9 + 7 \cdot 12 + \cdots + (3n - 2)(3n + 3) = 3n(n^2 + 2n - 1)$.

9. $1 \cdot 4 + 4 \cdot 7 + 7 \cdot 10 + \cdots + (3n - 2)(3n + 1) = n(3n^2 + 3n - 2)$.

10. $1 \cdot 2 \cdot 3 + 2 \cdot 3 \cdot 4 + 3 \cdot 4 \cdot 5 + \cdots + n(n + 1)(n + 2)$
$= \frac{1}{4}n(n + 1)(n + 2)(n + 3)$

11. $\dfrac{1}{1 \cdot 2} + \dfrac{1}{2 \cdot 3} + \dfrac{1}{3 \cdot 4} + \cdots + \dfrac{1}{n(n+1)} = \dfrac{n}{n+1}$.

12. $1^2 + 2^2 + 3^2 + \cdots + n^2 = \frac{1}{6}n(n + 1)(2n + 1)$.

13. $2^2 + 4^2 + 6^2 + \cdots + (2n)^2 = \frac{2}{3}n(n + 1)(2n + 1)$.

14. $4^2 + 7^2 + 10^2 + \cdots + (3n + 1)^2 = \frac{1}{2}n(6n^2 + 15n + 11)$.

15. $1^3 + 3^3 + 5^3 + \cdots + (2n - 1)^3 = n^2(2n^2 - 1)$.

16. $2^3 + 4^3 + 6^3 + \cdots + (2n)^3 = 2n^2(n + 1)^2$.

17. $2 + 2^2 + 2^3 + \cdots + 2^n = 2(2^n - 1)$.

18. $3 + 3^2 + 3^3 + \cdots + 3^n = \frac{3}{2}(3^n - 1)$.

19. $\dfrac{1}{2} + \dfrac{1}{2^2} + \dfrac{1}{2^3} + \cdots + \dfrac{1}{2^n} = 1 - \dfrac{1}{2^n}$.

20. $\dfrac{1}{1 \cdot 3} + \dfrac{1}{3 \cdot 5} + \dfrac{1}{5 \cdot 7} + \cdots + \dfrac{1}{(2n-1)(2n+1)} = \dfrac{n}{2n+1}$.

21. $\dfrac{1}{1 \cdot 2 \cdot 3} + \dfrac{1}{2 \cdot 3 \cdot 4} + \dfrac{1}{3 \cdot 4 \cdot 5} + \cdots + \dfrac{1}{n(n+1)(n+2)}$
$= \dfrac{n(n+3)}{4(n+1)(n+2)}$

22. Prove that if n is any positive integer, then $x^{2n} - y^{2n}$ is divisible by $x + y$

23. Prove that if n is any positive integer, then $x^{2n-1} + y^{2n-1}$ is divisible by $x + y$.

24. Prove that the sum of the interior angles of a polygon of n sides is

$$(n - 2) \cdot 180°.$$

25. Prove that if the theorem

$$1 + 3 + 5 + \cdots + (2n - 1) = n^2 + 4$$

were true for $n = k$, it would be true for $n = k + 1$. Is the theorem true or false?

26. Prove that if the theorem

$$1 + 5 + 9 + \cdots + (4n - 3) = 2n^2 - n + 3$$

were true for $n = k$, it would be true for $n = k + 1$. Is the theorem true?

27. Show that the equation

$$1 + 3 + 5 + \cdots + (2n - 1) = n^2 + (n - 1)(n - 2)(n - 3) \cdots (n - 10),$$

where n is a positive integer, is true for $n \leqq 10$ but is false for $n > 10$. Use the result of Prob. 2.

In each of the following cases prove that the given theorem is true or that it is false, n being any positive integer:

28. $\dfrac{1}{1 \cdot 3} + \dfrac{1}{3 \cdot 5} + \dfrac{1}{5 \cdot 7} + \cdots + \dfrac{1}{(2n - 1)(2n + 1)} = \dfrac{n}{2n + 1}.$

29. $2^0 + 2^1 + 2^2 + \cdots + 2^{n-1} = 2^n + 1.$

30. $\dfrac{4}{1 \cdot 2 \cdot 3} + \dfrac{5}{2 \cdot 3 \cdot 4} + \dfrac{6}{3 \cdot 4 \cdot 5} + \cdots + \dfrac{n + 3}{n(n + 1)(n + 2)}$

$$= \frac{n(5n + 11)}{4(n + 1)(n + 2)}.$$

31. Prove, by means of mathematical induction, the formula for the sum of an arithmetic progression of n terms.

32. Prove, by means of mathematical induction, the formula for the sum of a geometric progression of n terms.

33. Prove that if n is any natural number, then $\frac{1}{3}(n^3 + 2n)$ is a natural number.

34. Prove that if n is any natural number, then $\frac{1}{2}(n^4 + 5n)$ is a natural number.

35. Prove that if n is any natural number, then $\frac{1}{3}(n^3 + 6n^2 + 2n)$ is a natural number.

CHAPTER XII

THE BINOMIAL THEOREM

77. ***The binomial theorem for positive integral exponents.*** By direct multiplication we may verify the following results:

$$(a + b)^1 = a + b,$$
$$(a + b)^2 = a^2 + 2ab + b^2,$$
$$(a + b)^3 = a^3 + 3a^2b + 3ab^2 + b^3,$$
$$(a + b)^4 = a^4 + 4a^3b + 6a^2b^2 + 4ab^3 + b^4,$$
$$(a + b)^5 = a^5 + 5a^4b + 10a^3b^2 + 10a^2b^3 + 5ab^4 + b^5.$$

The letters a and b may denote any two numbers or expressions, and the right member of each identity is called the *expansion* of the expression on the left. By examining the above results, we seek to determine a general formula for the expansion of $(a + b)^n$. We observe that in all of the above cases:

(*a*) The first term is a^n and the second is $na^{n-1}b$.

(*b*) In each succeeding term the exponent of a is one less and that of b is one more than in the term immediately preceding, so that the sum of the exponents of a and b is equal to n in every term.

(*c*) If the coefficient in the second term is multiplied by the exponent of a in that term and divided by the number of the term, the result is the coefficient in the third term; if the coefficient in the third term is multiplied by the exponent of a in that term and divided by the number of the term, the result is the coefficient in the fourth term, and so on.

From these observations it appears that if n is any positive integer, the following formula may hold:

$$(a + b)^n = a^n + na^{n-1}b + \frac{n(n-1)}{1 \cdot 2} a^{n-2}b^2 + \frac{n(n-1)(n-2)}{1 \cdot 2 \cdot 3} a^{n-3}b^3 + \cdots + b^n.$$

This formula is called the *binomial formula* or the *binomial theorem* for positive integral exponents. That it does hold for such exponents will be proved by means of mathematical induction in Sec. 80.

Example 1

Write the first four terms and the last term of the expansion of $(x + y)^{10}$.

Solution

$$(x + y)^{10} = x^{10} + 10x^9y + \frac{10 \cdot 9}{1 \cdot 2}x^8y^2 + \frac{10 \cdot 9 \cdot 8}{1 \cdot 2 \cdot 3}x^7y^3 + \cdots + y^{10}$$
$$= x^{10} + 10x^9y + 45x^8y^2 + 120x^7y^3 + \cdots + y^{10}.$$

The student should observe, in connection with the next example, that it is best not to simplify the individual terms in the expansion until after the entire expansion has been written out, for one needs each term in its original form in order to get the next term from it.

Example 2

Expand $(2x - \sqrt{y})^5$.

Solution

We employ the usual formula, letting $a = 2x$, $b = -\sqrt{y}$, and $n = 5$. Then

$$(2x - \sqrt{y})^5 = (2x)^5 + 5(2x)^4(-\sqrt{y}) + 10(2x)^3(-\sqrt{y})^2 + 10(2x)^2(-\sqrt{y})^3 + 5(2x)(-\sqrt{y})^4 + (-\sqrt{y})^5$$
$$= 32x^5 - 80x^4y^{\frac{1}{2}} + 80x^3y - 40x^2y^{\frac{3}{2}} + 10xy^2 - y^{\frac{5}{2}}.$$

Example 3

Find the value of $(1.02)^8$ to four decimal places using the binomial formula.

Solution

We may write $(1.02)^8$ as $(1 + 0.02)^8$. Then

$$(1 + 0.02)^8 = 1^8 + 8(1)^7(0.02) + 28(1)^6(0.02)^2 + 56(1)^5(0.02)^3 + 70(1)^4(0.02)^4 + \cdots + (0.02)^8$$
$$= 1 + 8(0.02) + 28(0.0004) + 56(0.000008) + 70(0.00000016) + \cdots$$
$$= 1 + 0.16 + 0.0112 + 0.000448 + 0.0000112 + \cdots$$
$$= 1.1717.$$

Only the first five terms of the expansion were needed in order to obtain the result to four decimal places. The exact value of $(1.02)^8$ would of course be obtained if one used all nine terms.

78. *The definition of the symbol n!* By the symbol $n!$ where n is a positive integer, we mean the product of all the positive integers from 1 to n inclusive; *i.e.*,

$$n! = 1 \cdot 2 \cdot 3 \cdot 4 \cdots n.$$

Example

$$6! = 1 \cdot 2 \cdot 3 \cdot 4 \cdot 5 \cdot 6 = 720.$$

The fundamental property of $n!$ is contained in the equation

$$n! = n(n-1)! \qquad n \geqq 2.$$

For example,

$$6! = 6(5!), \quad 5! = 5(4!), \quad 4! = 4(3!), \quad 3! = 3(2!), \quad 2! = 2(1!).$$

If this relation were to hold for $n = 1$, we should have

$$1! = 1(0!).$$

We therefore *define* the symbol $0!$ by the equation

$$0! = 1.$$

With this definition of $0!$, the relation

$$n! = n(n-1)!$$

holds for *all* positive integral values of n.

79. *The general term in the expansion of* $(a + b)^n$. Sometimes one wishes to write down a particular term of the expansion of $(a + b)^n$ without writing the terms that precede it. This can easily be accomplished as in the following:

Example

Write down the sixth term of the expansion of $(a + b)^{14}$.

Solution

We observe from the binomial formula that the exponent of b in any term is one less than the number of the term, so in the sixth term we have b^5. Next, since the sum of the exponents of a

and b must be 14 (equal to n), the exponent of a must be 9. So the sixth term is some coefficient times a^9b^5.

We write down the coefficient by observing that its denominator is $1 \cdot 2 \cdot 3 \cdot 4 \cdot 5$ (the last factor being the same as the exponent of b), while its numerator has *the same number of factors* starting with 14 and going down; *i.e.*, $14 \cdot 13 \cdot 12 \cdot 11 \cdot 10$. Finally then, the sixth term is

$$\frac{14 \cdot 13 \cdot 12 \cdot 11 \cdot 10}{1 \cdot 2 \cdot 3 \cdot 4 \cdot 5} a^9b^5.$$

By using precisely the same observations in the general case, we may write down the following formula for the rth term of the expansion of $(a + b)^n$:

$$r\text{th term} = \frac{n(n-1)(n-2)\cdots(n-r+2)}{(r-1)!} a^{n-r+1}b^{r-1}.$$

This is a general formula from which one can get any term after the first by substituting for r the number of the desired term.

A similar general formula is that which gives the term involving b^r. This is the $(r + 1)$th term, and the formula for it can be derived from that for the rth term by replacing r by $r + 1$:

$$\text{Term involving } b^r = \frac{n(n-1)(n-2)\cdots(n-r+1)}{r!} a^{n-r}b^r.$$

This formula may also be used for finding any specified term. For example, if one wants the fifth term, he will recall that the fifth term involves b^4 and will then substitute 4 for r.

PROBLEMS

Expand each of the following by the binomial formula:

1. $(x + y)^6$.
2. $(a - 2b)^5$.
3. $(x + 2)^4$.
4. $(2x^2 - y)^5$.
5. $(\sqrt{x} + \frac{1}{2}y)^6$.
6. $(xy^2 - 2)^4$.
7. $(2a^2 - 3b^2)^3$.
8. $(1 - z^2)^6$.
9. $(\sqrt{a} + \sqrt{b})^6$.
10. $(y^3 - 3a)^4$.
11. $(2\sqrt{x} - 3\sqrt{y})^6$.
12. $\left(a\sqrt{2} + \frac{b}{2}\right)^5$.
13. $(1 + 2\sqrt{x})^8$.
14. $(a^2b - 2c^3)^6$.
15. $(\sqrt{x} + \sqrt{y})^{10}$.

Write out the first four terms of each of the following expansions:

16. $(a - 2b)^{12}$.
17. $(x + \frac{1}{2})^{14}$.
18. $\left(x - \frac{2}{x}\right)^{10}$.

19. $(1 + \sqrt{x})^{16}$. **20.** $\left(1 + \frac{1}{x}\right)^{16}$. **21.** $(a + 3\sqrt{b})^{12}$.

Using the binomial formula, find the value of each of the following expressions correct to four decimal places:

22. $(1.01)^{10}$. **23.** $(1.04)^6$. **24.** $(1.03)^9$.
25. $(0.98)^8$. **26.** $(0.97)^6$. **27.** $(1.1)^7$.

Find the indicated term in each of the following expansions without writing the preceding terms:

28. $(x + y)^{12}$, 7th term. **29.** $(a - b)^9$, 4th term.

30. $(y^2 - 2)^{10}$, 5th term. **31.** $\left(x^2 - \frac{1}{x}\right)^{12}$, 8th term.

32. $(2a^2 - b^2)^7$, 4th term. **33.** $(2a - \sqrt{b})^9$, 7th term.

34. Find the middle term of the expansion of $\left(x - \frac{2}{x}\right)^8$.

35. Find the term involving x^5 in the expansion of $(2x + 3y)^7$.

36. Find the term that does not involve x in the expansion of $\left(x^2 + \frac{1}{x^3}\right)^{10}$.

Find the value of each of the following:

37. $\frac{(9!)(3!)}{8!}$. **38.** $\frac{10!}{(7!)(3!)}$. **39.** $\frac{(12!)(6!)}{(8!)(10!)}$.

Simplify each of the following expressions:

40. $\frac{n!}{(n+1)!}$. **41.** $\frac{(k+2)!}{k!}$. **42.** $\frac{(n+1)!}{(n-1)!}$.

43. $\frac{(n+2)(n!)}{(n+2)!}$. **44.** $\frac{(n+1)\cdot(n-1)!}{n!}$. **45.** $\frac{(n+1)!}{n(n-2)!}$.

46. Show that the expression for the term involving b^r, namely,

$$\frac{n(n-1)(n-2)\cdots(n-r+1)}{r!}a^{n-r}b^r$$

can be written in the form

$$\frac{n!}{r!(n-r)!}a^{n-r}b^r.$$

80. *Proof of the binomial formula for positive integral exponents.* We shall now prove by means of mathematical induction that the formula

$$(1)\quad (a+b)^n = a^n + na^{n-1}b + \frac{n(n-1)}{2!}a^{n-2}b^2 + \cdots + \frac{n(n-1)\cdots(n-r+2)}{(r-1)!}a^{n-r+1}b^{r-1} + \frac{n(n-1)\cdots(n-r+1)}{r!}a^{n-r}b^r + \cdots + nab^{n-1} + b^n$$

is true for all positive integral values of n. The verification for the case $n = 1$ is trivial, and therefore we need prove only that *if the formula holds for* $\mathbf{n} = \mathbf{k}$, *it will hold for* $\mathbf{n} = \mathbf{k} + \mathbf{1}$. Our hypothesis is then:

$$(2)\quad (a+b)^k = a^k + ka^{k-1}b + \frac{k(k-1)}{2!}a^{k-2}b^2 + \cdots + \frac{k(k-1)\cdots(k-r+2)}{(r-1)!}a^{k-r+1}b^{r-1} + \frac{k(k-1)\cdots(k-r+1)}{r!}a^{k-r}b^r + \cdots + kab^{k-1} + b^k.$$

If we multiply both sides of this equation by $a + b$, the two members of the resulting equation will be equal. The left member will be $(a + b)^{k+1}$, and we must show that the right member is equivalent to the right member of (1) with n replaced by $k + 1$:

$$\begin{aligned}(3)\quad (a+b)^{k+1} = {} & a^{k+1} + ka^kb + \frac{k(k-1)}{2!}a^{k-1}b^2 + \cdots \\ & + \frac{k(k-1)\cdots(k-r+2)}{(r-1)!}a^{k-r+2}b^{r-1} \\ & + \frac{k(k-1)\cdots(k-r+1)}{r!}a^{k-r+1}b^r + \cdots \\ & + ka^2b^{k-1} + ab^k \\ & + a^kb + ka^{k-1}b^2 + \frac{k(k-1)}{2!}a^{k-2}b^3 + \cdots \\ & + \frac{k(k-1)\cdots(k-r+2)}{(r-1)!}a^{k-r+1}b^r \\ & + \frac{k(k-1)\cdots(k-r+1)}{r!}a^{k-r}b^{r+1} + \cdots \\ & + kab^k + b^{k+1}.\end{aligned}$$

When we combine like terms in the right member of (3), we see that the coefficient of $a^k b$ is $(k + 1)$. The coefficient of $a^{k-1}b^2$ is

$$\frac{k(k-1)}{2!} + k = k\left(\frac{k-1}{2} + 1\right) = k\left(\frac{k+1}{2}\right) = \frac{(k+1)k}{2!}.$$

. . . .

The coefficient of $a^{k-r+1}b^r$ is

$$\begin{aligned}
&\frac{k(k-1)\cdots(k-r+1)}{r!} + \frac{k(k-1)\cdots(k-r+2)}{(r-1)!} \\
&= \frac{k(k-1)\cdots(k-r+2)(k-r+1)}{r(r-1)!} + \frac{k(k-1)\cdots(k-r+2)}{(r-1)!} \\
&= \frac{k(k-1)\cdots(k-r+2)}{(r-1)!}\left(\frac{k-r+1}{r} + 1\right) \\
&= \frac{k(k-1)\cdots(k-r+2)}{(r-1)!}\left(\frac{k+1}{r}\right) \\
&= \frac{(k+1)(k)(k-1)\cdots(k-r+2)}{r!}.
\end{aligned}$$

. . . .

The coefficient of ab^k is $(k + 1)$.

We have then

$$\begin{aligned}
(4)\quad (a+b)^{k+1} &= a^{k+1} + (k+1)a^k b + \frac{(k+1)k}{2!}a^{k-1}b^2 + \cdots \\
&+ \frac{(k+1)(k)(k-1)\cdots(k-r+2)}{r!}a^{k-r+1}b^r \\
&+ \cdots + (k+1)ab^k + b^{k+1}.
\end{aligned}$$

This equation is what (1) becomes for the case $n = k + 1$. The proof is therefore complete.

81. ***The binomial series.*** By the binomial formula we have the following expansion of $(1 + x)^n$ for positive integral values of n:

$$\begin{aligned}
(1+x)^n = 1 + nx &+ \frac{n(n-1)}{2!}x^2 + \frac{n(n-1)(n-2)}{3!}x^3 \\
&+ \frac{n(n-1)(n-2)(n-3)}{4!}x^4 + \cdots.
\end{aligned}$$

The expansion contains $n + 1$ terms.

If n is any real number that is not a positive integer, the corresponding expansion can be written down in a purely formal fashion, and its meaning can then be investigated. Thus for the case $n = \frac{1}{2}$, we may tentatively write

$$(1 + x)^{\frac{1}{2}} = 1 + \frac{1}{2}x + \frac{\frac{1}{2}(-\frac{1}{2})}{2!}x^2 + \frac{\frac{1}{2}(-\frac{1}{2})(-\frac{3}{2})}{3!}x^3 + \frac{\frac{1}{2}(-\frac{1}{2})(-\frac{3}{2})(-\frac{5}{2})}{4!}x^4 + \cdots$$

$$= 1 + \frac{1}{2}x - \frac{1}{2^2 \cdot 2!}x^2 + \frac{1 \cdot 3}{2^3 \cdot 3!}x^3 - \frac{1 \cdot 3 \cdot 5}{2^4 \cdot 4!}x^4 + \cdots.$$

We observe immediately that this expansion does not terminate as it does if n is a positive integer—the terms continue indefinitely. This happens whenever n is not a positive integer (or 0). The result is an infinite series which, since it arises from the binomial formula, is called the *binomial series.*

The question that now arises is similar to the one that confronted us when we considered the case of the sum of an infinite geometric progression: *Does the sum of the first* **n** *terms approach a limit as* **n** *becomes larger and larger, and if so is this limit equal to the value of* $(1 + x)^n$? The answer to both questions is *yes* if x is any number between -1 and $+1$, but the proof cannot be given here. Thus if we put $x = \frac{1}{2}$ in the above expansion, we have

$$\left(1 + \frac{1}{2}\right)^{\frac{1}{2}} = 1 + \frac{1}{4} - \frac{1}{2^4 \cdot 2!} + \frac{1 \cdot 3}{2^6 \cdot 3!} - \frac{1 \cdot 3 \cdot 5}{2^8 \cdot 4!} + \cdots.$$

The sum of the first two terms on the right is 1.2500; the sum of the first three terms is 1.2188; the sum of the first four terms is 1.2266; and the sum of the first five terms is 1.2241. As more and more terms are taken, the sum approaches nearer and nearer to the precise value of $(1 + \frac{1}{2})^{\frac{1}{2}}$ or $\sqrt{1.5}$. Since $\sqrt{1.5} = 1.22474$, the use of five terms in the above expansion gives the result correct to three significant figures.

We conclude with another example of the use of the binomial series in computation.

Example

Compute $\sqrt[3]{130}$.

Solution

$$\sqrt[3]{130} = \sqrt[3]{125 + 5} = 5\sqrt[3]{1 + \frac{5}{125}} = 5\left(1 + \frac{1}{25}\right)^{\frac{1}{3}}.$$

$$\left(1 + \frac{1}{25}\right)^{\frac{1}{3}} = 1 + \frac{1}{3}\left(\frac{1}{25}\right) + \frac{\frac{1}{3}(-\frac{2}{3})}{2!}\left(\frac{1}{25}\right)^2 + \frac{\frac{1}{3}(-\frac{2}{3})(-\frac{5}{3})}{3!}\left(\frac{1}{25}\right)^3 + \cdots$$

$$= 1 + \frac{1}{3}\left(\frac{1}{25}\right) - \frac{1 \cdot 2}{3^2 \cdot 2!}\left(\frac{1}{25}\right)^2 + \frac{1 \cdot 2 \cdot 5}{3^3 \cdot 3!}\left(\frac{1}{25}\right)^3 + \cdots$$

$$= 1 + 0.013333 - 0.000178 + 0.000004 - \cdots$$

$$= 1.01316.$$

Hence $\sqrt[3]{130} = 5(1.01316) = 5.0658$ (approximately).

It may be observed that the use of only two terms would give the result correct to four significant digits.

PROBLEMS

Write out the first five terms of each of the following expansions:

1. $\sqrt{1 - x}$.
2. $\sqrt[3]{1 + x}$.
3. $\sqrt[3]{1 - x}$.
4. $(1 + x)^{-2}$.
5. $\dfrac{1}{1 - x}$.
6. $\dfrac{1}{(1 - x^2)^2}$.
7. $\dfrac{1}{\sqrt{1 + x}}$.
8. $\dfrac{1}{\sqrt{1 - x}}$.

9. Divide 1 by $1 - x^2$ and compare the result with that obtained by expanding $(1 - x^2)^{-1}$ by the binomial formula.

10. Divide 1 by $1 + x$ and compare the result with that obtained by expanding $(1 + x)^{-1}$ by the binomial formula.

11. Show that $(a + b)^n$ can be written in the form $a^n\left(1 + \frac{b}{a}\right)^n$. Write down the first five terms of the expansion of each of these and show that they are equivalent.

Use the binomial series to find the value of each of the following quantities correct to four significant digits:

12. $\sqrt{50}$.
13. $\sqrt{98}$.
14. $\sqrt{84}$.
15. $\sqrt[3]{29}$.
16. $\sqrt[4]{260}$.
17. $\sqrt[3]{25}$.
18. $\sqrt{1.04}$.
19. $\sqrt[3]{1.03}$.
20. $\sqrt[4]{1.02}$.
21. $(1.04)^{-5}$.
22. $(1.03)^{-6}$.
23. $(1.02)^{-8}$.
24. $(1.03)^{-4}$.
25. $(1.04)^{-6}$.
26. $(1.03)^{-7}$.
27. $(1.05)^{-10}$.

CHAPTER XIII

COMPOUND INTEREST AND ANNUITIES

82. *Compound interest.* Money paid by one person or organization for the use of another's money is called *interest.* The amount of money whose use is involved is called the *principal.* The *rate of interest* is the ratio of the amount of interest earned in one unit of time to the principal. In expressing interest rates, the unit of time is understood to be 1 year unless otherwise specified.

Example

If \$100 earns \$4.50 in 1 year, the rate of interest is 0.045 or $4\frac{1}{2}$ per cent. In most of our work we shall express the rate in decimal form.

In many types of investment, interest is payable at the end of specified equal periods of time, such as annually, semiannually, or quarterly. Often, however, the interest due at the end of the first period is added to the principal, and this sum serves as the principal for the second period; the interest on this new principal for the second period is added to this principal to make a new principal for the third period; and so on. When this is done, the total amount of interest earned by the original principal in a given length of time is called the *compound interest* for that length of time. The sum of the original principal and the compound interest is called the *compound amount* or simply the *amount* at the end of that time. The time between successive additions of interest to the principal is called the *conversion period* or *interest period.* If the conversion period is 1 year, 6 months, or 3 months, we say that the interest is compounded annually, semiannually, or quarterly, respectively.

Example

If \$100 were invested at 4 per cent, compounded quarterly, then the rate per period would be 1 per cent. The principal for the first period would be \$100, and the interest for that period would be \$1. The principal for the second period would be \$101, and the interest for that period would be \$1.01; the principal for the third period would then be \$102.01; etc. The student may show that the compound amount at the end of four periods (1 year) is \$104.06.

We wish now to derive a formula for the compound amount A in terms of the original principal P, the interest rate i per period, and the number n of periods. The derivation is as follows:

The principal for the first period is the original principal P so that the interest for this period is Pi. The amount at the end of the first period is then $P + Pi$ or $P(1 + i)$. This is the principal for the second period. The interest for the second period is then $i \cdot P(1 + i)$, so the amount at the end of two periods is

$$P(1 + i) + iP(1 + i) = P(1 + i)(1 + i) = P(1 + i)^2.$$

The interest for the third period is $iP(1 + i)^2$ so the amount at the end of three periods is

$$P(1 + i)^2 + iP(1 + i)^2 = P(1 + i)^2(1 + i) = P(1 + i)^3$$

and so on. We see that at the end of n periods the compound amount would be

$$\mathbf{A = P(1 + i)^n}.$$

(The student may supply the formal details of the proof by mathematical induction that is obviously indicated here.)

Example

If a principal of \$100 is invested at 4 per cent, compounded semiannually, for 8 years, the number of periods is 16 and the rate per period is 0.02. The compound amount is then

$$\begin{aligned} A = 100(1 + 0.02)^{16} &= 100(1.02)^{16} \\ &= 100(1.3728) = \$137.28. \end{aligned}$$

If the interest were compounded quarterly, then $n = 32$ and the rate per period is 0.01. In this case

$$A = 100(1 + 0.01)^{32} = 100(1.01)^{32}$$
$$= 100(1.3749) = \$137.49.$$

If $P = 1$, the above compound interest formula becomes $A = (1 + i)^n$. Thus the factor $(1 + i)^n$ gives the compound amount corresponding to a principal of 1, at the end of n conversion periods at rate i per period.

Very elaborate tables giving the value of $(1 + i)^n$ for various values of i and n are in common use. Table VI (page 313) in this text is a brief table of this kind. When a table is not available, one can compute the value of $(1 + i)^n$ by using logarithms or by means of the binomial formula.

83. *Present value.* The *present value* of a sum of money due at the end of n periods at rate i per period, is the principal which, if invested, would accumulate to the specified sum by the end of the specified time. It can be found by solving the compound interest formula for P in terms of A, i, and n:

$$P = \frac{A}{(1 + i)^n} = A(1 + i)^{-n}.$$

Example

If A is to pay B the sum of \$800 at the end of 5 years, and if money is worth 4 per cent compounded annually, then the present value would be

$$P = 800(1.04)^{-5} = 800(0.82193) = \$657.54.$$

If A should take \$657.54 and invest it at the above rate, it would accumulate to the required \$800 by the end of 5 years. Of course B could accept the payment of \$657.54 now as equivalent to \$800 at the end of 5 years and invest it himself.

The factor $(1 + i)^{-n}$ gives the present value of 1 due at the end of n periods at rate i per period. Tables giving the value of $(1 + i)^{-n}$ for various values of i and n are available. A brief one is Table VII (page 316) in this text.

If any three of the four numbers in the equation

$$A = P(1 + i)^n \qquad \text{or} \qquad P = A(1 + i)^{-n}$$

are given, one can of course solve for the remaining one.

Example

The holder of a bond that will be worth $2,000 6 years from now offers to sell it for $1,500 cash. What interest rate, compounded annually, would the purchaser be earning?

Solution

In the formula $A = P(1 + i)^n$ we may let $A = 2{,}000, P = 1{,}500$, and $n = 6$. Then we have

$$\begin{aligned} 1{,}500(1 + i)^6 &= 2{,}000, \\ (1 + i)^6 &= \frac{2{,}000}{1{,}500} = \frac{4}{3}, \\ 6 \log (1 + i) &= \log 4 - \log 3, \\ \log (1 + i) &= \tfrac{1}{6}(\log 4 - \log 3) = 0.02082, \\ 1 + i &= 1.049, \\ i &= 0.049 = 4.9 \text{ per cent.} \end{aligned}$$

PROBLEMS

Use logarithms in solving each of the following problems for the unknown quantity:

1. $(1 + i)^{14} = 1.782$.
2. $(1 + i)^8 = 1.524$.
3. $(1 + i)^{36} = 2.238$.
4. $(1 + i)^{17} = 1.864$.
5. $(1.04)^n = 2.614$.
6. $(1.02)^n = 1.500$.

In solving the following problems use either logarithms or tables giving the values of $(1 + i)^n$ and $(1 + i)^{-n}$:

7. Find the compound amount of $250 at the end of 8 years at 4 per cent compounded semiannually.
8. Find the compound amount of $600 at the end of 6 years at 5 per cent compounded quarterly.
9. Find the compound amount of $4,000 at the end of 12 years at 3 per cent compounded semiannually.
10. Find the compound amount of $480 at the end of 10 years at $4\frac{1}{2}$ per cent compounded annually.
11. Find the compound amount of $800 at the end of $7\frac{1}{2}$ years at 4 per cent compounded semiannually.

12. At what interest rate, compounded annually, will a given sum of money double itself in 15 years?

13. In how many years will a given sum double itself if the interest rate is $3\frac{1}{2}$ per cent compounded semiannually?

14. Find the present value of $600 due at the end of 4 years if money is worth $3\frac{1}{2}$ per cent compounded annually.

15. Find the present value of $1,000 due at the end of 9 years if money is worth 4 per cent compounded semiannually.

16. Find the present value of $750 due at the end of $6\frac{1}{2}$ years if money is worth 4 per cent compounded semiannually.

17. The present value of a bond that will be worth $100 at the end of 10 years is $75. This corresponds to what interest rate, compounded annually?

18. A house can be purchased for $20,000 cash, or by paying $7,000 cash, $7,000 at the end of 1 year, and $7,000 at the end of 2 years. What interest rate, compounded annually, is involved in the second plan?

19. Using the binomial theorem, check the value given for $(1.01)^{10}$ in Table VI (page 313).

20. Using the binomial theorem, check the value given for $(1.02)^{20}$ in Table VI (page 313).

21. Using the binomial theorem, check the value given for $(1.01)^{-10}$ in Table VII (page 316).

22. Using the binomial theorem, check the value given for $(1.04)^{-20}$ in Table VII (page 317).

84. ***Annuities.*** An *annuity* is a sequence of equal payments made at equal intervals of time. The time between two successive payments is called the *payment period*, and the total time from the beginning of the first payment period to the end of the last one is called the *term* of the annuity. Payments may be made at the beginning or at the end of each payment period, but we shall consider only the case in which they are made at the *end* of the period. Thus by an annuity of $100 per year for 3 years we shall mean one in which the first $100 is paid at the end of the first year, the second at the end of the second year, and the third at the end of the third year

85. ***Amount of an annuity.*** Let the amount of each payment be denoted by R and the number of payment periods by n. Assume that each payment draws compound interest from the date of payment at rate i per period. Let S denote the total amount that has accumulated at the end of the term of the annuity. We can derive a formula for S in terms of R, i, and n, as follows:

The first payment, being made at the end of the first period, draws compound interest for $n - 1$ periods and therefore amounts to $R(1 + i)^{n-1}$ at the end of the term; the second payment draws interest for $n - 2$ periods and hence amounts to $R(1 + i)^{n-2}$; and so on. The last payment is made at the end of the last period and draws no interest. It then amounts to R. The total accumulation is then

$$S = R + R(1 + i) + R(1 + i)^2 + \cdots + R(1 + i)^{n-1}.$$

The terms of this sum form a geometric progression of n terms with first term R and common ratio $1 + i$. Hence, using the formula on page 159, we have

$$S = \frac{R - R(1 + i)^n}{1 - (1 + i)}.$$

This reduces immediately to

$$S = R\frac{(1 + i)^n - 1}{i}. \tag{1}$$

If $R = 1$, this formula becomes

$$s = \frac{(1 + i)^n - 1}{i}. \tag{2}$$

Formula (2) then gives the accumulated amount of an annuity of 1 per period for n periods at interest rate i per period, each payment being made at the *end* of the period. Formula (2) simply states that if each payment is R instead of 1, the accumulation is R times as much.

Examples

The amount of an annuity of \$100 per year for 10 years at 4 per cent compounded annually is

$$S = 100\frac{(1.04)^{10} - 1}{0.04}$$
$$= 100(12.0061) = \$1{,}200.61.$$

The amount of an annuity of \$25 per month for 36 months at $\frac{1}{2}$ per cent per month is

$$S = 25\frac{(1.005)^{36} - 1}{0.005}$$
$$= 25(39.3361) = \$983.40.$$

Elaborate tables giving the value of the factor

$$\frac{(1 + i)^n - 1}{i}$$

for various values of i and n are in common use. A brief table of this sort is Table VIII (page 319) in this text.

The term *sinking fund* is usually applied to a fund created to meet a future obligation or to make a future purchase. Frequently such a fund is built up by setting aside equal amounts at the ends of equal periods of time, and the problem then is one to which the above formula applies.

Example

A company estimates that \$15,000 will be needed at the end of 8 years to replace a machine. How much should be set aside at the end of each year for this purpose if the money can be invested at 4 per cent compounded annually?

Solution

We are to determine the annual payment R so that the accumulated amount is \$15,000.

$$R\frac{(1.04)^8 - 1}{0.04} = 15{,}000;$$

$$R(9.214226) = 15{,}000;$$

$$R = \frac{15{,}000}{9.214226} = \$1{,}627.92.$$

86. *Present value of an annuity.* The sum of the present values of the payments of an annuity, at the beginning of the term of the annuity, is called the *present value* of the annuity.

Example

The amount that one must pay now, in cash, for an annuity of \$100 per year for the next 10 years is the *present value* of this annuity. It is the sum of the present values of the ten \$100 payments, these being computed at whatever interest rate is specified.

A formula for the present value P of an annuity of R per period for n periods, at interest rate i per period, can be derived as follows:

The first payment is to be made at the end of the first period so its present value is $R(1+i)^{-1}$; the second payment is to be made at the end of the second period so its present value is $R(1+i)^{-2}$; and so on. The last payment is to be made at the end of the last period so its present value is $R(1+i)^{-n}$. The sum is

$$P = R(1+i)^{-1} + R(1+i)^{-2} + \cdots + R(1+i)^{-n}.$$

The terms form a geometric progression with first term $R(1+i)^{-1}$ and common ratio $(1+i)^{-1}$. Hence

$$P = \frac{R(1+i)^{-1} - R(1+i)^{-1}(1+i)^{-n}}{1-(1+i)^{-1}}.$$

If we multiply numerator and denominator of the right member by $(1+i)$, we have

$$P = \frac{R - R(1+i)^{-n}}{i},$$

or

$$P = R\,\frac{1-(1+i)^{-n}}{i}. \tag{1}$$

For the case in which $R = 1$, this formula becomes

$$p = \frac{1-(1+i)^{-n}}{i}. \tag{2}$$

Formula (2) gives the present value of an annuity of 1 per period for n periods at interest rate i per period, and formula (1) simply states that for an annuity of R per period the present value is R times as much.

Tables giving the value of the factor

$$\frac{1-(1+i)^{-n}}{i}$$

for various values of i and n are available. A brief table of this kind is Table IX (page 322) in this text.

Example

A man buys a house for \$16,000. He pays \$4,000 in cash and agrees to pay the balance, principal and interest, in 20 equal installments, one at the end of each 6 months for the next 10 years.

What should be the amount of each payment if the interest rate is 5 per cent compounded semiannually?

Solution

The present value of the unpaid balance is $12,000 and the interest rate is $2\frac{1}{2}$ per cent per period for 20 periods. Hence

$$R\frac{1 - (1.025)^{-20}}{0.025} = 12{,}000;$$
$$R(15.5892) = 12{,}000;$$
$$R = \frac{12{,}000}{15.5892} = \$769.76.$$

It is customary in practice to add other items such as taxes and insurance to these payments.

PROBLEMS

Find the value of each of the following to six significant digits by using the binomial formula. Obtain an approximate check by means of logarithms:

1. $\dfrac{(1.02)^8 - 1}{0.02}$.

2. $\dfrac{(1.03)^6 - 1}{0.03}$.

3. $\dfrac{1 - (1.04)^{-3}}{0.04}$.

4. $\dfrac{1 - (1.02)^{-6}}{0.02}$.

Find the amount of each of the following annuities:

5. $500 at the end of each year for 12 years at 4 per cent compounded annually.

6. $100 at the end of each 6 months for 15 years at 3 per cent compounded semiannually.

7. $80 at the end of each 6 months for $6\frac{1}{2}$ years at 4 per cent compounded semiannually.

8. $75 at the end of each 3 months for 12 years at 4 per cent compounded quarterly.

9. A father decides at the birth of a son to provide for the son's college education by setting aside a certain amount on each of the son's first 18 birthdays. What must this sum be if the accumulated amount available on the eighteenth birthday is to be $4,000, assuming an interest rate of 3 per cent compounded annually?

10. What sum placed in a savings account on the date of a boy's birth would provide him with $800 on each of his eighteenth, nineteenth, twentieth, and twenty-first birthdays if the interest rate were 4 per cent compounded annually?

11. A trucker estimates that he will have to buy a new truck in 5 years and that it will cost $4,600. What amount should he put into a savings account at the end of each 6 months in order to buy the new truck if the interest rate is 3 per cent compounded semiannually? Assume that he can get \$600 for his old truck.

12. A man buys a house for \$15,000. He pays \$5,000 on the date of purchase and is to pay the balance in 10 equal installments at intervals of 6 months over the next 5 years. What should be the amount of these payments if money is worth 4 per cent compounded semiannually?

13. A man buys a house on which he is to pay \$200 in principal and interest every 3 months for 15 years (60 payments in all). On the day that he makes the twenty-fourth payment he wishes to pay off the entire balance because he is selling the house. How much should he pay, in addition to the twenty-fourth payment, in order to settle the debt if the interest rate is 4 per cent compounded quarterly?

14. A house is offered for sale on the following terms: \$4,000 cash and \$800 at the end of each 6 months for 8 years. What would be the corresponding cash price for the house if the interest rate were 6 per cent compounded semiannually?

15. The beneficiary of a \$12,000 life insurance policy is to receive 10 equal annual payments (instead of \$12,000 in cash), the first payment being made at the time of death of the insured. What is the amount of each payment if the interest rate is 3 per cent compounded annually?

16. The beneficiary of a life insurance policy is to receive 10 annual payments of \$1,200 each, the first payment to be made at the time of death of the insured. What would be the equivalent cash settlement at the time of death of the insured if money is worth 3 per cent compounded annually?

17. Show that, if each payment is made at the *beginning* instead of at the end of the payment period, the formula for the sum is

$$S = R(1+i)\left[\frac{(1+i)^n - 1}{i}\right].$$

18. Show that, if each payment is made at the *beginning* instead of at the end of the payment period, the formula for the present value is

$$P = R(1+i)\left[\frac{1-(1+i)^{-n}}{i}\right].$$

CHAPTER XIV

THEORY OF EQUATIONS

87. *Introduction.* A *polynomial* in x is a function of x that can be written in the form

$$a_0x^n + a_1x^{n-1} + a_2x^{n-2} + \cdots + a_{n-1}x + a_n$$

where the a's are constants and n in a positive integer. It is of degree n, if $a_0 \neq 0$.* We may now agree to use the symbol $P(x)$ to mean a polynomial in x, and when we wish to indicate the degree, we may use $P_n(x)$ to mean a polynomial of degree n.

On page 50 we defined a rational integral equation in x, of degree n, to be an equation that can be written in the form

$$a_0x^n + a_1x^{n-1} + a_2x^{n-2} + \cdots + a_{n-1}x + a_n = 0 \qquad (a_0 \neq 0)$$

in which n is a positive integer. Thus a rational integral equation in x is the result of equating to zero a polynomial in x of degree one or more. We may agree to denote such an equation by writing

$$P_n(x) = 0,$$

it being understood that $n \geqq 1$.

We have seen that if the equation is of degree one (linear), it has precisely one root, and if it is of degree two (quadratic), it has two roots, and we have developed the following formulas which give these roots in terms of the coefficients:

$$a_0x + a_1 = 0; \qquad \text{root: } x = -\frac{a_1}{a_0};$$

$$a_0x^2 + a_1x + a_2 = 0; \qquad \text{roots: } x = \frac{-a_1 \pm \sqrt{a_1^2 - 4a_0a_2}}{2a_0}.$$

Some progress has been made on the case in which the equation is of degree three or more. On page 71 we stated, but did not

* A constant is regarded as a polynomial of degree zero.

prove, the fundamental theorem of algebra which assures us that such an equation has a root within the field of complex numbers. On page 72 we solved the equation

$$x^3 + 8 = 0$$

by factoring the left member and thus writing the equation in the equivalent form

$$(x + 2)(x^2 - 2x + 4) = 0.$$

From this we could be assured that the equation has the following three roots and no others:

$$x = -2; \qquad x = 1 + \sqrt{3}\, i; \qquad x = 1 - \sqrt{3}\, i.$$

In the set of problems beginning on page 74 we similarly solved several equations of higher degree, but we did not develop any general method for solving such equations.

In Sec. 35 of Chap. IV on Functions and Graphs, we became familiar with the idea of locating, at least approximately, the real roots of the equation $P_n(x) = 0$ by making the graph of the equation $y = P_n(x)$ and finding the points where it crosses, or touches, the x-axis. Thus Fig. 14 (page 85) exhibits the fact that the equation

$$x^4 - \tfrac{5}{2}x^3 - 13x^2 + 10x + 36 = 0$$

has both $+2$ and -2 as roots, and it shows that there is another real root between 4 and 5. We called -2 a *double* root because the factor $x + 2$ appeared twice when we factored the left member of the equation.

In the present chapter we shall continue our study of rational integral equations. We shall prove that the equation $P_n(x) = 0$ has exactly n roots. We shall prove that if the coefficients are real numbers, then the imaginary roots, if there are any, occur in *conjugate pairs*, so that, for example, if $2 + 3i$ is a root then $2 - 3i$ must also be a root. We shall develop various other theorems in regard to the roots, some of which will be useful in connection with the problem of determining the roots. Finally, we shall indicate some methods that may be employed in computing the real roots to any desired number of decimal places.

88. *Synthetic division.* In the work of this chapter we shall frequently wish to divide a given polynomial $P_n(x)$ by the first degree polynomial $x - r$, where r is a constant. We take up in this section a short method of performing this division. The method is called *synthetic division*, and we shall illustrate it by first dividing the polynomial

(1) $$5x^3 - 16x^2 + 19x + 4$$

by $x - 2$ using ordinary long division. We shall then see how the process can be shortened:

$$\begin{array}{r|l}
 & 5x^2 - 6x + 7 \quad (\textit{quotient}) \\
\hline
x - 2 & 5x^3 - 16x^2 + 19x + 4 \\
 & 5x^3 - 10x^2 \\
 & \quad - 6x^2 + 19x \\
 & \quad - 6x^2 + 12x \\
 & \qquad + 7x + 4 \\
 & \qquad + 7x - 14 \\
 & \qquad\qquad + 18 \quad (\textit{remainder})
\end{array}$$

The quotient is thus $5x^2 - 6x + 7$, and the remainder is 18.

Our first step in shortening this process is to write only the coefficients. We then have:

(2)
$$\begin{array}{r|l}
 & 5 - 6 + 7 \\
\hline
1 - 2 & 5 - 16 + 19 + 4 \\
 & 5 - 10 \\
 & \quad - 6 + 19 \\
 & \quad - 6 + 12 \\
 & \qquad + 7 + 4 \\
 & \qquad + 7 - 14 \\
 & \qquad\qquad + 18
\end{array}$$

Observe now that the numbers 5, -6, and 7, which are the coefficients in the quotient, have each been written down three times. We can avoid this repetition. Also, we can dispense with "bringing down" the 19 and 4, and thus avoid another repetition. Furthermore, instead of writing $1 - 2$ as the divisor we can write only the -2, it being understood that the coefficient of x is 1.

We can then compress all the essentials of the above results into the following form:

$$
(3)\qquad \begin{array}{r|l}
-2 & 5 - 16 + 19 + \ \ 4 \\
\hline
 & \quad\ - 10 + 12 - 14 \\
\hline
 & 5 - \ \ 6 + \ \ 7 + 18
\end{array}
$$

In this form the numbers 5, -6, and $+7$ in the bottom line are the coefficients in the quotient, and the last number, 18, is the remainder.

We finally make one more minor modification. In form (3) above, we *subtract* the numbers of the second row from the corresponding ones of the first row in order to get those in the third row. If we change the sign of the indicated divisor (replace -2 by 2), we automatically change the sign of every number of the second row; then we could *add* instead of *subtract*. Also, it is customary to put the divisor on the right instead of on the left. The final form is then

$$
(4)\qquad \begin{array}{rrrr|l}
5 & -16 & +19 & +\ 4 & 2 \\
 & +10 & -12 & +14 & \\
\hline
5 & -\ \ 6 & +\ \ 7 & +18 &
\end{array}
$$

The synthetic process of dividing a given polynomial $P(x)$ by $x - r$ is therefore carried out by writing down three rows of numbers as illustrated by form (4), as follows:

STEP 1. As row one, write down the coefficients of the polynomial, including signs, in order of decreasing powers of x. Supply a zero coefficient for each missing power. At the right end of this row, place the number r as shown. (If the divisor is $x - 2$, then $r = 2$; if it is $x + 4$, then $r = -4$; etc.)

STEP 2. Copy down the first number of row one as the first number of row three.

STEP 3. Multiply this first number of row three by r and place the result under the second number of row one (as the first number of row two). Add these numbers to obtain the second number of row three. (In the illustration above we multiply 5 by 2 and place the product, 10, under -16. Then we add -16 and $+10$ to get -6.)

STEP 4. Multiply the second number of row three by r and place the result under the third number in row one. Add to obtain the third number of row three. Continue this until the end of row one is reached.

The numbers of the third row, except for the last one, will be the coefficients in the quotient whose degree will be less by one than that of the dividend $P_n(x)$. The last number will be the remainder.

Example

Divide $3x^4 - 14x^2 + 5$ by $x + 2$.

Solution

$$\begin{array}{rrrrr|l} 3 & +0 & -14 & +0 & +5 & -2 \\ & -6 & +12 & +4 & -8 & \\ \hline 3 & -6 & -2 & +4 & -3 & \end{array}$$

The quotient is $3x^3 - 6x^2 - 2x + 4$ and the remainder is -3. We may therefore write

$$\frac{3x^4 - 14x^2 + 5}{x + 2} = (3x^3 - 6x^2 - 2x + 4) + \frac{-3}{x + 2}$$

or $$3x^4 - 14x^2 + 5 = (x + 2)(3x^3 - 6x^2 - 2x + 4) - 3.$$

This last result is in the familiar form

$$\textit{Dividend} = \textit{divisor} \cdot \textit{quotient} + \textit{remainder}.$$

While it is of course possible to stop the division at any stage, we have in both of the examples of this section carried it to the point where the remainder is a constant. This is what we shall imply throughout this chapter when we speak of dividing a polynomial $P_n(x)$ by $x - r$.

89. *The remainder theorem and the factor theorem.* Let the polynomial $P(x)$ be divided by $x - r$. Let the quotient be denoted by $Q(x)$ and the remainder by R. We then have the identity

$$P(x) = (x - r)Q(x) + R$$

which holds for all values of x.* In particular, for $x = r$, we have

$$P(r) = R$$

* The student should read again the discussion on p. 33.

which means that the value of the polynomial $P(x)$ when x has the value r is equal to the remainder R. This is the remainder theorem which may be stated formally as follows:

Remainder theorem. *If a polynomial* $\mathbf{P(x)}$ *is divided by* $\mathbf{x - r}$, *the remainder is equal to* $\mathbf{P(r)}$; *i.e.*, the remainder is equal to the value of the polynomial when $x = r$.

Example

Find the value of $x^4 - 2x^3 + 36x + 5$ when $x = -4$.

Solution

We may divide the polynomial by $x - (-4)$ or $x + 4$ and take the remainder:

$$\begin{array}{rrrrr|l} 1 & -2 & +0 & +36 & +5 & -4 \\ & -4 & +24 & -96 & +240 & \\ \hline 1 & -6 & +24 & -60 & +245 & \end{array}$$

The required value is 245. The usual way of finding it would of course be that of substituting -4 for x. Thus we have

$$\begin{aligned} (-4)^4 - 2(-4)^3 + 36(-4) + 5 &= 256 + 128 - 144 + 5 \\ &= 245. \end{aligned}$$

As a corollary to the remainder theorem we have the following:

Factor theorem. *If* $\mathbf{r}$ *is a root of the equation* $\mathbf{P(x)} = 0$, *then* $\mathbf{x - r}$ *is a factor of the polynomial* $\mathbf{P(x)}$. The proof is as follows:

If r is a root of $P(x) = 0$, then of course $P(r) = 0$ which implies that the remainder $R = 0$. The identity

$$P(x) = (x - r)Q(x) + R$$

then becomes

$$P(x) = (x - r)Q(x).$$

Thus $x - r$ is a factor of $P(x)$, since of course $Q(x)$ is a polynomial.

The converse of the factor theorem is also true; *i.e.*, *if* $\mathbf{x - r}$ *is a factor of* $\mathbf{P(x)}$, *then* $\mathbf{r}$ *is a root of the equation* $\mathbf{P(x)} = 0$. The student may supply the proof.

Example

If we divide the left member of the equation

$$x^4 - \tfrac{5}{2}x^3 - 13x^2 + 10x + 36 = 0 \qquad (1)$$

by $x - 2$, we get the following result:

$$\begin{array}{rrrrr|l} 1 & -\frac{5}{2} & -13 & +10 & +36 & 2 \\ & +2 & -1 & -28 & -36 & \\ \hline 1 & -\frac{1}{2} & -14 & -18 & +0 & \end{array}$$

The remainder is *zero.* This means that 2 is a root of (1) and $x - 2$ is a factor of its left member. The remaining numbers in the third row are the coefficients of the other factor so that (1) is equivalent to

$$(x - 2)(x^3 - \tfrac{1}{2}x^2 - 14x - 18) = 0. \tag{2}$$

To find the other roots of (1) we may then seek the roots of the equation

$$x^3 - \tfrac{1}{2}x^2 - 14x - 18 = 0. \tag{3}$$

(See Prob. 9 of the next set.)

PROBLEMS

In each of the following problems find the quotient and the remainder using synthetic division. Check your result by multiplication:

1. $(4x^3 + 5x^2 - 23x + 6) \div (x - 2)$.
2. $(7x^3 - 14x^2 + 23x - 12) \div (x + 1)$.
3. $(2y^3 + 12y^2 + 16y - 9) \div (y + 3)$.
4. $(3x^4 - 8x^2 + 5x - 7) \div (x - 1)$.
5. $(x^4 + 8x + 2) \div (x + 1)$.
6. $(x^5 - 10x^3 + 24x^2 + 75x + 20) \div (x + 3)$.
7. $(2y^4 + 12y^3 + 13y^2 - 75) \div (y + 5)$.
8. $(x^5 - 1) \div (x - 1)$.

9. Let r_1 be a root of the equation $P_n(x) = 0$. Then, from the factor theorem,

$$P_n(x) = (x - r_1)Q(x),$$

and $Q(x)$ is obviously a polynomial of degree $n - 1$. Prove that, if any number different from r_1 is another root of $P_n(x) = 0$, it is a root of the equation $Q(x) = 0$.

10. Show that 7 is a root of the equation $6x^3 - 31x^2 - 87x + 70 = 0$. Find two more roots by solving a quadratic equation.

11. Show that -3 is a root of the equation $2x^3 - x^2 - 36x - 45 = 0$. Find two more roots by solving a quadratic equation.

12. Show that $\frac{2}{3}$ is a root of the equation

$$12x^3 + 16x^2 - 7x - 6 = 0.$$

Find two more roots by solving a quadratic equation.

13. Show that $-\frac{3}{2}$ is a root of the equation

$$2x^3 - 5x^2 + 14x + 39 = 0.$$

Find two more roots by solving a quadratic equation.

14. Show that $x - 3$ is a factor of the polynomial

$$x^5 - 17x^3 + 75x - 9$$

and find the other factor.

15. Show that $x + 4$ is a factor of the polynomial

$$x^5 + 4x^4 + x^2 - 3x - 28$$

and find the other factor.

16. Show that $x + 2$ is a factor of the polynomial $x^5 + 32$ and find the other factor.

In each of the following problems use synthetic division to find the value of the given polynomial for every integral value of x in the specified interval. Use these function values (and obtain additional values wherever necessary) to draw the corresponding graph. Find, or locate between consecutive integers, the real values of x in the interval for which the value of the polynomial is zero:

17. $4x^3 - 16x^2 + 9x + 9$; $\quad -2 \leqq x \leqq 4$.
18. $x^3 - x^2 - 5x + 5$; $\quad -3 \leqq x \leqq 3$.
19. $6x^3 - 13x^2 - 36x + 28$; $\quad -3 \leqq x \leqq 4$.
20. $3x^3 - 10x^2 + 15x - 50$; $\quad -1 \leqq x \leqq 4$.
21. $2x^3 - 11x^2 + 10x + 18$; $\quad -2 \leqq x \leqq 5$.
22. $-2x^3 - 7x^2 + 7x + 28$; $\quad -4 \leqq x \leqq 3$.
23. $x^4 - 2x^3 - 5x^2 + 6x + 5$; $\quad -2 \leqq x \leqq 3$.
24. $x^4 - 7x^3 + 12x^2 + 7x - 11$; $\quad -2 \leqq x \leqq 4$.
25. $2x^4 + 17x^3 + 37x^2 - 8x - 46$; $\quad -5 \leqq x \leqq 2$.
26. $x^5 - 12x^2 + 5x - 7$; $\quad -2 \leqq x \leqq 3$.

27. Prove the converse of the factor theorem; *i.e.*, prove that if $x - r$ is a factor of $P(x)$, then r is a root of the equation $P(x) = 0$.

28. By making a graph of the equation $y = x^4 - 9x^3 + 18x^2 - 3$, show that the equation $x^4 - 9x^3 + 18x^2 - 3 = 0$ is satisfied by four real values of x. Locate each of these roots between two consecutive integers.

29. By making a graph of the equation $y = 2x^3 - 15x^2 + 24x + 9$, show that the equation $2x^3 - 15x^2 + 24x + 9 = 0$ is satisfied by three real values of x. Find these roots.

30. Show that $2x - 3$ is a "double" factor of the left side of the equation $4x^4 - 20x^3 + 17x^2 + 30x - 36 = 0$. Find four roots of this equation. Sketch the corresponding graph of the polynomial.

31. Show that $2x + 1$ is a "double" factor of the left side of the equation $4x^4 - 20x^3 + x^2 + 18x + 6 = 0$. Find four roots of this equation. Sketch the corresponding graph of the polynomial.

90. ***Number of roots.*** The fundamental theorem of algebra (page 71) assures us that every rational integral equation,

See 90 & 92

$P_n(x) = 0$, has a root. More precisely, it assures us that if the coefficients are any complex numbers, there is a value of x *within this same field of numbers* for which the equation is true.

This theorem is of course far from trivial. An equation that is not a rational integral equation may not have a root; for example, the exponential equation $2^x = 0$ has no root at all, real or imaginary. Neither has the equation $(x - 3)/(x + 5) = 1$. Even a rational integral equation may not have a root within the field of numbers to which its coefficients are restricted. Thus if we restrict a and b, in the equation $ax + b = 0$, to the field of integers, the equation may or may not have a solution in that same field. If we restrict a and b to the field of rational numbers, then the equation has a solution in this same field if $a \neq 0$. The fundamental theorem guarantees that every rational integral equation whose coefficients are elements of the field of complex numbers has a solution within this same field.

We now prove the following:

Theorem. *Every rational integral equation of degree* **n** *has exactly* **n** *roots.*

Proof: Let the equation be $P_n(x) = 0$, by which we mean

$$a_0x^n + a_1x^{n-1} + \cdots + a_{n-1}x + a_n = 0, \qquad a_0 \neq 0, \tag{1}$$

n being a positive integer. Let r_1 denote a root, the existence of which is assured by the fundamental theorem. Then $(x - r_1)$ is a factor of $P_n(x)$ and

$$P_n(x) = (x - r_1)P_{n-1}(x) \tag{2}$$

where $P_{n-1}(x)$ is a polynomial of degree $n - 1$ having a_0x^{n-1} as its term of highest degree. Consider now the equation

$$P_{n-1}(x) = 0.$$

The fundamental theorem assures us that it has a root, which we may denote by r_2. Then

$$P_{n-1}(x) = (x - r_2)P_{n-2}(x) \tag{3}$$

where $P_{n-2}(x)$ is a polynomial of degree $n - 2$ having a_0x^{n-2} as its term of highest degree. If we combine (2) and (3), we have

$$P_n(x) = (x - r_1)(x - r_2)P_{n-2}(x). \tag{4}$$

After n such applications of the fundamental theorem, we have

$$(5) \qquad P_n(x) = (x - r_1)(x - r_2) \cdots (x - r_n)P_0(x)$$

where $P_0(x)$ is a polynomial of degree zero, namely, the constant a_0. This means that the polynomial $P_n(x)$ can be expressed as a product of the constant a_0 and n linear factors of the type $(x - r_i)$. Thus

$$(6) \qquad P_n(x) = a_0(x - r_1)(x - r_2) \cdots (x - r_n).$$

It follows immediately that the equation $P_n(x) = 0$ has the n roots $r_1, r_2, \cdots, r_n$, and no others.

It should be emphasized that the numbers $r_1, r_2, \cdots, r_n$ need not be real. Also, they are not necessarily all distinct. If a certain factor $(x - r_i)$ appears m times in the right member of (6), then r_i is called a *root of multiplicity m*—a double root, triple root, etc. A root corresponding to a linear factor that occurs only once is often called a *simple root.*

Examples

The rational integral equation $x^3 - x^2 - 7x + 15 = 0$ is equivalent to

$$(x + 3)(x - 2 - i)(x - 2 + i) = 0.$$

It has the three roots -3, $2 + i$, $2 - i$. The above theorem does not tell us how these factors may be found; it merely guarantees their existence.

The equation

$$2x^3 + (1 + 2i)x^2 + (4 + i)x + 2 = 0$$

is equivalent to

$$2(x + \tfrac{1}{2})(x - i)(x + 2i) = 0.$$

It has the roots $-\frac{1}{2}$, i, and $-2i$.

91. ***Identical polynomials.*** An important corollary of the above theorem is the following:

Theorem. *If two polynomials in* **x**, *of degree not greater than* **k**, *are equal in value for more than* **k** *distinct values of* **x** *then the polynomials are identical term by term; i.e., the coefficients of like powers of* **x** *are equal.*

Proof: Let the two polynomials be

$$a_0x^k + a_1x^{k-1} + \cdots + a_{k-1}x + a_k,$$
$$b_0x^k + b_1x^{k-1} + \cdots + b_{k-1}x + b_k.$$

Then, by hypothesis, the equation

$$(1) \quad a_0x^k + a_1x^{k-1} + \cdots + a_{k-1}x + a_k = b_0x^k + b_1x^{k-1} + \cdots + b_{k-1}x + b_k$$

or

$$(2) \quad (a_0 - b_0)x^k + (a_1 - b_1)x^{k-1} + \cdots + (a_{k-1} - b_{k-1})x + (a_k - b_k) = 0$$

is satisfied by more than k distinct values of x. If any of the numbers $(a_0 - b_0)$, $(a_1 - b_1)$, $\cdots$, $(a_{k-1} - b_{k-1})$ were different from zero, then (2) would be a rational integral equation of degree k or less with more than k roots. Since this is impossible, we must have $a_0 = b_0$, $a_1 = b_1$, $\cdots$, $a_{k-1} = b_{k-1}$. Then we must also have $a_k = b_k$ because otherwise (2) could not be true for any value of x. This completes the proof.

Example 1

The polynomials $x^3 - 7x^2 + 9$ and $4x^2 + 6x - 13$ cannot have equal values for more than three distinct values of x. This means that their graphs cannot intersect at more than three points.

Example 2

For what values of A, B, and C, is the equation

$$A(x - 2) + (Bx + C)(2x + 5) = 25x + 76$$

an identity?

Solution

When the indicated operations are performed on the left member, the equation becomes

$$2Bx^2 + (A + 5B + 2C)x + (-2A + 5C) = 25x + 76.$$

These two polynomials will be equal for all values of x if and only if the coefficients of like powers are equal; *i.e.*, if and only if

$$\begin{aligned} 2B &= 0, \\ A + 5B + 2C &= 25, \\ -2A + 5C &= 76. \end{aligned}$$

This system has the unique solution $A = -3$, $B = 0$, $C = 14$. If A, B, and C have these values, the two sides of the given equation are equal for *all* values of x. For any other values of A, B, and C, the two sides cannot be equal for more than *two* distinct values of x.

92. ***Formation of an equation whose roots are given.*** Let it be specified that the roots of a rational integral equation of degree n are to be the given numbers $r_1, r_2, \cdots, r_n$, and let it be required to find the equation. Obviously the solution to this problem is

$$a_0(x - r_1)(x - r_2) \cdots (x - r_n) = 0$$

in which a_0 is an arbitrary constant different from zero.

Example 1

The quadratic equation whose roots are $\frac{3}{2}$ and -5 is

$$a_0(x - \tfrac{3}{2})(x + 5) = 0,$$

where a_0 may be any constant except zero. We may obtain an equation with integral coefficients by choosing $a_0 = 2$. We then have

$$(2x - 3)(x + 5) = 0$$

or

$$2x^2 + 7x - 15 = 0.$$

Example 2

The rational integral equation that has $2 + i$, $2 - i$, and 0, as simple roots, has -1 as a double root, and has no other roots, is

$$a_0(x - 2 - i)(x - 2 + i)(x)(x + 1)^2 = 0,$$

or

$$a_0(x^5 - 2x^4 - 2x^3 + 6x^2 + 5x) = 0.$$

If we choose $a_0 = 1$, we get the simplest such equation with integral coefficients.

PROBLEMS

In each of the following problems find the values of A and B or of A, B, and C for which the given equation is an identity:

1. $A(3x - 5) + B(x + 7) = 38 - 2x$.
2. $A(7x + 2) + B(2x - 3) = -5(x - 4)$.
3. $A(2x + 5) + B(3x - 1) = 85$.
4. $A(2x + 4) + B(7 - x) = 378$.

5. $8x^2 + 41x - 105 = A(x - 3)(x + 6) + B(x - 1)(x + 6) + C(x - 1)(x - 3)$.

6. $5x^2 - 45x - 80 = A(2x + 1)(x + 3) + B(x - 7)(x + 3) + C(2x + 1)(x - 7)$.

7. $(Ax + B)(2x + 1) + C(x^2 - 1) = -2x - 7$.

8. $(Ax + B)(x - 2) + C(x^2 + x + 1) = 8x^2 + 3$.

9. $A(x - 3)(x - 1) + B(x - 1) + C(x - 3)^2 = x^2 + 6x - 1$.

Find the roots of each of the following equations. Give the multiplicity of each multiple root:

10. $7(x + 1)^2(x - 3) = 0$.

11. $\frac{1}{2}x^2(x^2 - 4) = 0$.

12. $8(x^2 - 4)(x^3 + 8) = 0$.

13. $x^4 - 2x^3 + 10x^2 = 0$.

14. $3(x + 1)^3(x^2 - 4x + 5)^2 = 0$.

15. $x^5 + 27x^2 = 0$.

16. $5(x^4 - 1)^2 = 0$.

17. $(x^2 - x - 2)(x^2 - 7x + 10) = 0$.

18. Prove that zero is a root of the cubic equation

$$ax^3 + bx^2 + cx + d = 0$$

if and only if $d = 0$. What is the corresponding statement involving the case in which $c = 0$ and $d = 0$?

In each of the following problems form the rational integral equation with the smallest possible integral coefficients that has the given roots and no others:

19. -1 and 0 as simple roots; 2 as a double root.

20. 0 and $-\frac{2}{3}$ as simple roots; $-\frac{1}{2}$ as a double root.

21. $\frac{4}{3}$, $-\frac{2}{3}$, and -1 as simple roots.

22. 2, $2i$, -2, and $-2i$ as simple roots.

23. $\sqrt{3}$, $-\sqrt{3}$, $1 + i$, and $1 - i$ as simple roots.

24. $-2 + \sqrt{3}$, $-2 - \sqrt{3}$, i, and $-i$ as simple roots.

25. $-\frac{2}{3}$ as a triple root; -1 as a double root.

26. $-1 + 2i$, $-1 - 2i$, and 0 as double roots.

In each of the following problems form a rational integral equation that has the given roots and no others:

27. i, $2i$, and $-3i$ as simple roots.

28. $1 + i$, $1 - i$, and $-i$ as simple roots.

29. $1 + i$, $1 - i$, i, and $-i$ as simple roots.

30. $-3i$, i, and $-4i$ as simple roots.

31. Using synthetic division, show that i is a root of the equation

$$x^3 + 7x - 6i = 0.$$

Show also that $2i$ is a root.

32. Using synthetic division, show that $2 + i$ is a root of the equation

$$x^5 - 2x^4 - 2x^3 + 6x^2 + 5x = 0.$$

33. Show that the quadratic equation whose roots are r_1 and r_2 can be written in the form

$$a_0[x^2 - (r_1 + r_2)x + r_1r_2] = 0,$$

where a_0 is any constant except zero. Hence imply that if the coefficient of x^2 is taken as 1, the coefficient of x is equal to minus the sum of the roots, and the constant term is equal to the product of the roots. For example, the sum of the numbers $2 + 3i$ and $2 - 3i$ is 4, and their product is 13. The quadratic equation having the roots $2 \pm 3i$ can then be written as $x^2 - 4x + 13 = 0$.

34. Show that the cubic equation whose roots are r_1, r_2, and r_3 can be written in the form

$$a_0[x^3 - (r_1 + r_2 + r_3)x^2 + (r_1r_2 + r_1r_3 + r_2r_3)x - r_1r_2r_3] = 0,$$

where a_0 is any constant except zero. Hence imply that if the coefficient of x^3 is taken as 1, the coefficient of x^2 is equal to minus the sum of the roots, the coefficient of x is equal to the sum of all the products that can be made from the roots by taking them two at a time, and the constant term is equal to minus the product of the roots.

35. Show that 2 is a double root of the equation

$$x^4 - x^3 - 7x^2 + 8x + 4 = 0$$

and find the other roots.

36. Show that 3 is a double root of the equation

$$2x^4 + x^3 - 54x^2 + 81x + 54 = 0$$

and find the other roots.

37. Show that -1 is a triple root of the equation

$$2x^4 + 3x^3 - 3x^2 = 7x + 3$$

and find the other root.

38. Show that 2 is a triple root of the equation

$$2x^5 - 7x^4 - 4x^3 + 32x^2 = 16(x + 1)$$

and find the other roots.

39. Show that 1.5 is a double root of the equation

$$4x^4 + 25x^2 + 36 = 12x(x^2 + 4)$$

and find the other roots.

40. Show that both $+0.5$ and -0.5 are double roots of the equation

$$8x^4(2x^2 + 1) = 7x^2 - 1.$$

Find the other roots.

93. ***Imaginary roots.*** In this section we wish to prove that, in the case of a rational integral equation whose coefficients are real numbers, any imaginary roots that occur are in conjugate pairs; *i.e.*, if $p + qi$ is a root then $p - qi$ is also a root. This will mean that if the coefficients are real numbers, then either the roots are all real, or there are an *even* number of imaginary roots. First we prove a preliminary theorem:

Theorem. *If the value of $\mathbf{x}^m$ ($\mathbf{m}$ a positive integer) when $\mathbf{x} = \mathbf{a} + \mathbf{bi}$ is $\mathbf{C} + \mathbf{Di}$, then the value of $\mathbf{x}^m$ when $\mathbf{x} = \mathbf{a} - \mathbf{bi}$ is $\mathbf{C} - \mathbf{Di}$.*

The proof by mathematical induction is as follows:

PART I. *Hypothesis:*

$$(a + bi)^k = C + Di \qquad \text{and} \qquad (a - bi)^k = C - Di.$$

Conclusion:

$$(a + bi)^{k+1} = C_1 + D_1 i \qquad \text{and} \qquad (a - bi)^{k+1} = C_1 - D_1 i.$$

Proof: By hypothesis,

$$(a + bi)^k = C + Di \qquad \text{and} \qquad (a - bi)^k = C - Di.$$

Multiplying both sides of the first equation by $a + bi$, and both sides of the second by $a - bi$, we have:

$$(a + bi)^{k+1} = (aC - bD) + (bC + aD)i = C_1 + D_1 i;$$
$$(a - bi)^{k+1} = (aC - bD) - (bC + aD)i = C_1 - D_1 i.$$

Thus if the theorem were true for $m = k$, it would be true for $m = k + 1$.

PART II. For the case in which $m = 1$, $x^m = x$; its value for $x = a + bi$ is $a + bi$, and its value for $x = a - bi$ is $a - bi$. This completes the proof.

Example

The value of x^3 for $x = 6 + 2i$ is $144 + 208i$; its value for $x = 6 - 2i$ is $144 - 208i$.

It is obvious that the theorem obtained by replacing x^m by kx^m in the above theorem is also true, k being any real number. Finally, we have the following situation regarding any polynomial $P(x)$ with real coefficients:

If the value of a polynomial $\mathbf{P(x)}$ whose coefficients are real numbers is $\mathbf{C} + \mathbf{Di}$ when $\mathbf{x} = \mathbf{a} + \mathbf{bi}$, then its value is $\mathbf{C} - \mathbf{Di}$ when $\mathbf{x} = \mathbf{a} - \mathbf{bi}$.

Example

The value of the polynomial $x^3 + 7x - 3$ when $x = 2 + i$ is $13 + 18i$. When $x = 2 - i$, it is $13 - 18i$.

We are now in a position to prove very easily the following basic theorem:

Theorem. *If a rational integral equation* $\mathbf{P(x)} = 0$ *with real coefficients has the root* $\mathbf{a + bi}$*, then it has also the root* $\mathbf{a - bi}$.

Proof: By hypothesis, $a + bi$ is a root of the equation $P(x) = 0$, so that $P(a + bi) = 0$:

$$P(a + bi) = C + Di = 0.$$

But this means that both $C = 0$ and $D = 0$, since a complex number is zero only if its real and imaginary parts are both zero. It follows immediately that

$$P(a - bi) = C - Di = 0,$$

which means that $a - bi$ is also a root.

Example

If $1 + i$ is a root of the equation

$$x^4 - 2x^3 - 2x^2 + 8x - 8 = 0,$$

what are the other roots?

Solution

If $1 + i$ is a root, then $1 - i$ is also a root. This means that $(x - 1 - i)$ and $(x - 1 + i)$ are factors and that consequently the product

$$(x - 1 - i)(x - 1 + i) = x^2 - 2x + 2$$

is a factor. The other factor is found by ordinary long division to be $x^2 - 4$; *i.e.*,

$$x^4 - 2x^3 - 2x^2 + 8x - 8 = (x^2 - 2x + 2)(x^2 - 4).$$

The remaining two roots are of course found by setting the factor $x^2 - 4$ equal to zero. They are 2 and -2.

The following theorem concerning the occurrence of a certain kind of irrational root is closely related to the above theorem regarding imaginary roots.

Theorem. *If a rational integral equation with rational coefficients has the root* $\boldsymbol{a} + \sqrt{\boldsymbol{b}}$, *where* $\boldsymbol{a}$ *and* $\boldsymbol{b}$ *are rational numbers and* $\boldsymbol{b}$ *is not a perfect square, then it has also the root* $\boldsymbol{a} - \sqrt{\boldsymbol{b}}$.

The proof may be supplied by the student.

Example

Find all the roots of the equation

$$2x^3 - 5x^2 - 14x - 3 = 0,$$

given that $2 - \sqrt{5}$ is a root.

Solution

If $2 - \sqrt{5}$ is a root, then $2 + \sqrt{5}$ is also a root. Then $(x - 2 + \sqrt{5})$ and $(x - 2 - \sqrt{5})$ are factors, and their product

$$(x - 2 + \sqrt{5})(x - 2 - \sqrt{5}) = x^2 - 4x - 1$$

is a factor. The other factor is found by long division to be $2x + 3$. The third root is then $-\frac{3}{2}$.

The theorem does not imply that we always have an even number of irrational roots, for an irrational root may not be a number of the form $a + \sqrt{b}$. For example, the equation $x^3 - 2 = 0$ has one irrational root, $\sqrt[3]{2}$, and a pair of conjugate imaginary roots. The theorem implies only that a *certain kind* of irrational root can appear only in pairs if the coefficients are rational.

94. *Variations of sign.* Let $P(x)$ be a polynomial with real coefficients whose terms are arranged in the usual order of descending powers of x. It is understood for our present purpose that missing powers are not to be supplied with zero coefficients. A *variation of sign* is said to occur among the coefficients wherever the coefficients of two consecutive terms have opposite signs.

Example

The signs of the coefficients in the polynomial

$$3x^5 - 7x^4 + 9x^2 + 6x - 5$$

are

$$+ \quad - \quad + \quad + \quad - \,.$$

There are *three* variations of sign.

95. *Descartes' rule of signs.* It is obvious that the equation

$$x^2 + 4x + 5 = 0,$$

in which the left member has no variation of sign, cannot have any positive root, because for any positive value of x every term on the left is positive and the sum cannot be zero. On the other hand the equation

$$x^2 - 7x + 10 = 0,$$

in which the left member has two variations in sign, has the two positive roots 2 and 5. A connection between the number of positive roots and the number of variations in sign is given by the following rule:

DESCARTES' RULE OF SIGNS. *The number of positive roots of the rational integral equation* $\boldsymbol{P(x)} = 0$ *with real coefficients is either equal to the number of variations of sign in the polynomial* $\boldsymbol{P(x)}$, *or is less than that number by a positive even integer.*

The proof of this theorem will not be given.*

Example

The left member of the equation

$$x^3 - 4x^2 + 6x + 9 = 0$$

has two variations of sign. There are then either two positive roots or no positive roots.

Corresponding information concerning the number of negative roots can be obtained by considering the number of variations of sign in the polynomial $P(-x)$. This polynomial is written down from the polynomial $P(x)$ by replacing x by $-x$. The rule is:

The number of negative roots of the rational integral equation $\boldsymbol{P(x)} = 0$ *with real coefficients is either equal to the number of variations of sign in the polynomial* $\boldsymbol{P(-x)}$, *or is less than that number by a positive even integer.*

Example

In the case of the equation

$$x^3 - 4x^2 + 6x + 9 = 0,$$

* See, for example, Uspensky, "Theory of Equations," New York, McGraw-Hill Book Company, Inc., 1948.

the polynomial $P(-x)$ is

$$(-x)^3 - 4(-x)^2 + 6(-x) + 9,$$

or

$$-x^3 - 4x^2 - 6x + 9.$$

There is one variation of sign in this polynomial. It follows that the given equation has *exactly one negative root.*

In the previous example we found that this same equation has either two positive roots or no positive roots. We can finally conclude that it has either two positive roots and one negative root, or a pair of conjugate imaginary roots and one negative root.

PROBLEMS

1. Find the value of the polynomial $4x^2 - 6x + 2$ for $x = 1 + i$ and for $x = 1 - i$.

2. Find the value of the polynomial $2x^3 - 9x - 7$ for $x = 2 + 3i$ and for $x = 2 - 3i$.

3. Find the value of the polynomial $2x^5 + 16x^2 + 9$ for $x = 1 + 2i$ and for $x = 1 - 2i$.

In each of the following cases discuss the nature of the roots of the given equation, taking into account the degree of the equation and the theorems of the preceding three sections:

4. $4x^3 - 3x^2 + 2x - 7 = 0.$

5. $x^3 + 2x^2 + x - 4 = 0.$

6. $3x^3 + 5x^2 - 2x + 8 = 0.$

7. $8x^3 - 6x^2 - 7 = 0.$

8. $2x^3 + x + 6 = 0.$

9. $x^4 - 5x^3 - 6x - 9 = 0.$

10. $2x^4 - 6x^2 + 5 = 0.$

11. $2x^4 + 3x^2 = 5x + 20.$

12. $4x^5 + x^3 = x(4x + 1).$

13. $x^5 + 32 = 0.$

14. $x^5 + 7 = 5x.$

15. $x^6 - 64 = 0.$

16. $x^6 + 4x^4 = x^2 + 8.$

17. It is desired to find a number x such that

$$x^2(2x^2 + 5) = 7(x + 5).$$

Show that there are exactly two real numbers, one positive and one negative, that satisfy this condition.

18. By considering the equation $x^3 + 1 = 0$, show that there are three cube roots of -1, and that only one of them is a real number. Find these three numbers.

19. By considering the equation $x^4 - 1 = 0$, show that there are four fourth roots of 1, two of which are real. Find all four of these numbers.

20. By considering the equation $x^5 - 32 = 0$, show that there are five fifth roots of 32, only one of which is real. Find this real root, and by using synthetic division obtain an equation of degree 4 whose roots are the other four fifth roots of 32.

21. How many real roots has the equation $x^n - 1 = 0$, n being a positive even integer? How many if n is a positive odd integer?

22. How many real roots has the equation $x^n + 1 = 0$, n being a positive even integer? How many if n is a positive odd integer?

In each of the following cases verify that the given number is a root of the given equation, and find the other roots:

23. $3x^3 - 10x^2 + 7x + 10 = 0$; $2 + i$.
24. $2x^3 - 9x^2 + 44x - 85 = 0$; $1 - 4i$.
25. $3x^3 - 7x^2 + 27x - 63 = 0$; $3i$.
26. $x^4 - 2x^3 + 2x^2 + 4x - 8 = 0$; $\sqrt{2}$.
27. $x^4 - 6x^3 + 8x^2 = 2x + 1$; $2 + \sqrt{5}$.
28. $x^4 + 8x^2 + 18x = 3(2x^3 + 11)$; $3 + \sqrt{2}\,i$.
29. $x^4 - 4x^2 + 12x = 9$; $1 - \sqrt{2}\,i$.

30. Verify that $2 + 2i$ is one of the four fourth roots of -64, and find the other three.

96. *Integral roots.* The following theorem is useful in determining whether or not a given rational integral equation whose coefficients are integers has any root that is an integer.

Theorem. *If a rational integral equation with integral coefficients has an integer for a root, this root must be a factor of the constant term.*

Proof: Let the equation be

$$a_0x^n + a_1x^{n-1} + \cdots + a_{n-1}x + a_n = 0 \qquad (a_0 \neq 0).$$

Let it also be assumed that $a_n \neq 0$. (The theorem is true if $a_n = 0$, since any integer may be regarded as a factor of zero.)

If an integer r ($r \neq 0$)* is a root, then

$$a_0r^n + a_1r^{n-1} + \cdots + a_{n-1}r + a_n = 0,$$

or

$$a_0r^{n-1} + a_1r^{n-2} + \cdots + a_{n-1} = -\frac{a_n}{r}.$$

Now the left member of this equality is an integer because it is a sum of several terms, each of which is an integer under the conditions of our hypothesis. Hence the right member must also be an integer, but this means that r must be a factor of a_n, which was to be proved.

* Zero is a root if and only if $a_n = 0$. We have already disposed of this case.

Example

The only integers that could possibly be roots of the equation

$$5x^3 - 9x + 14 = 0$$

are ± 1, ± 2, ± 7, and ± 14. When we try each of these, using synthetic division, we find that none of them are roots. We can conclude that no integer is a root of the equation.

97. *Rational roots.* A useful extension of the preceding theorem enables us to determine rather quickly whether or not a given equation whose coefficients are integers has any roots that are rational numbers.

Theorem. *If the rational integral equation*

$$a_0x^n + a_1x^{n-1} + \cdots + a_{n-1}x + a_n = 0 \qquad (a_0 \neq 0),$$

whose coefficients are integers, has the number $\boldsymbol{p/q}$ *as a root, where* $\boldsymbol{p}$ *and* $\boldsymbol{q}$ *are integers and the fraction is in lowest terms, then* $\boldsymbol{p}$ *is a factor of* $\boldsymbol{a_n}$ *and* $\boldsymbol{q}$ *is a factor of* $\boldsymbol{a_0}$.

Proof: As in the preceding theorem, we need consider only the case in which $a_n \neq 0$. Let p/q be a root where p and q are integers different from zero, and relatively prime. Then

$$(1) \qquad a_0\left(\frac{p}{q}\right)^n + a_1\left(\frac{p}{q}\right)^{n-1} + \cdots + a_{n-1}\left(\frac{p}{q}\right) + a_n = 0,$$

or, multiplying both members by q^n,

$$(2) \qquad a_0p^n + a_1p^{n-1}q + \cdots + a_{n-1}pq^{n-1} + a_nq^n = 0.$$

If we subtract a_nq^n from both members of this equation and then divide both members by p, we have

$$(3) \qquad a_0p^{n-1} + a_1p^{n-2}q + \cdots + a_{n-1}q^{n-1} = -\frac{a_nq^n}{p}.$$

Now the left member of this equation is an integer under the conditions of our hypothesis, so the right member must also be an integer. This implies that p is a factor of a_nq^n, and since p and q are relatively prime, p is a factor of a_n. This proves the first part of the theorem.

In order to prove the second part, we return to equation (2), subtract a_0p^n from both members, and then divide both members by q. We thus get

$$(4) \qquad a_1p^{n-1} + \cdots + a_{n-1}pq^{n-2} + a_nq^{n-1} = -\frac{a_0p^n}{q}.$$

The left member of (4) is an integer, and consequently the right member must be an integer, and since q is not a factor of p, it must be a factor of a_0. This completes the proof.

Example

The only rational numbers that could possibly be roots of the equation

$$2x^3 + 7x^2 + 2x - 6 = 0$$

are those that have ± 1, ± 2, ± 3, or ± 6 as numerators, and ± 1 or ± 2 as denominators. These are the numbers

$$\pm 1, \qquad \pm 2, \qquad \pm 3, \qquad \pm 6, \qquad \pm \tfrac{1}{2}, \qquad \pm \tfrac{3}{2}.$$

When these are tried out, using synthetic division, it is found that $-\frac{3}{2}$ is the only one that is a root:

$$\begin{array}{rrrr|l} 2 & +7 & +2 & -6 & -\frac{3}{2} \\ & -3 & -6 & +6 & \\ \hline 2 & +4 & -4 & +0 & \end{array}$$

Descartes' rule of signs assures us that there is one positive root, and we know now that it is an irrational number. Also there must be another negative root, and this also must be irrational.

We can of course find the other two roots in this case by solving the quadratic equation

$$2x^2 + 4x - 4 = 0.$$

They are $-1 + \sqrt{3}$ or 0.732, and $-1 - \sqrt{3}$ or -2.732.

If the coefficient a_0 of the highest power of x is 1, then q must be either $+1$ or -1, and consequently any rational root p/q must be an integer. In fact we have the following:

Corollary. *Any rational root of the equation*

$$x^n + a_1x^{n-1} + \cdots + a_{n-1}x + a_n = 0,$$

in which the coefficient of $\mathbf{x}^n$ *is one and all the remaining coefficients are integers, is an integer that is a divisor of* $\mathbf{a}_n$.

Example

The only rational numbers that could be roots of the equation

$$x^5 + 12x^2 - 7x + 4 = 0$$

are the integers ± 1, ± 2, and ± 4. When these are tried out, using synthetic division, it is found that none of them are roots. We can conclude that *any real roots that exist must be irrational numbers.*

Descartes' rule of signs assures us that there is exactly one negative real root, and also tells us that there are either two positive roots or no positive roots.

98. *Bounds for the roots.* In connection with the preliminary problems of finding the rational roots of a given equation and locating the other real roots between consecutive integers, the following theorem is often useful:

Theorem. *If in the synthetic division of the polynomial*

$$P(x) \equiv a_0x^n + a_1x^{n-1} + \cdots + a_{n-1}x + a_n$$

by $x - r$, *where* a_0 *and* r *are both positive, all the numbers in the third row are positive, then the equation* $P(x) = 0$ *has no root larger than* r.

Example

One of the numbers that might be a rational root of the equation

$$x^3 - 4x^2 - 4x + 12 = 0$$

is 6. When we try out this number we have:

$$\begin{array}{rrrr|l} 1 & -4 & -4 & +12 & \underline{6} \\ & +6 & +12 & +48 & \\ \hline 1 & +2 & +8 & +60 & \end{array}$$

We of course conclude that 6 is not a root, since the remainder is not zero. And we can further conclude, in accordance with the above theorem, that there is no real root larger than 6; *i.e.*, we have already passed the last real root and we need not try any larger numbers. The reason is obvious. If we replace 6 by a larger number, the effect will be to increase all the numbers in the third row except the first one. The remainder will be greater than 60 for every "divisor" greater than 6.

A corresponding lower bound for the negative roots is established by the following statement:

If r *is negative,* a_0 *being positive, and the numbers in the third row are alternately positive and negative, then there is no negative root whose absolute value is greater than that of* r.

Example

When we use -2 as the indicated divisor in the problem of the previous example, the result is:

$$\begin{array}{rrrr|r} 1 & -4 & -4 & +12 & -2 \\ & -2 & +12 & -16 & \\ \hline 1 & -6 & +8 & -4 & \end{array}$$

We can conclude not only that -2 is not a root, but also that there is no root "to the left" of -2; there is no need to try -3, -4, etc. Again the reason is obvious. If the divisor were a negative number with larger absolute value, then the absolute values of all the numbers in the third row, except the first one, would be larger.

The two results together indicate that all the real roots of the equation lie between -2 and $+6$.

PROBLEMS

In each of the following cases find any integral roots that the given equation may have:

1. $2x^4 - 6x^3 + 11x^2 - 27x + 9 = 0.$
2. $3x^3 - 13x^2 + 6x - 8 = 0.$
3. $3x^3 + 4x^2 - 15x = 20.$
4. $5x^3 + 11x^2 - 2x = 8.$
5. $x^5 + 3x^4 - 7x^2 = 19x - 6.$

In each of the following cases find any rational roots that the given equation may have. Find the irrational and imaginary roots also when it is possible to do so by solving a quadratic equation:

6. $2x^3 + 3x^2 - 14x - 21 = 0.$ **7.** $4x^3 + 14x^2 - 6x = 21.$
8. $x^5 + 5x^3 - 7x^2 - 35 = 0.$ **9.** $2x^3 - 11x^2 - 7x + 6 = 0.$
10. $3x^4 - 11x^3 + 9x^2 + 13x - 10 = 0.$
11. $4x^4 + 16x^3 + x^2 + 6x + 8 = 0.$

12. $3x^4 + 8x^3 - 39x^2 - 96x + 36 = 0.$
13. $x^4 - x^3 + 6x^2 - 2x + 8 = 0.$
14. $x^6 - 21x^3 - 58x + 12 = 0.$
15. $2x^5 + x^4 + 3x^3 + 5x^2 - 2x + 6 = 0.$
16. $3x^5 - 5x^4 - 9x^3 + 15x^2 - 12x + 20 = 0.$
17. $4x^4 - 16x^3 - 13x^2 + 4x + 3 = 0.$
18. $4x^4 - 13x^3 - 7x^2 + 41x - 14 = 0.$

19. Show that the equation

$$x^4 + x^3 - 2x^2 - 7x - 35 = 0$$

has exactly one positive root and that this root is an irrational number. Find two consecutive integers between which it lies.

20. Show that the equation

$$2x^5 + 3x^3 - 18x^2 - 27 = 0$$

has exactly one real root and that this root is an irrational number. Find two consecutive integers between which it lies.

21. Find all rational roots of the equation

$$2x^5 - x^4 - 12x^3 - 13x^2 - 14x = 12,$$

and locate each of the other real roots between two consecutive integers.

22. Using synthetic division, verify that i is a root of the equation

$$2x^5 - x^4 - 24x^3 + 6x^2 - 26x + 7 = 0.$$

Find the remaining roots.

23. Using synthetic division, verify that $1 + i$ is a root of the equation

$$x^5 - 2x^4 - 9x^3 + 8x^2 + 6x - 28 = 0.$$

Find the remaining roots.

In each of the following cases sketch the graph of the polynomial over the interval that contains the real roots of the given equation. Locate each real root between two consecutive integers:

24. $x^3 - 4x^2 - 7x + 12 = 0.$
25. $2x^4 + 3x^3 - 10x^2 + 6x - 20 = 0.$
26. $2x^3 - 5x^2 - 5x - 7 = 0.$
27. $x^4 - 14x^2 - 8 = 0.$
28. $4x^3 + 24x^2 + 27x - 12 = 0.$
29. $x^4 - 6x^3 + 20 = 0.$
30. $x^4 - 4x^3 - 9x^2 + 16x + 25 = 0.$

99. ***Irrational roots.*** We have proved that the rational integral equation $P_n(x) = 0$ has exactly n roots. We have seen how any rational roots that may exist can be found, and how each irrational root can be located between two consecutive integers. Our next task is that of developing a method by which we can determine these irrational roots to any desired degree of accuracy.

In the next few pages two such methods will be given. The first is called the *method of successive linear approximations*, and the second is known as *Horner's method* (after W. G. Horner who published it in 1819).

It should be mentioned again that if a given equation has one or more rational roots it is usually best to remove the corresponding factors and thus obtain an equation of lower degree for the computation of the irrational roots. For example, by trying out the rational numbers that might be roots of the equation

$$2x^4 + 3x^3 - 14x^2 - 15x + 9 = 0, \tag{1}$$

we find that $-\frac{3}{2}$ is a root:

$$\begin{array}{rrrrr|r} 2 & +3 & -14 & -15 & +9 & -\frac{3}{2} \\ & -3 & +0 & +21 & -9 & \\ \hline 2 & +0 & -14 & +6 & +0 & \end{array}$$

Equation (1) can then be written in the form

$$(x + \tfrac{3}{2})(2x^3 - 14x + 6) = 0$$

or

$$(2x + 3)(x^3 - 7x + 3) = 0. \tag{2}$$

The remaining roots of (1) are of course the three roots of the equation

$$x^3 - 7x + 3 = 0. \tag{3}$$

We would then use (3) instead of (1) in finding these roots.*

Our methods of finding the irrational roots are based upon the fact that a polynomial is a *continuous function*. This means essentially that the value of a polynomial $P(x)$ changes gradually as x changes, and that the graph is consequently a continuous curve within the usual meaning of the word continuous; there are no breaks in the graph, and in fact there are no sudden changes in direction—no sharp corners. These assertions can be proved

* It should be mentioned that outside the realm of textbooks on mathematics one seldom meets an equation of higher degree that has a rational root. In many cases the coefficients are given in decimal form, and one proceeds immediately to the task of finding roots in decimal form.

but we shall not give the proofs here. It is because of the important property of continuity of a polynomial that we can make the following assertion:

If the value of a polynomial $P(x)$ is positive for $x = x_1$ and negative for $x = x_2$, or vice versa, then there is at least one root of the equation $P(x) = 0$ between x_1 and x_2. More briefly, *if* $\mathbf{P}(\mathbf{x}_1)$ *and* $\mathbf{P}(\mathbf{x}_2)$ *have opposite signs, then* $\mathbf{P}(\mathbf{x})$ *equals zero for at least one value of* $\mathbf{x}$ *between* $\mathbf{x}_1$ *and* $\mathbf{x}_2$.

Example

In the case of the polynomial whose graph is shown in Fig. 27, we have

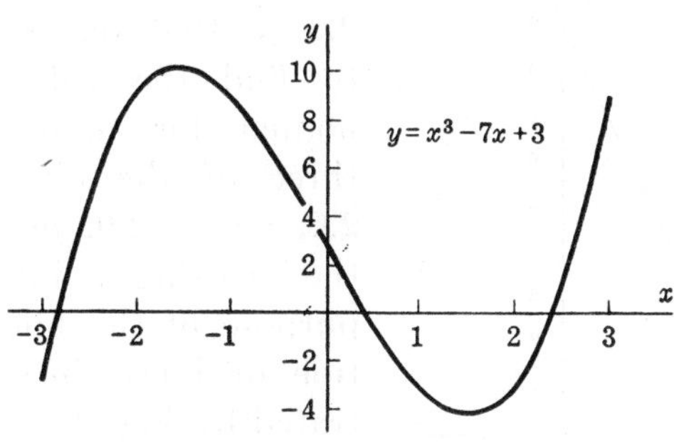

Figure 27

$$P(2) = -3 \qquad \text{and} \qquad P(3) = +9.$$

There must then be a root of the equation

$$x^3 - 7x + 3 = 0$$

between 2 and 3. Similarly, there is a root between 0 and 1, and another between -2 and -3. These roots are certainly irrational. Why?

100. ***The method of successive linear approximations.*** We shall illustrate the method by finding the root of the equation

$$x^3 - 7x + 3 = 0$$

that is between 2 and 3 (Fig. 27).

We could locate the root between two successive tenths by trying 2.1, 2.2, 2.3, etc. In this way we would find that the value of the polynomial at $x = 2.3$ is -0.933; at $x = 2.4$, it is $+0.024$. We would then know that the root is between 2.3 and 2.4. Thus

$$\begin{array}{rrrrl} 1 & +0.0 & -7.00 & +3.000 & \underline{|2.3} \\ & +2.3 & +5.29 & -3.933 & \\ \hline 1 & +2.3 & -1.71 & -\mathbf{0.933} & \quad [P(\mathbf{2.3})] \end{array}$$

$$\begin{array}{rrrrl} 1 & +0.0 & -7.00 & +3.000 & \underline{|2.4} \\ & +2.4 & +5.76 & -2.976 & \\ \hline 1 & +2.4 & -1.24 & +\mathbf{0.024} & \quad [P(\mathbf{2.4})] \end{array}$$

It would be advantageous to be able to guess in advance that the root is in the neighborhood of 2.3 so that one would not have to find the value of the polynomial for more than two or three of the nine numbers 2.1, 2.2, $\cdots$, 2.9, in order to locate the crossing point. This is the purpose of the linear approximation or linear interpolation illustrated by Fig. 28:

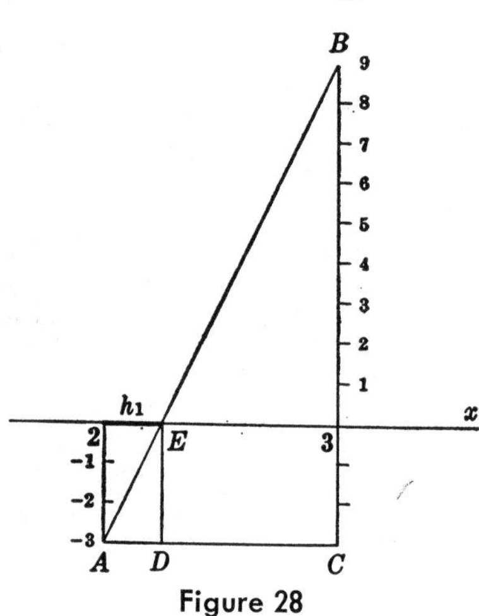

Figure 28

At $x = 2$ the value of the polynomial is -3 and at $x = 3$ it is $+9$. We draw a straight line joining $A(2,-3)$ and $B(3,9)$, and find by proportion the distance h_1 from 2 to the point E where this line crosses the x-axis. Thus

$$\frac{AD}{AC} = \frac{DE}{CB} \qquad \text{or} \qquad \frac{h_1}{1} = \frac{3}{12} \qquad \text{or} \qquad h_1 = 0.25.$$

The curve crosses the x-axis a little to the right of E, so we are led to try 2.3 and 2.4 with the result indicated above.*

The next step is to locate the root between two successive hundredths. Again, we could try 2.31, 2.32, 2.33, etc., until we find the place where the change in sign occurs. Much com-

* It will be observed that this process is identical with that of interpolation as used, for example, in connection with tables of logarithms. Note also that the straight line serves as an approximation to the curve.

putation can be saved by making a second linear approximation as illustrated by Fig. 29.

$$\frac{AD}{AC} = \frac{DE}{CB} \quad \text{or} \quad \frac{h_2}{0.1} = \frac{0.933}{0.933 + 0.024} = \frac{0.933}{0.957} = 0.9+.$$

This gives $h_2 = 0.09+$ so we guess that the crossing point is between 2.39 and 2.40. We then compute $P(2.39)$ and $P(2.40)$:

$$\begin{array}{rrrr|l}
1 & +0.00 & -7.0000 & +3.000000 & 2.39 \\
 & +2.39 & +5.7121 & -3.078081 & \\
\hline
1 & +2.39 & -1.2879 & \mathbf{-0.078081} & [\boldsymbol{P}(\mathbf{2.39})]
\end{array}$$

$$\begin{array}{rrrr|l}
1 & +0.0 & -7.00 & +3.000 & 2.40 \\
 & +2.4 & +5.76 & -2.976 & \\
\hline
1 & +2.4 & -1.24 & \mathbf{+0.024} & [\boldsymbol{P}(\mathbf{2.40})]
\end{array}$$

This shows that the root is between 2.39 and 2.40. To locate it between successive thousandths we make another linear interpolation:

$$\frac{h_3}{0.01} = \frac{0.078081}{0.078081 + 0.024};$$

$$h_3 = \frac{0.078081}{0.102081}(0.01) = 0.007+.$$

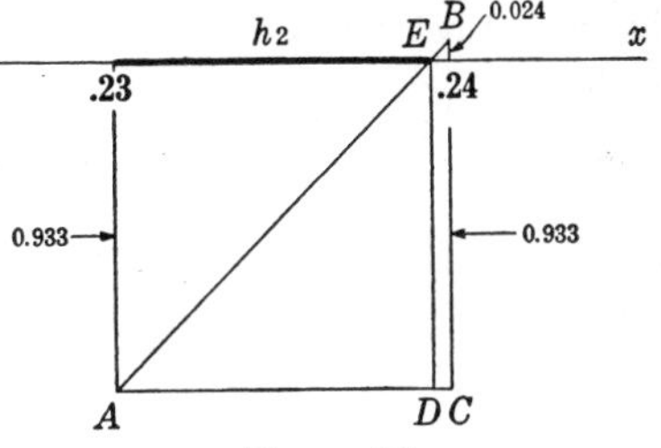

Figure 29

This leads us to believe that the crossing point is near 2.397, so we compute $P(2.397)$ and $P(2.398)$:

$$\begin{array}{rrrr|l}
1 & +0.000 & -7.000000 & +3.000000000 & 2.397 \\
 & +2.397 & +5.745609 & -3.006775227 & \\
\hline
1 & +2.397 & -1.254391 & \mathbf{-0.006775227} & [\boldsymbol{P}(\mathbf{2.397})]
\end{array}$$

$$\begin{array}{rrrr|l}
1 & +0.000 & -7.000000 & +3.000000000 & 2.398 \\
 & +2.398 & +5.750404 & -2.996531208 & \\
\hline
1 & +2.398 & -1.249596 & \mathbf{+0.003468792} & [\boldsymbol{P}(\mathbf{2.398})]
\end{array}$$

This assures us that the root is between 2.397 and 2.398.

The process can obviously be carried on to obtain the root to any desired number of decimal places. If in the above case we want the root correct to three decimal places, we must determine whether the digit in the fourth decimal place is more or less than 5.

It turns out that $P(2.3975)$ is negative, so the root lies between 2.3975 and 2.3980. Correct to three decimal places the root is then 2.398.

This method can be applied to equations other than rational integral equations, but the function values must in general be found by direct substitution instead of by synthetic division. If $f(x)$ is any function that is continuous in the sense described above, over the interval from x_1 to x_2, then there is at least one root (and in any case an odd number of roots) of the equation $f(x) = 0$ between x_1 and x_2 if $f(x_1)$ and $f(x_2)$ have opposite signs.

PROBLEMS

1. Let $f(x) = \dfrac{5x - 2}{3x - 8}$. Then $f(2) = -4$ and $f(3) = +13$. Does it follow that the equation

$$\frac{5x - 2}{3x - 8} = 0$$

has a root between 2 and 3? Explain the situation using a graph of the function $f(x)$.

2. Show that the equation

$$2^x = \tfrac{1}{2}x + 10$$

has a root between 3 and 4. What assumption are you making in regard to the function 2^x? Can you state whether this root is rational or irrational?

3. Find, correct to three decimal places, the root of the equation

$$2x^3 - 15x^2 + 26x - 12 = 0$$

that lies between 5 and 6, using the above method. Check your result by finding that there is a rational root and hence finding all the roots.

4. Find, correct to three decimal places, the root of the equation

$$x^3 - 10x - 12 = 0$$

that lies between 3 and 4, using the above method. Check your result by finding that there is a rational root and hence finding all the roots.

5. Show that the equation

$$x^3 + x = 28$$

has only one real root and that it is an irrational number. Compute it, correct to two decimal places.

6. Show that the equation

$$x^3 + 3x + 27 = 0$$

has only one real root and that this is an irrational number. Compute it, correct to two decimal places.

7. After removing the rational root from the equation

$$2x^4 - x^3 + 20x^2 - 46x + 18 = 0$$

find the irrational root, correct to two decimal places.

8. Compute, correct to three decimal places, the larger of the two irrational roots of the equation

$$x^4 - 4x^3 + 5x^2 - 16x + 4 = 0.$$

9. Evaluate $\sqrt[3]{50}$ to three decimal places by finding the real root of the equation $x^3 - 50 = 0$. Check your result by using logarithms.

10. Compute, correct to two decimal places, the real root of the equation

$$x^3 + 4.72x - 56.4 = 0.$$

11. Compute, correct to two decimal places, the real root of the equation

$$x^5 - 1.54x^2 = 36.7.$$

Check your result by direct substitution, using logarithms.

12. Compute, correct to two decimal places, the positive root of the equation

$$x^4 - 1.87x^3 = 12.5.$$

Check your result by direct substitution, using logarithms.

13. A rectangular box with open top is made from a sheet of tin that is 18 in. long and 12 in. wide by cutting a square from each corner and turning up the sides. What size square should be cut out if the volume of the box is to be 180 cu. in.?

14. Show that the problem of solving the system

$$x^2 + y^2 = 5,$$
$$xy - 6y = 8,$$

for x, reduces to that of solving the equation

$$x^4 - 12x^3 + 31x^2 + 60x - 116 = 0.$$

15. Show that the problem of solving the system

$$x^2 + xy + 3y^2 = 4,$$
$$x^2 + y^2 = 2x,$$

for x, reduces to that of solving the equation

$$5x^4 - 26x^3 + 52x^2 - 48x + 16 = 0.$$

16. Show that the problem of solving the system

$$x^2 + 3xy - 2y^2 = 8,$$
$$x^2 + 2y^2 = 3x,$$

for x, reduces to that of solving the equation

$$17x^4 - 51x^3 - 46x^2 + 96x + 128 = 0.$$

Show that $x = 2$ is a rational root, and that the only other positive root is an irrational number between 2 and 3.

101. ***Transformation to diminish the roots of an equation by a constant h.*** Before we can take up Horner's method of computing irrational roots, we must develop a simple method of solving the following problem:

Given a rational integral equation $P_n(x) = 0$ having roots $r_1, r_2, \cdots r_n$, find a rational integral equation $T_n(x) = 0$ having as its roots the numbers $r_1 - h, r_2 - h, \cdots, r_n - h$, where h is a given real number.

In some cases in which the roots of the given equation are known, the required equation may be easily determined.

Example

The student may verify that the roots of the equation

$$2x^3 - 19x^2 + 59x - 60 = 0$$

are $2\frac{1}{2}$, 3, and 4. If we want an equation each of whose roots is less by 2 than the corresponding root of this equation, then these roots must be $\frac{1}{2}$, 1, and 2. The required equation is then

$$(2x - 1)(x - 1)(x - 2) = 0$$

or

$$2x^3 - 7x^2 + 7x - 2 = 0.$$

Our problem is not quite so simple as this. We must be able to find the coefficients for the required equation *directly from those of the given equation*, without knowing the roots. We shall think of the process as that of transforming the given equation into a new equation, each of whose roots is less by h than the corresponding root of the given equation. The method that we shall develop below makes use of synthetic division.

Let the given equation be

$$(1) \qquad a_0x^n + a_1x^{n-1} + \cdots + a_{n-1}x + a_n = 0.$$

Let us tentatively form a new equation by taking (1) and replacing x by $x' + h$:

$$(2) \quad a_0(x' + h)^n + a_1(x' + h)^{n-1} + \cdots + a_{n-1}(x' + h) + a_n = 0.$$

Now if r is any root of (1), *i.e.*, if r is any number that satisfies (1), then $r - h$ will satisfy (2), for substituting $r - h$ for x' in (2) gives precisely the same thing as substituting r for x in (1). Thus (2) is the desired equation. In order to reduce it to the usual standard form, it is necessary to expand each term and then collect like powers of x. This would result in an equation of the form

$$(3) \qquad A_0x'^n + A_1x'^{n-1} + \cdots + A_{n-1}x' + A_n = 0$$

in which the A's denote the new coefficients.

Example

If we replace x by $x' + 2$ in the equation

$$2x^3 - 19x^2 + 59x - 60 = 0$$

we have

$$2(x' + 2)^3 - 19(x' + 2)^2 + 59(x' + 2) - 60 = 0.$$

If we expand each term and then collect like powers of x, we get

$$2x'^3 - 7x'^2 + 7x' - 2 = 0.$$

The roots of the given equation are $2\frac{1}{2}$, 3, and 4. Those of the new equation are $\frac{1}{2}$, 1, and 2, each being less by 2 than the corresponding root of the original equation.

It will be observed that the coefficient in the term of highest degree is the same in the new equation as in the original one ($A_0 = a_0$). The student should see immediately that this would be true in all cases.

In order to obtain an easier way of finding the A's, we may take (3) and replace x' by $x - h$. The result is

$$(4) \quad A_0(x - h)^n + A_1(x - h)^{n-1} + \cdots + A_{n-1}(x - h) + A_n = 0.$$

Now if each of the terms in (4) were expanded and if like powers of x were then collected, the resulting equation would be identical with (1); *i.e.*, (4) is simply equation (1) in disguise. We may use this fact to find an easy way of computing the A's from (1) as follows:

If the left member of (4) were to be divided by $x - h$, the remainder would obviously be A_n. But (4) is identical with (1), so if the left member of (1) were divided by $x - h$, the remainder would be A_n. We then have an easy way of finding A_n—simply divide the left member of (1) by $x - h$ and take the remainder as A_n.

Example

In the case of the example that we have been using we have:

$$\begin{array}{rrrr|l} 2 & -19 & +59 & -60 & \underline{2} \\ & +4 & -30 & +58 & \\ \hline 2 & -15 & +29 & -\mathbf{2} & (A_n) \end{array}$$

We observe next that the quotient obtained when the left member of (4) is divided by $x - h$ is

$$A_0(x - h)^{n-1} + A_1(x - h)^{n-2} + \cdots + A_{n-2}(x - h) + A_{n-1}.$$

If this is divided again by $x - h$, the remainder is obviously A_{n-1}. This means that, if the quotient obtained by dividing the left member of (1) by $x - h$ were again divided by $x - h$, the remainder would be A_{n-1}.

Example

In the above example the coefficients in the quotient are 2, -15, $+29$. We may divide this quotient again by $x - 2$ as follows:

$$\begin{array}{rrr|l} 2 & -15 & +29 & \underline{2} \\ & +4 & -22 & \\ \hline 2 & -11 & +\mathbf{7} & (A_{n-1}) \end{array}$$

It is left for the student to continue this analysis of the situation and show that the work can be arranged as in the following examples:

Example 1

Transform the equation

$$2x^3 - 19x^2 + 59x - 60 = 0$$

into an equation each of whose roots is less by 2 than the corresponding root of this equation.

Solution

$$\begin{array}{rrrrrl}
 & 2 & -19 & +59 & -60 & \underline{|2} \\
 & & +4 & -30 & +58 & \\
\hline
 & 2 & -15 & +29 & -2 & (A_3) \\
 & & +4 & -22 & & \\
\hline
 & 2 & -11 & +7 & (A_2) & \\
 & & +4 & & & \\
\hline
(A_0) & 2 & -7 & (A_1) & &
\end{array}$$

The required equation is

$$2x^3 - 7x^2 + 7x - 2 = 0.$$

The student should be careful to observe that the "indicated divisor" for each of the successive divisions is the *same number*, in this case 2. It is *not* 2 for the first division, -2 for the second, 7 for the third, etc.

In most of our applications of this idea we shall be dealing with decimal fractions as in the following example:

Example 2

Find the equation each of whose roots is less by 0.6 than the corresponding root of the equation

$$5x^4 - 3x^2 + 7x - 4 = 0.$$

Solution

$$\begin{array}{rrrrrrl}
 & 5 & +0 & -3.00 & +7.000 & -4.000 & \underline{|0.6} \\
 & & +3 & +1.80 & -0.720 & +3.768 & \\
\hline
 & 5 & +3 & -1.20 & +6.280 & \mathbf{-0.232} & (A_4) \\
 & & +3 & +0.36 & -0.504 & & \\
\hline
 & 5 & +6 & -0.84 & \mathbf{+5.776} & (A_3) & \\
 & & +3 & +5.40 & & & \\
\hline
 & 5 & +9 & \mathbf{-4.56} & (A_2) & & \\
 & & +3 & & & & \\
\hline
(A_0) & \mathbf{5} & \mathbf{+12} & (A_1) & & &
\end{array}$$

The required equation is

$$5x^4 + 12x^3 - 4.56x^2 + 5.776x - 0.232 = 0.$$

By changing the sign of the indicated divisor one obtains an equation whose roots are *greater* than those of the given equation.

Example

The equation whose roots are greater by 1 (or less by -1) than those of the equation $x^3 - 8 = 0$ is found as follows:

$$
\begin{array}{rrrr|r}
1 & +0 & +0 & -8 & -1 \\
 & -1 & +1 & -1 & \\
\hline
1 & -1 & +1 & -9 & \\
 & -1 & +2 & & \\
\hline
1 & -2 & +3 & & \\
 & -1 & & & \\
\hline
1 & -3 & & &
\end{array}
$$

The equation is $x^3 - 3x^2 + 3x - 9 = 0$. The student should verify that the roots of the given equation are 2, $-1 + \sqrt{3}i$, and $-1 - \sqrt{3}i$, and that those of the new equation are 3, $\sqrt{3}i$, and $-\sqrt{3}i$.*

PROBLEMS

1. Verify that the roots of the equation

$$x^3 - 6x^2 + 11x - 6 = 0$$

are 1, 2, and 3. Then find, in two different ways, a cubic equation whose roots are -1, 0, and 1.

2. Verify that the roots of the equation

$$2x^3 - 5x^2 - 14x + 8 = 0$$

are -2, $\frac{1}{2}$, and 4. Then find, in two different ways, a cubic equation whose roots are -3, $-\frac{1}{2}$, and 3.

3. Verify that the roots of the equation

$$6x^3 - 13x^2 - 13x + 20 = 0$$

are $-\frac{4}{3}$, 1, and $\frac{5}{2}$. Then find, in two different ways, a cubic equation whose roots are 0, $\frac{7}{3}$, and $\frac{23}{6}$.

* Strictly speaking, one cannot say that $\sqrt{3}i$ is greater by 1 than $-1 + \sqrt{3}i$, since one imaginary number is neither greater than nor less than another. It would be better to say that the one root is equal to the other root *plus one.*

4. Verify that the roots of the equation

$$x^3 - 4x^2 + x + 26 = 0$$

are -2, $3 + 2i$, and $3 - 2i$. Then find, in two different ways, a cubic equation whose roots are 0, $5 + 2i$, and $5 - 2i$.

5. Find an equation each of whose roots is equal to the corresponding root of the following equation minus one:

$$x^4 - 4x^3 + 5x^2 - 2x - 20 = 0.$$

Note that this new equation can be solved easily. By solving it, find the roots of the given equation.

6. Find an equation each of whose roots is equal to the corresponding root of the equation

$$x^4 + 8x^3 + 16x^2 = 25$$

plus 2. Note that this new equation can be solved easily. By solving it, find the roots of the given equation.

7. The equation $x^4 - 9 = 0$ has the roots $\pm\sqrt{3}$ and $\pm\sqrt{3}\,i$. Find, in two different ways, an equation of fourth degree whose roots are $2 \pm \sqrt{3}$ and $2 \pm \sqrt{3}\,i$.

8. By replacing x in a given equation by $x' + h$ we obtained an equation in x' each of whose roots is less by h than the corresponding root of the given equation. If we should replace x by $1/x'$, what relation would each root of the resulting equation bear to the corresponding root of the given equation? What if we replace x by $-x'$?

9. Find a cubic equation each of whose roots is the reciprocal of the corresponding root of the equation

$$2x^3 - 7x^2 + 2x + 3 = 0.$$

Find also a cubic equation each of whose roots is the negative of the corresponding root of this equation. (See Prob. 8.)

10. Find a cubic equation each of whose roots is the reciprocal of the corresponding root of the equation

$$6x^3 - 5x^2 - 3x + 2 = 0.$$

Find also a cubic equation each of whose roots is the negative of the corresponding root of this equation. (See Prob. 8.)

11. Show that if one should find the positive roots of the equation

$$x^4 - 16x^2 - 8x - 32 = 0 \tag{1}$$

and change their signs, he would have the negative roots of the equation

$$x^4 - 16x^2 + 8x - 32 = 0. \tag{2}$$

Hence imply that one can find all the real roots of (2) by finding its positive roots and then finding the positive roots of (1). (See Prob. 8.)

12. After the positive roots of the equation

$$x^4 + 2x^3 - 8x^2 - 10x + 15 = 0$$

are found, it is desired to find the negative roots by finding the positive roots of another equation. What is this other equation? (See Prob. 8.)

Transform each of the following equations into an equation each of whose roots is less than the corresponding root of the given equation by the specified amount h:

13. $2x^3 - x^2 + 7x + 5 = 0;\quad h = 2.$
14. $x^3 + 6x^2 - 4x - 9 = 0;\quad h = 1.$
15. $2x^4 - 6x^3 + 7x - 12 = 0;\quad h = -1.$
16. $x^5 - 4x^3 + 8x + 8 = 0;\quad h = 2.$
17. $x^4 - 10x^3 + 4x^2 - 8 = 0;\quad h = 0.5.$
18. $x^3 + 6x^2 + 12x - 4 = 0;\quad h = 0.2.$
19. $x^3 + 9x^2 + 27x - 4 = 0;\quad h = 0.1.$
20. $3x^4 + x^3 - 8x^2 = 3x + 3;\quad h = 1.$

21. Show that the equation

$$x^4 - 2x^3 + x^2 + 4x - 6 = 0$$

has an irrational root between 1 and 2. Then transform the equation so as to decrease the roots by 1. Between what two integers does the corresponding root of this new equation lie? Show that it is between 0.4 and 0.5, and hence imply that the original equation has a root between 1.4 and 1.5.

22. Show that the equation

$$2x^4 + x^3 - 26x^2 - 14x - 28 = 0$$

has only one positive root and that this is an irrational number between 3 and 4. Transform the equation so as to decrease the roots by 3. Between what two integers does the corresponding root of this new equation lie? Show that it is between 0.4 and 0.5, and hence imply that the original equation has a root between 3.4 and 3.5.

23. Show that the equation

$$3x^4 - 5x^3 - 5x^2 - 8x - 6 = 0$$

has a root between 2 and 3. Transform the equation so as to decrease the roots by 2. Between what two integers does the corresponding root of this new equation lie? Show that it is between 0.7 and 0.8, and hence imply that the original equation has a root between 2.7 and 2.8.

24. Show that the equation

$$3x^4 = 9x^3 + 13x^2 + 11x + 2$$

has a root between 4 and 5. Transform the equation so as to decrease the roots by 4. Between what two integers does the corresponding root of this new equation lie? Show that it is between 0.2 and 0.3 and hence imply that the original equation has a root between 4.2 and 4.3.

25. Show that the equation

$$x^4 - 3x^3 - 12x^2 - 18x - 20 = 0$$

has a root between -1 and -2. Transform the equation so as to increase the roots by 2. Between what two integers does the corresponding root of this new equation lie? Show that it is between 0.2 and 0.3, and hence imply that the original equation has a root -1.7 and -1.8.

26. Show that the equation

$$x^4 - 9x^3 + 14x^2 + 13x + 23 = 0$$

has a root between -3 and -4. Transform the equation so as to increase the roots by 4. Between what two integers does the corresponding root of this new equation lie? Show that it is between 0.4 and 0.5, and hence imply that the original equation has a root between -3.5 and -3.6.

27. Show that the equation

$$2x^4 - 5x^3 = 29x^2 - 14x + 16$$

has a root between -3 and -4. Transform the equation so as to increase the roots by 3. Between what two integers does the corresponding root of this new equation lie? Show that it is between -0.1 and -0.2, and hence imply that the original equation has a root between -3.1 and -3.2.

102. ***Horner's method.*** We have seen that a preliminary step in the computation of an irrational root is that of locating the root between two consecutive integers. If we find, for example, that such a root lies between 2 and 3, we may try 2.1, 2.2, 2.3, etc., until we have located it between two consecutive tenths. If it turns out that the root is between 2.3 and 2.4, we may try 2.31, 2.32, 2.33, etc., until we have located it between two successive hundredths, and so on.

Horner's method is a somewhat simplified procedure for doing this. If the root is between 2 and 3, we do not immediately try 2.1, 2.2, etc. Instead, we transform the equation into a new equation whose roots are less by 2 than those of the given equation. This new equation has a root between 0 and 1, and we may try

0.1, 0.2, etc., on it, instead of trying 2.1, 2.2, etc., on the original equation. This simplifies the arithmetic.

If we find that the root of this new equation is between 0.3 and 0.4, we again do not immediately try 0.31, 0.32, etc. Instead, we transform the equation into another new one whose roots are less by 0.3. This equation will have the root between 0 and 0.1, and on it we can try 0.01, 0.02, etc. The indicated divisors are thus always numbers of one digit. We never have to use two- or three-digit divisors like 2.3 or 2.36.

We shall illustrate this method by computing the root of the equation

$$x^3 - 7x + 3 = 0$$

that lies between 2 and 3. The left member of the equation may be denoted by $P(x)$.

After making certain that there is a root between 2 and 3 by finding, using synthetic division, that $P(2) = -3$ and $P(3) = +9$, we obtain a new equation having roots less by 2 than those of the given equation:

$$\begin{array}{rrrr|l}
1 & +0 & -7 & +3 & 2 \\
 & +2 & +4 & -6 & \\
\hline
1 & +2 & -3 & -3 & \\
 & +2 & +8 & & \\
\hline
1 & +4 & +5 & & \\
 & +2 & & & \\
\hline
1 & +6 & & &
\end{array}$$

The new equation is $x^3 + 6x^2 + 5x - 3 = 0$,* and the root that we seek is between 0 and 1. By trying 0.1, 0.2, etc., we find that the remainder is still negative at 0.3 but positive at 0.4, so that the root lies between 0.3 and 0.4. We then proceed to transform

* A simple check results from the fact that this new polynomial should have the same value at $x = 0$ as the old one had at $x = 2$, and the same value at $x = 1$ as the old one had at $x = 3$:

$$P(x) = x^3 - 7x + 3: \qquad P(2) = -3;\ P(3) = 9.$$
$$P_1(x) = x^3 + 6x^2 + 5x - 3: \qquad P_1(0) = -3;\ P_1(1) = 9.$$

This check, which can be applied at each successive step, is valuable because numerical errors enter so readily.

this equation into a new one whose roots are less by 0.3:

1	+6.0	+5.00	−3.000	0.3
	+0.3	+1.89	+2.067	
1	+6.3	+6.89	−0.933	
	+0.3	+1.98		
1	+6.6	+8.87		
	+0.3			
1	+6.9			

The new equation is $x^3 + 6.9x^2 + 8.87x - 0.933 = 0$, and the root that we seek is between 0 and 0.1. If we try 0.01, 0.02, etc., we find that the remainder is negative at 0.09 and positive at 0.10 so that the root of the original equation is between 2.39 and 2.40. We then transform to an equation whose roots are less by 0.09:

1	+6.90	+ 8.8700	−0.933000	0.09
	+0.09	+ 0.6291	+0.854919	
1	+6.99	+ 9.4991	−0.078081	
	+0.09	+ 0.6372		
1	+7.08	+10.1363		
	+0.09			
1	+7.17			

The new equation is $x^3 + 7.17x^2 + 10.1363x - 0.078081 = 0$, and the root that we seek is between 0 and 0.01 so we may try 0.001, 0.002, etc. Work can be saved by estimating the next digit in the following way: For x between 0 and 0.01, the terms involving x^3 and x^2 are small compared to the other terms. Thus the value of x that satisfies the last equation above does not differ much from the value of x that satisfies the *linear* equation

$$10.1363x - 0.078081 = 0.$$

This is about 0.007, so instead of trying 0.001, 0.002, etc., we start by trying 0.007, and we find quickly that the root is between 0.007 and 0.008. (This same method of estimating the root could be used, with somewhat less accuracy, in the preceding steps.)

Finally, it is convenient to arrange the work in the following way:

1	+0	−7	+3	2
	+2	+4	−6	
1	+2	−3	−3	
	+2	+8		
1	+4	+5		
	+2			
1	+6.0	+5.00	−3.000	0.3
	+0.3	+1.89	+2.067	
1	+6.3	+6.89	−0.933	
	+0.3	+1.98		
1	+6.6	+8.87		
	+0.3			
1	+6.90	+ 8.8700	−0.933000	0.09
	+0.09	+ 0.6291	+0.854919	
1	+6.99	+ 9.4991	−0.078081	
	+0.09	+ 0.6372		
1	+7.08	+10.1363		
	+0.09			
1	+7.170	+10.136300	−0.078081000	0.007
	+0.007	+ 0.050239	+0.071305773	
1	+7.177	+10.186539	−0.006775227	
	+0.007	+ 0.050288		
1	+7.184	+10.236827		
	+0.007			
1	+7.191			

Estimation of next digit: $\dfrac{0.00678}{10.2} = 0.0006+.$

Root, correct to three decimal places: **2.398.**

PROBLEMS

1. Evaluate $\sqrt[3]{42}$, correct to three decimal places, by finding the real root of the equation $x^3 - 42 = 0$ using Horner's method.

2. Evaluate $\sqrt[3]{120}$, correct to three decimal places, by finding the real root of the equation $x^3 - 120 = 0$ using Horner's method.

3. Evaluate $\sqrt[3]{21}$, correct to three decimal places, by finding the real root of the equation $x^3 - 21 = 0$ using Horner's method.

In each of the following problems use Horner's method to find the specified root of the given equation, correct to three decimal places:

4. $x^4 - 5x^3 + 2x^2 + x + 7 = 0$; root between 1 and 2.
5. $3x^3 - 7x^2 - 17x + 5 = 0$; root between 0 and 1.
6. $2x^3 + 7x^2 = 6(x + 3)$; root between 1 and 2.
7. $4x^3 = 37x + 3$; root between 3 and 4.
8. $2x^3 + 33 = 7x^2 + 16x$; root between -2 and -3.
9. $x^4 - 3x^3 - 27x^2 = 4(7x + 6)$; root between -3 and -4.
10. $2x^4 - 15x^3 - 12x^2 - 19x = 6$; root between 0 and -1.

11. Find, correct to three decimal places, the positive number x such that

$$2x^4 = 11x^3 + 8 + 7x(3x + 2).$$

12. Find, correct to three decimal places, the negative number x such that

$$2x(x^3 + 12) = 5x^3 + 49x^2 + 26.$$

13. Find, correct to three decimal places, the larger of the two real roots of the equation

$$4x^4 - 8x^3 = 13x^2 + 10x - 22.$$

14. Find, correct to two decimal places, the smaller of the two real roots of the equation of Prob. 13.

15. Find, correct to two decimal places, the larger of the two real roots of the equation

$$3x^4 - 14x^3 - 21x^2 + 16x - 10 = 0.$$

16. Show that the problem of solving for x in the system

$$x^2 + y^2 = 3,$$
$$xy - 4y = 5,$$

reduces to that of solving the equation

$$x^4 - 8x^3 + 13x^2 + 24x - 23 = 0.$$

17. Show that the problem of solving for x in the system

$$x^2 + 2xy + 3y^2 = 4,$$
$$x^2 + y^2 = 4x,$$

reduces to that of solving the equation

$$x^4 - 8x^3 + 20x^2 - 12x + 2 = 0.$$

18. Show that the problem of solving for x in the system

$$x^2 + 3xy - y^2 = 4,$$
$$2x^2 + y^2 = 3x.$$

reduces to that of solving the equation

$$27x^4 - 45x^3 - 15x^2 + 24x + 16 = 0.$$

103. ***Algebraic solution of the cubic equation.*** We have seen that the linear equation $ax + b = 0$ has the unique solution $x = -b/a$. We have also been able to solve the general quadratic equation $ax^2 + bx + c = 0$ in terms of the coefficients, the result being

$$x = \frac{-b \pm \sqrt{b^2 - 4ac}}{2a}.$$

We have mentioned (page 72) that the solutions of rational integral equations of degree three and four can also be expressed as algebraic functions of the coefficients, but that methods such as those just described are usually preferable for computing the real roots of a given equation with numerical coefficients. We have also mentioned (page 72) that for equations of degree higher than four it is impossible to express the roots as algebraic functions of the coefficients, except in special cases.

We shall conclude this chapter by presenting the algebraic solution of the cubic equation

$$x^3 + ax^2 + bx + c = 0.*$$

As a preliminary step, we solve the equation $x^3 = 1$ as follows:

$$x^3 = 1;$$
$$x^3 - 1 = 0;$$
$$(x - 1)(x^2 + x + 1) = 0.$$

Roots: $x = 1$; $x = -\frac{1}{2} + \frac{1}{2}\sqrt{3}\,i$; $x = -\frac{1}{2} - \frac{1}{2}\sqrt{3}\,i$.

The student may verify that the cube of each of the numbers

$$1, \quad -\tfrac{1}{2} + \tfrac{1}{2}\sqrt{3}\,i, \quad -\tfrac{1}{2} - \tfrac{1}{2}\sqrt{3}\,i,$$

is 1, so that each is a cube root of 1. He may further verify that if we denote the second of these numbers by the greek letter ω

* There is no loss in generality if we take the coefficient of x^3 as 1. If it is not equal to 1, we can divide both sides of the equation by this coefficient, and the resulting equation is equivalent to the given one.

(omega) then the third one is equal to ω^2. Thus the three cube roots of 1 are

$$1, \omega, \text{ and } \omega^2, \qquad \text{where } \omega = -\tfrac{1}{2} + \tfrac{1}{2}\sqrt{3}\, i.$$

We now consider the equation $x^3 = k$ where k is any constant. If r is any root, so that $r^3 = k$, then $r\omega$ and $r\omega^2$ are the other two roots. For if $r^3 = k$, then

$$(r\omega)^3 = r^3\omega^3 = r^3 \cdot 1 = k.$$
$$(r\omega^2)^3 = r^3(\omega^3)^2 = r^3 \cdot 1^2 = k.$$

In particular if k is a real number and $\sqrt[3]{k}$ is its principal cube root, then $\sqrt[3]{k} \cdot \omega$ and $\sqrt[3]{k} \cdot \omega^2$ are the other two cube roots.

Example

The principal cube root of -8 is -2. The others are -2ω and $-2\omega^2$. These numbers are

$$-2(-\tfrac{1}{2} + \tfrac{1}{2}\sqrt{3}\, i) \qquad \text{or} \qquad 1 - \sqrt{3}\, i,$$

and

$$-2(-\tfrac{1}{2} - \tfrac{1}{2}\sqrt{3}\, i) \qquad \text{or} \qquad 1 + \sqrt{3}\, i.$$

We are now in a position to solve the general cubic equation

$$x^3 + ax^2 + bx + c = 0. \tag{1}$$

Let us first make the substitution $x = y + k$ where k is a constant whose value is to be decided upon later. We get

$$(y + k)^3 + a(y + k)^2 + b(y + k) + c = 0,$$

which reduces to

$$y^3 + (3k + b)y^2 + (3k^2 + 2bk + c)y + (k^3 + bk^2 + ck + d) = 0. \tag{2}$$

We now observe that we can eliminate the term involving y^2 if we take $k = -\frac{1}{3}b$. For this value of k, (2) becomes

$$y^3 + \left(c - \frac{b^2}{3}\right)y + \left(d - \frac{bc}{3} + \frac{2b^3}{27}\right) = 0, \tag{3}$$

or

$$y^3 + py + q = 0, \tag{4}$$

where $\quad p = c - \dfrac{b^2}{3} \quad$ and $\quad q = d - \dfrac{bc}{3} + \dfrac{2b^3}{27}.$

Equation (4) is called the *reduced cubic.* Any cubic of the form (1) can be reduced to the form (4) by letting $x = y - \frac{1}{3}b$. We proceed then to solve (4), the next step being that of letting

$$y = z - \frac{p}{3z}. \tag{5}$$

When this substitution is made in (4), we get

$$\left(z - \frac{p}{3z}\right)^3 + p\left(z - \frac{p}{3z}\right) + q = 0,$$

which reduces to

$$z^3 - \frac{p^3}{27z^3} + q = 0. \tag{6}$$

If we multiply both sides of this equation by z^3, we have

$$z^6 + qz^3 - \frac{p^3}{27} = 0. \tag{7}$$

Equation (7) can be regarded as a quadratic equation in z^3:

$$(z^3)^2 + q(z^3) - \frac{p^3}{27} = 0. \tag{8}$$

By using the quadratic formula, we obtain the following solutions of (8):

$$z^3 = -\frac{q}{2} + \sqrt{R}, \quad \text{and} \quad z^3 = -\frac{q}{2} - \sqrt{R}, \tag{9}$$

where $$R = \tfrac{1}{27}p^3 + \tfrac{1}{4}q^2.$$

Each of the equations in (9) is of the form $z^3 = k$. Hence if z_1 is any root of the first of these equations, then $z_1\omega$ and $z_1\omega^2$ are the other two. If we substitute each of these into (5), we get the three roots of (4):

$$\begin{aligned} y_1 &= z_1 - \frac{p}{3z_1}; \\ y_2 &= z_1\omega - \frac{p}{3z_1\omega} = z_1\omega - \frac{p\omega^2}{3z_1}; \\ y_3 &= z_1\omega^2 - \frac{p}{3z_1\omega^2} = z_1\omega^2 - \frac{p\omega}{3z_1}. \end{aligned} \tag{10}$$

Substituting the three values of y given by (10) into the relation $x = y - (b/3)$, we finally obtain the three roots of (1):

$$
\begin{aligned}
x_1 &= z_1 - \frac{p}{3z_1} - \frac{b}{3}; \\
x_2 &= z_1\omega - \frac{p\omega^2}{3z_1} - \frac{b}{3}; \\
x_3 &= z_1\omega^2 - \frac{p\omega}{3z_1} - \frac{b}{3}.
\end{aligned} \tag{11}
$$

It can be shown that if instead of the roots z_1, $z_1\omega$, and $z_1\omega^2$ of the first of equations (9), we had used the roots z_2, $z_2\omega$, and $z_2\omega^2$ of the second of these equations, we would arrive at the same three values of x. Hence we may regard z_1 in (11) as denoting any one of the three roots of either of the two equations in (9).

Example

Solve the equation

$$x^3 - 4x^2 + 8x - 8 = 0$$

using the above formulas.

Solution

In this case $b = -4$, $c = 8$, and $d = -8$. We first compute p and q:

$$
\begin{aligned}
p &= c - \frac{b^2}{3} = 8 - \frac{16}{3} = \frac{8}{3}; \\
q &= d - \frac{bc}{3} + \frac{2b^3}{27} = -8 + \frac{32}{3} - \frac{128}{27} = -\frac{56}{27}.
\end{aligned}
$$

We next compute R and write down one of the two equations in (9):

$$
\begin{aligned}
R &= \frac{p^3}{27} + \frac{q^2}{4} = \frac{8^3}{27^2} + \frac{56^2}{27^2(4)} = \frac{8^2}{27^2} \cdot \frac{81}{4}; \\
\sqrt{R} &= \tfrac{36}{27}.
\end{aligned}
$$

The first of equations (9) is then

$$z^3 = -\frac{q}{2} + \sqrt{R} = \frac{28}{27} + \frac{36}{27} = \frac{64}{27}.$$

We may take $z_1 = \frac{4}{3}$ and write down the three roots of the given equation from (11):

$$x_1 = \frac{4}{3} - \frac{\frac{8}{3}}{4} + \frac{4}{3} = 2;$$

$$x_2 = \frac{4}{3}\,\omega - \frac{\frac{8}{3}\omega^2}{4} + \frac{4}{3}$$

$$= \frac{4}{3}\,\omega - \frac{2}{3}\,\omega^2 + \frac{4}{3} = 1 + \sqrt{3}\,i;$$

$$x_3 = \frac{4}{3}\,\omega^2 - \frac{\frac{8}{3}\omega}{4} + \frac{4}{3}$$

$$= \frac{4}{3}\,\omega^2 - \frac{2}{3}\,\omega + \frac{4}{3} = 1 - \sqrt{3}\,i.$$

The student may check these results by verifying that 2 is a rational root and then finding the remaining roots by solving a quadratic equation.

PROBLEMS

Solve each of the following equations by carrying through each step of the above-described method. Then solve it by using the formulas as in the above illustrative example:

1. $x^3 + 2x^2 + 2x + 1 = 0.$

2. $x^3 + 9x^2 + 30x + 36 = 0.$

3. $x^3 + 3x^2 - 6x - 36 = 0.$

4. $x^3 - 3x^2 - x + 3 = 0.$

5. $x^3 - 3x^2 + 6x - 4 = 0.$

6. $x^3 + 3x^2 + 21x + 19 = 0.$

7. $x^3 - 6x^2 - 3x + 18 = 0.$ (A cube root of $2 + 11i$ is $2 + i$.)

8. $x^3 - x^2 - 4x - 6 = 0.$ (A cube root of $100 + 51\sqrt{3}$ is $4 + \sqrt{3}$.)

CHAPTER XV

DETERMINANTS

104. ***Determinants of second order.*** In Chap. V (pages 90 to 94) we considered the following system of two linear equations in the unknowns x and y:

$$a_1x + b_1y = c_1, \tag{1}$$
$$a_2x + b_2y = c_2. \tag{2}$$

We eliminated y by multiplying (1) by b_2 and (2) by b_1 and then subtracting. In a similar fashion we eliminated x. The resulting equations were

$$(a_1b_2 - a_2b_1)x = c_1b_2 - c_2b_1, \tag{3}$$
$$(a_1b_2 - a_2b_1)y = a_1c_2 - a_2c_1. \tag{4}$$

We proceeded to show that if $a_1b_2 - a_2b_1 \neq 0$, the given system has the unique solution

$$x = \frac{c_1b_2 - c_2b_1}{a_1b_2 - a_2b_1}; \qquad y = \frac{a_1c_2 - a_2c_1}{a_1b_2 - a_2b_1}. \tag{5}$$

We then discussed the case in which $a_1b_2 - a_2b_1 = 0$, finding that in this case the equations are either inconsistent or dependent (page 93).

We can put the above results into a convenient and easily remembered form by introducing a new way of writing the number $a_1b_2 - a_2b_1$:

The symbol

$$\begin{vmatrix} a_1 & b_1 \\ a_2 & b_2 \end{vmatrix}$$

in which any four numbers are arranged in a square array consisting of two rows and two columns, and flanked on the sides by vertical bars, is called a *determinant of second order.* The symbol stands for the number $a_1b_2 - a_2b_1$; *i.e., by definition,*

$$\begin{vmatrix} a_1 & b_1 \\ a_2 & b_2 \end{vmatrix} \equiv a_1b_2 - a_2b_1.$$

The right member of this identity is called the *expansion* of the determinant on the left. The four individual numbers in the determinant are called its *elements.* The (horizontal) rows are numbered from top to bottom and the (vertical) columns are numbered from left to right.

Example

$$\begin{vmatrix} 2 & -3 \\ 5 & 7 \end{vmatrix} = 2(7) - 5(-3) = 14 + 15 = 29.$$

In this determinant the elements 2 and -3 form the *first row.* The *second row* consists of the elements 5 and 7. The *first column* has the elements 2 and 5, and the *second column* is that containing the elements -3 and 7.

Using the determinant notation, we can write formulas (5) as follows:

$$(5) \qquad x = \frac{\begin{vmatrix} c_1 & b_1 \\ c_2 & b_2 \end{vmatrix}}{\begin{vmatrix} a_1 & b_1 \\ a_2 & b_2 \end{vmatrix}}; \qquad y = \frac{\begin{vmatrix} a_1 & c_1 \\ a_2 & c_2 \end{vmatrix}}{\begin{vmatrix} a_1 & b_1 \\ a_2 & b_2 \end{vmatrix}}.$$

These are easily remembered for the following reason: In each case the denominator is the determinant whose elements are the coefficients of x and y, arranged just as they appear in the given equations (1) and (2). This determinant is called the *determinant of the coefficients* and may be denoted by the letter D:

$$D = \begin{vmatrix} a_1 & b_1 \\ a_2 & b_2 \end{vmatrix}.$$

The numerator in the solution for x is this same determinant D with the coefficients of x replaced by the constants c_1 and c_2; *i.e.,* with a_1 and a_2 replaced by c_1 and c_2, respectively. The numerator in the solution for y similarly has c_1 and c_2 in place of the corresponding coefficients of y. We may denote these determinants by D_x and D_y:

$$D_x = \begin{vmatrix} c_1 & b_1 \\ c_2 & b_2 \end{vmatrix}; \qquad D_y = \begin{vmatrix} a_1 & c_1 \\ a_2 & c_2 \end{vmatrix}.$$

We can then write formulas (5) in the form

$$\text{(6)} \qquad x = \frac{D_x}{D}; \qquad y = \frac{D_y}{D}.$$

Example

Solve the system

$$4x + 15y = 8,$$
$$2x - 6y = 13.$$

Solution

$$D = \begin{vmatrix} 4 & 15 \\ 2 & -6 \end{vmatrix} = 4(-6) - 2(15) = -24 - 30 = -54.$$

$$D_x = \begin{vmatrix} 8 & 15 \\ 13 & -6 \end{vmatrix} = 8(-6) - 13(15) = -48 - 195 = -243.$$

$$D_y = \begin{vmatrix} 4 & 8 \\ 2 & 13 \end{vmatrix} = 4(13) - 2(8) = 52 - 16 = 36.$$

$$x = \frac{D_x}{D} = \frac{-243}{-54} = 4\frac{1}{2};$$

$$y = \frac{D_y}{D} = \frac{36}{-54} = -\frac{2}{3}.$$

105. ***Determinants of third order.*** The symbol

$$\begin{vmatrix} a_1 & b_1 & c_1 \\ a_2 & b_2 & c_2 \\ a_3 & b_3 & c_3 \end{vmatrix}$$

in which nine numbers are arranged in a square array consisting of three rows and three columns, and flanked on the sides by vertical bars, is called a *determinant of third order*. It is defined to denote the number that is obtained as follows:

$$\begin{vmatrix} a_1 & b_1 & c_1 \\ a_2 & b_2 & c_2 \\ a_3 & b_3 & c_3 \end{vmatrix} = a_1\begin{vmatrix} b_2 & c_2 \\ b_3 & c_3 \end{vmatrix} - b_1\begin{vmatrix} a_2 & c_2 \\ a_3 & c_3 \end{vmatrix} + c_1\begin{vmatrix} a_2 & b_2 \\ a_3 & b_3 \end{vmatrix}$$
$$= a_1(b_2c_3 - b_3c_2) - b_1(a_2c_3 - a_3c_2) + c_1(a_2b_3 - a_3b_2)$$
$$= a_1b_2c_3 + a_2b_3c_1 + a_3b_1c_2 - a_1b_3c_2 - a_2b_1c_3 - a_3b_2c_1.$$

As in the case of a determinant of second order, the individual numbers in the determinant are called its *elements*. Any of the expressions on the right is called an *expansion* of the determinant.

Example

$$\begin{vmatrix} 2 & 3 & -1 \\ 4 & 0 & 5 \\ 3 & -2 & 2 \end{vmatrix} = 2\begin{vmatrix} 0 & 5 \\ -2 & 2 \end{vmatrix} - 3\begin{vmatrix} 4 & 5 \\ 3 & 2 \end{vmatrix} + (-1)\begin{vmatrix} 4 & 0 \\ 3 & -2 \end{vmatrix}$$

$$= 2[0 - (-10)] - 3[8 - 15] - 1[-8 - 0]$$
$$= 2(10) - 3(-7) - 1(-8)$$
$$= 20 + 21 + 8 = 49.$$

By the *minor* of an element we mean the determinant that remains when the row and column in which that element appears are omitted. For example, the minor of the element a_2 (below) is the determinant that remains when the first column and second row are omitted:

$$\begin{vmatrix} a_1 & b_1 & c_1 \\ a_2 & b_2 & c_2 \\ a_3 & b_3 & c_3 \end{vmatrix} \quad \text{minor of } a_2 \text{ is } A_2 \equiv \begin{vmatrix} b_1 & c_1 \\ b_3 & c_3 \end{vmatrix}.$$

As indicated here, we often denote the minor of an element by the corresponding capital letter. Thus the minor of a_1 is denoted by A_1, the minor of b_3 by B_3, etc.

Using this notation, we can write the above expansion of a determinant D of third order as follows:

$$D = \begin{vmatrix} a_1 & b_1 & c_1 \\ a_2 & b_2 & c_2 \\ a_3 & b_3 & c_3 \end{vmatrix} = a_1A_1 - b_1B_1 + c_1C_1.$$

This is called the *expansion of the determinant using the minors of the elements of the first row.* Precisely the same result is obtained if one uses either of the following expansions using the minors of the elements of the second and third rows, respectively:

$$D = -a_2A_2 + b_2B_2 - c_2C_2$$
$$= a_3A_3 - b_3B_3 + c_3C_3.$$

Finally, the same result is obtained from any of the following expansions using the minors of the elements of the first, second, and third columns, respectively:

$$\begin{aligned} D &= a_1A_1 - a_2A_2 + a_3A_3 \\ &= -b_1B_1 + b_2B_2 - b_3B_3 \\ &= c_1C_1 - c_2C_2 + c_3C_3. \end{aligned}$$

The proofs are left to the exercises. The following points concerning each expansion should be observed: It is the sum of the three products formed by multiplying each element of one row or one column by its minor and attaching a plus or minus sign. *This sign is plus if the sum of the number of the row and the number of the column in which the element lies is even, minus if odd.* Consider, for example, the term involving the product c_2C_2. The element c_2 is in the *second row* and *third column*. The sum of the number of the row and the number of the column is 5—an odd number. Therefore the term is $-c_2C_2$ instead of c_2C_2.

Example

The determinant of the above illustrative example may be expanded using the minors of the elements of the third column as follows:

$$\begin{vmatrix} 2 & 3 & -1 \\ 4 & 0 & 5 \\ 3 & -2 & 2 \end{vmatrix} = -1\begin{vmatrix} 4 & 0 \\ 3 & -2 \end{vmatrix} - 5\begin{vmatrix} 2 & 3 \\ 3 & -2 \end{vmatrix} + 2\begin{vmatrix} 2 & 3 \\ 4 & 0 \end{vmatrix}$$

$$= -1(-8) - 5(-13) + 2(-12)$$

$$= 8 + 65 - 24 = 49.$$

If we use the minors of the elements of the *second row* we have

$$\begin{vmatrix} 2 & 3 & -1 \\ 4 & 0 & 5 \\ 3 & -2 & 2 \end{vmatrix} = -4\begin{vmatrix} 3 & -1 \\ -2 & 2 \end{vmatrix} + 0\begin{vmatrix} 2 & -1 \\ 3 & 2 \end{vmatrix} - 5\begin{vmatrix} 2 & 3 \\ 3 & -2 \end{vmatrix}$$

$$= -4(4) + 0 - 5(-13)$$

$$= -16 + 65 = 49.$$

Consider now the following system of three linear equations in the three unknowns x, y, and z:

(1) $$a_1x + b_1y + c_1z = d_1,$$

(2) $$a_2x + b_2y + c_2z = d_2,$$

(3) $$a_3x + b_3y + c_3z = d_3.$$

If we eliminate z between (1) and (2), and also eliminate z between (1) and (3), we have the following two equations in x and y:

(4) $$(a_1c_2 - a_2c_1)x + (b_1c_2 - b_2c_1)y = d_1c_2 - d_2c_1,$$
(5) $$(a_1c_3 - a_3c_1)x + (b_1c_3 - b_3c_1)y = d_1c_3 - d_3c_1.$$

The result of eliminating y between (4) and (5) is

(6) $$(a_1b_2c_3 + a_2b_3c_1 + a_3b_1c_2 - a_1b_3c_2 - a_2b_1c_3 - a_3b_2c_1)x$$
$$= d_1b_2c_3 + d_2b_3c_1 + d_3b_1c_2 - d_1b_3c_2 - d_2b_1c_3 - d_3b_2c_1.$$

If the coefficient of x is different from zero, we can divide by it and we then have:

(7) $$x = \frac{d_1b_2c_3 + d_2b_3c_1 + d_3b_1c_2 - d_1b_3c_2 - d_2b_1c_3 - d_3b_2c_1}{a_1b_2c_3 + a_2b_3c_1 + a_3b_1c_2 - a_1b_3c_2 - a_2b_1c_3 - a_3b_2c_1}.$$

The denominator of (7) is precisely the expansion of the determinant D of the coefficients in the given equations (1), (2), and (3); the numerator is the same thing with the constants d_1, d_2, and d_3 in place of the corresponding coefficients of x. If we denote this determinant by D_x, we can write (6) in the form $xD = D_x$, and (7) as $x = D_x/D$.

When we proceed similarly to solve for y and z, we arrive at similar results. We are finally able to say that if (1), (2), and (3) are to be true, then we must have

(8) $$xD = D_x; \qquad yD = D_y; \qquad zD = D_z.$$

If $D \neq 0$, the only possible solution of the system is given by

(9) $$x = \frac{D_x}{D}; \qquad y = \frac{D_y}{D}; \qquad z = \frac{D_z}{D}.$$

One can show by direct substitution that the values of x, y, and z given by (9) satisfy the given system, so this is the solution for the case in which $D \neq 0$.

If $D = 0$, the system has no solution if any one of the determinants D_x, D_y, D_z is different from zero. This can be proved as follows: Suppose that $D = 0$ and say $D_y \neq 0$. Assume that there is a solution $x = x_0$, $y = y_0$, $z = z_0$ of the system. Then from (8) it must be true that

$$y_0(D) = D_y \qquad \text{or} \qquad y \cdot 0 = D_y.$$

This relation cannot be true if $D_y \neq 0$, so the assumption that there is a solution is false. The equations are said to be *inconsistent* if there is no solution.

If $D = 0$, and also $D_x = D_y = D_z = 0$, the system may or may not have a solution. We shall not go into the details of the various cases. See Probs. 30 to 32 of the next set.

Example

In order to solve the system

$$\begin{aligned} 2x + y + 3z &= 5, \\ x - \tfrac{1}{2}y - 2z &= 2, \\ 4x + 2y - z &= 3, \end{aligned}$$

we evaluate the following determinants:

$$D = \begin{vmatrix} 2 & 1 & 3 \\ 1 & -\frac{1}{2} & -2 \\ 4 & 2 & -1 \end{vmatrix} = 2(4\tfrac{1}{2}) - 1(7) + 3(4) = 14.$$

$$D_x = \begin{vmatrix} 5 & 1 & 3 \\ 2 & -\frac{1}{2} & -2 \\ 3 & 2 & -1 \end{vmatrix} = 5(4\tfrac{1}{2}) - 1(4) + 3(5\tfrac{1}{2}) = 35.$$

$$D_y = \begin{vmatrix} 2 & 5 & 3 \\ 1 & 2 & -2 \\ 4 & 3 & -1 \end{vmatrix} = 2(4) - 5(7) + 3(-5) = -42.$$

$$D_z = \begin{vmatrix} 2 & 1 & 5 \\ 1 & -\frac{1}{2} & 2 \\ 4 & 2 & 3 \end{vmatrix} = 2(-5\tfrac{1}{2}) - 1(-5) + 5(4) = 14.$$

We have then

$$x = \frac{D_x}{D} = \frac{35}{14} = \frac{5}{2}; \quad y = \frac{D_y}{D} = \frac{-42}{14} = -3; \quad z = \frac{D_z}{D} = \frac{14}{14} = 1.$$

PROBLEMS

Evaluate each of the following determinants:

1. $\begin{vmatrix} 2 & -3 \\ 4 & 0 \end{vmatrix}$ **2.** $\begin{vmatrix} -1 & 2 \\ 2 & 1 \end{vmatrix}$ **3.** $\begin{vmatrix} \frac{7}{2} & \frac{3}{4} \\ 6 & 2 \end{vmatrix}$

4. $\begin{vmatrix} 1 & 0 & -2 \\ -3 & 1 & 1 \\ 2 & 4 & 3 \end{vmatrix}$ **5.** $\begin{vmatrix} 3 & 6 & 7 \\ -2 & -1 & 4 \\ 2 & 5 & -1 \end{vmatrix}$ **6.** $\begin{vmatrix} 1 & 2 & -5 \\ 0 & 6 & 0 \\ -2 & -1 & 4 \end{vmatrix}$

7. $\begin{vmatrix} 14 & 0 & -2 \\ 3 & -1 & 0 \\ 0 & 5 & 1 \end{vmatrix}$ **8.** $\begin{vmatrix} 1 & 3 & 1 \\ -2 & -1 & 5 \\ 3 & 7 & 0 \end{vmatrix}$ **9.** $\begin{vmatrix} 0 & 6 & -2 \\ 1 & 5 & 1 \\ -2 & -6 & 7 \end{vmatrix}$

Solve each of the following systems using determinants:

10. $3x - y = 10,$
$6x + 5y = 6.$

11. $4x + 5y = 5,$
$6x - y = -18.$

12. $4x + y = 2.5,$
$5x - 2y = 14.5.$

13. $\frac{6}{x} + \frac{7}{y} = -\frac{9}{4},$
$\frac{1}{x} + \frac{3}{y} = \frac{1}{12}.$

14. $\frac{12}{x} - \frac{7}{y} = \frac{40}{3},$
$\frac{5}{x} + \frac{2}{y} = -1.$

15. $1.6x - 0.7y = 1.14,$
$2.5x + 1.5y = 6.45.$

16. $3x + y + 2z = 3,$
$x - 5y - z = 5,$
$2x + 3y + 2z = 0.$

17. $x + y + 2z = 10,$
$3x - y - 3z = -3,$
$5x + 3y - z = 1.$

18. $4x + 2y + z = 6,$
$2x + \frac{1}{2}y + 2z = 12,$
$x - y + \frac{3}{4}z = \frac{1}{2}.$

19. $4x + 8y - 3z = 8,$
$x - \frac{5}{2}y + z = -1,$
$12x + y + \frac{1}{2}z = 7.$

20. $7x - y + 3z = -9,$
$2x + \frac{1}{7}y - 2z = 2,$
$x + y + \frac{5}{7}z = 3.$

21. $2x + 8y + z = 6,$
$3x - 6y + z = 18,$
$x + y + \frac{2}{3}z = 6.$

22. $3x - 5y = -21,$
$5x - 8z = \frac{3}{2},$
$x + y + z = -6.$

23. $y + 3z = -4,$
$x + 5y = 2,$
$\frac{2}{3}x + z = 0.$

Show that each of the following systems of equations is inconsistent:

24. $x - 2y + 3z = 7,$
$2x + y - 7z = 20,$
$3x - 6y + 9z = 12.$

25. $x - y + 3z = 5,$
$2x + 6y - z = 4,$
$5x + 11y + z = 10.$

26. $3x + y + 2z = 8,$
$x - y + 5z = 2,$
$9x - y + 19z = 18.$

27. $x + 3z = 10,$
$y - 2z = 4,$
$2x + 3y = 5.$

28. We have defined the value of a determinant of third order by the equation:

$$D = \begin{vmatrix} a_1 & b_1 & c_1 \\ a_2 & b_2 & c_2 \\ a_3 & b_3 & c_3 \end{vmatrix} = a_1A_1 - b_1B_1 + c_1C_1.$$

Prove that
(a) $D = -a_2A_2 + b_2B_2 - c_2C_2.$
(b) $D = a_3A_3 - b_3B_3 + c_3C_3.$

29. Referring to Prob. 28, prove that
(a) $D = a_1A_1 - a_2A_2 + a_3A_3.$
(b) $D = -b_1B_1 + b_2B_2 - b_3B_3.$
(c) $D = c_1C_1 - c_2C_2 + c_3C_3.$

106 – 107 – 109

30. Show that for the following system $D = 0$ and also $D_x = D_y = D_z = 0$:

$$\begin{aligned} x + 2y - z &= 3, \\ 3x - 4y + 2z &= 9, \\ x + 12y - 6z &= 3. \end{aligned}$$

Show that the system has the solution $x = 3$, $y = a$, $z = 2a$, where a is any number.

31. Show that for the following system $D = 0$ and also $D_x = D_y = D_z = 0$:

$$\begin{aligned} x - y - 4z &= 10, \\ 2x - 2y + z &= 2, \\ 3x - 3y - z &= 8. \end{aligned}$$

Show that the system has the solution $x = a + 1$, $y = a - 1$, $z = -2$, where a is any number.

32. Show that for the following system $D = 0$ and also $D_x = D_y = D_z = 0$:

$$\begin{aligned} 2x + y - z &= 6, \\ x + 5y - 2z &= 6, \\ 2x - 5y + z &= 2. \end{aligned}$$

Show that the system has the solution $x = a + 1$, $y = a - 1$, $z = 3a - 5$, where a is any number.

106. *Determinants of order n.* The symbol

$$\begin{vmatrix} a_1 & b_1 & c_1 & d_1 \\ a_2 & b_2 & c_2 & d_2 \\ a_3 & b_3 & c_3 & d_3 \\ a_4 & b_4 & c_4 & d_4 \end{vmatrix}$$

is a determinant of fourth order, and of course the notation can be extended to determinants of still higher order. It would be difficult, however, to use this notation to indicate a determinant of order n, where n may be any integer greater than one, because of the limited number of letters at our disposal. We therefore devise a new notation as follows:

Observe that in the above notation the letter used indicates the column, and its subscript indicates the row, in which the element is located. Thus c_2 is in the *third column* (third letter of alphabet) and *second row* (subscript 2). In our new notation we employ a single letter, say a, and *two subscripts*. The first subscript indicates the row and the second subscript indicates the column in which the element lies. Thus a_{11} (read "a one one" not "a eleven") denotes the element in the first row, first column; a_{53}

(read "a five three") denotes the element in the fifth row, third column, etc.

Using this notation, we indicate a determinant of order n, where n is an integer greater than one,* as follows:

$$\begin{vmatrix} a_{11} & a_{12} & a_{13} & \cdots & a_{1n} \\ a_{21} & a_{22} & a_{23} & \cdots & a_{2n} \\ a_{31} & a_{32} & a_{33} & \cdots & a_{3n} \\ \cdot & \cdot & \cdot & \cdots & \cdot \\ \cdot & \cdot & \cdot & \cdots & \cdot \\ a_{n1} & a_{n2} & a_{n3} & \cdots & a_{nn} \end{vmatrix}.$$

For the case in which $n = 2$, we have defined the symbol as follows:

$$\begin{vmatrix} a_{11} & a_{12} \\ a_{21} & a_{22} \end{vmatrix} = a_{11}a_{22} - a_{21}a_{12}.$$

For the case in which $n = 3$, we have given the following definition:

$$\begin{vmatrix} a_{11} & a_{12} & a_{13} \\ a_{21} & a_{22} & a_{23} \\ a_{31} & a_{32} & a_{33} \end{vmatrix} = a_{11}\begin{vmatrix} a_{22} & a_{23} \\ a_{32} & a_{33} \end{vmatrix} - a_{12}\begin{vmatrix} a_{21} & a_{23} \\ a_{31} & a_{33} \end{vmatrix} + a_{13}\begin{vmatrix} a_{21} & a_{22} \\ a_{31} & a_{32} \end{vmatrix}$$
$$= a_{11}A_{11} - a_{12}A_{12} + a_{13}A_{13},$$

where the minors, denoted by the capital letters, are to be expanded in accordance with the above definition of a determinant of order two.

The corresponding definition for a determinant of higher order is entirely analogous. If D denotes the above determinant of order n, where $n > 2$, then, by definition,

$$\boldsymbol{D = a_{11}A_{11} - a_{12}A_{12} + a_{13}A_{13} - \cdots + (-1)^{1+n}a_{1n}A_{1n}}.$$

We have seen that in the special case in which $n = 3$ this same number D results from the similar expansion using the elements of any other row, or any column. This is true in the general case. The term in the expansion that involves the element from the ith row and jth column is

$$(-1)^{i+j}a_{ij}A_{ij}.$$

* We could of course include $n = 1$, but this case is trivial. A determinant of order one would consist of one element and its value would be simply that number. It would be confusing to use the symbol $|a|$ to denote such a determinant because this symbol is already employed for a different purpose.

Example

The value of the determinant

$$\begin{vmatrix} 2 & 0 & -1 & 3 \\ 4 & 5 & 2 & -1 \\ 3 & -6 & 0 & 5 \\ -1 & 2 & 3 & 1 \end{vmatrix}$$

is

$$2\begin{vmatrix} 5 & 2 & -1 \\ -6 & 0 & 5 \\ 2 & 3 & 1 \end{vmatrix} - 0\begin{vmatrix} 4 & 2 & -1 \\ 3 & 0 & 5 \\ -1 & 3 & 1 \end{vmatrix} + (-1)\begin{vmatrix} 4 & 5 & -1 \\ 3 & -6 & 5 \\ -1 & 2 & 1 \end{vmatrix} - 3\begin{vmatrix} 4 & 5 & 2 \\ 3 & -6 & 0 \\ -1 & 2 & 3 \end{vmatrix}$$

$$= 2(-25) - 0 + (-1)(-104) - 3(-117) = 405.$$

We have used here the minors of the elements of the *first row*. If we use the minors of the elements of the *second row*, we have

$$-4\begin{vmatrix} 0 & -1 & 3 \\ -6 & 0 & 5 \\ 2 & 3 & 1 \end{vmatrix} + 5\begin{vmatrix} 2 & -1 & 3 \\ 3 & 0 & 5 \\ -1 & 3 & 1 \end{vmatrix} - 2\begin{vmatrix} 2 & 0 & 3 \\ 3 & -6 & 5 \\ -1 & 2 & 1 \end{vmatrix} + (-1)\begin{vmatrix} 2 & 0 & -1 \\ 3 & -6 & 0 \\ -1 & 2 & 3 \end{vmatrix}$$

$$= -4(-70) + 5(5) - 2(-32) - 1(-36) = 405.$$

PROBLEMS

1. Write down a general determinant of fourth order using the double subscript notation.

2. Write down a general determinant of fifth order using the double subscript notation.

3. If we denote by D the determinant

$$\begin{vmatrix} a_1 & b_1 & c_1 & d_1 \\ a_2 & b_2 & c_2 & d_2 \\ a_3 & b_3 & c_3 & d_3 \\ a_4 & b_4 & c_4 & d_4 \end{vmatrix}$$

then, by definition,

$$D = a_1A_1 - b_1B_1 + c_1C_1 - d_1D_1.$$

Show that

$$D = -a_2A_2 + b_2B_2 - c_2C_2 + d_2D_2.$$

4. Referring to Prob. 3, show that

$$D = c_1C_1 - c_2C_2 + c_3C_3 - c_4C_4.$$

5. Referring to Prob. 3, show that

$$D = -d_1D_1 + d_2D_2 - d_3D_3 + d_4D_4.$$

In each of the following problems evaluate the given determinant using the minors of the elements of some row. Then check by evaluating it using the minors of the elements of some column. Observe that less work is required if one can use a row or column containing several zeros:

6. $\begin{vmatrix} 1 & 0 & -1 & 2 \\ 1 & 1 & -3 & -2 \\ 3 & 0 & 1 & 0 \\ 0 & 2 & -2 & 1 \end{vmatrix}$

7. $\begin{vmatrix} 3 & 1 & 1 & 0 \\ -2 & -2 & 4 & 0 \\ 3 & 0 & 0 & 5 \\ 2 & 0 & 1 & 3 \end{vmatrix}$

8. $\begin{vmatrix} 1 & -1 & 2 & 3 \\ 3 & 0 & 1 & 4 \\ 2 & 1 & 3 & 0 \\ 0 & 0 & -2 & -1 \end{vmatrix}$

9. $\begin{vmatrix} 5 & 0 & 0 & 5 \\ 0 & 0 & 0 & 1 \\ 3 & 0 & 1 & -2 \\ -2 & 1 & 6 & -1 \end{vmatrix}$

10. $\begin{vmatrix} 2 & 2 & 1 & -1 \\ -2 & 2 & 0 & 1 \\ 0 & 0 & 1 & 1 \\ 2 & 0 & 0 & 1 \end{vmatrix}$

11. $\begin{vmatrix} 1 & 0 & -1 & 0 \\ 3 & 2 & 0 & 1 \\ 0 & 1 & 5 & 0 \\ -2 & 0 & 0 & 0 \end{vmatrix}$

12. $\begin{vmatrix} 12 & 8 & -2 & 0 \\ 0 & 0 & 0 & 1 \\ -3 & -2 & 4 & 0 \\ 3 & 0 & 7 & 0 \end{vmatrix}$

13. $\begin{vmatrix} -7 & -2 & 0 & 1 \\ 0 & 0 & 1 & 1 \\ 2 & -2 & 0 & 0 \\ 3 & 0 & -1 & 1 \end{vmatrix}$

14. $\begin{vmatrix} -3 & -2 & 1 & 1 \\ 0 & 3 & 1 & 1 \\ 2 & 1 & 1 & 1 \\ 0 & 2 & 0 & 0 \end{vmatrix}$

15. $\begin{vmatrix} 2 & 1 & 0 & 3 \\ -3 & 0 & 2 & 4 \\ 3 & 5 & 6 & 0 \\ 4 & 2 & 0 & 6 \end{vmatrix}$

16. $\begin{vmatrix} 2 & 3 & 0 & 0 & 1 \\ -2 & 1 & 0 & 3 & 0 \\ -3 & 4 & 2 & 0 & 0 \\ 1 & 7 & 0 & 5 & 0 \\ 0 & 0 & 0 & 3 & 1 \end{vmatrix}$

17. $\begin{vmatrix} 1 & -1 & 1 & 0 & 0 \\ 2 & -2 & 3 & 0 & 3 \\ 0 & 0 & 1 & 2 & -1 \\ 3 & 3 & 3 & 0 & 0 \\ 0 & 0 & -1 & 0 & 1 \end{vmatrix}$

18. $\begin{vmatrix} 0 & 0 & 1 & 1 & 3 \\ 2 & -6 & 5 & 0 & 0 \\ 0 & 1 & 1 & 2 & -1 \\ 0 & 0 & 0 & 1 & 1 \\ 0 & 1 & -1 & 1 & 2 \end{vmatrix}$

19. $\begin{vmatrix} -1 & 1 & 2 & 1 & 0 \\ -3 & -1 & 2 & 0 & 0 \\ 0 & 0 & 1 & 0 & 0 \\ 2 & -1 & 1 & 0 & 0 \\ 0 & 0 & 1 & 3 & 0 \end{vmatrix}$

20. $\begin{vmatrix} 1 & 0 & 1 & 1 & 2 & -1 \\ 2 & 0 & 0 & 1 & 1 & -1 \\ -1 & -1 & 0 & 0 & 0 & 0 \\ 0 & 0 & 0 & 1 & -2 & 3 \\ 0 & 0 & 0 & 1 & 0 & 0 \\ 0 & 0 & 1 & 0 & 0 & 0 \end{vmatrix}$

107. ***Some properties of determinants.*** In this section we list several properties of a determinant D. Some of these are useful in connection with the problem of evaluating a determinant.

***Theorem* I.** *If a determinant* $\boldsymbol{D'}$ *is constructed from a determinant* $\boldsymbol{D}$ *by taking each row of* $\boldsymbol{D}$ *as the corresponding column of* $\boldsymbol{D'}$, *then* $\boldsymbol{D' = D}$. For example,

$$\begin{vmatrix} a_1 & b_1 & c_1 \\ a_2 & b_2 & c_2 \\ a_3 & b_3 & c_3 \end{vmatrix} = \begin{vmatrix} a_1 & a_2 & a_3 \\ b_1 & b_2 & b_3 \\ c_1 & c_2 & c_3 \end{vmatrix}.$$

This property is sometimes expressed by saying that *the value of a determinant is unchanged if its rows and columns are interchanged.* The details of the proof will be left for the student. HINT: First show directly that the theorem holds for the case $n = 2$. Then, for $n = 3$, compare the expansion of D using the elements of the first row with that of D' using the elements of the first column. This will point the way to a proof by induction.

From this property it follows that for every theorem concerning the rows of a determinant there is a corresponding theorem concerning the columns, and vice versa.

***Theorem* II.** *If every element of any row (or column) is zero, then* $\boldsymbol{D} = 0$. The proof is obvious; consider the expansion using the minors of the elements of this row or column.

***Theorem* III.** *If every element of any one row (or column) is multiplied by the same number* $\boldsymbol{k}$, *the value of the determinant is multiplied by* $\boldsymbol{k}$. The proof is left to the student. HINT: Compare the expansions of the determinants

$$\begin{vmatrix} a_1 & b_1 & c_1 \\ a_2 & b_2 & c_2 \\ a_3 & b_3 & c_3 \end{vmatrix} \quad \text{and} \quad \begin{vmatrix} ka_1 & kb_1 & kc_1 \\ a_2 & b_2 & c_2 \\ a_3 & b_3 & c_3 \end{vmatrix}$$

using the minors of the elements of the first row.

Observe that this theorem authorizes one to "factor out" any number that is a common factor of all the elements of any row (or column). Thus

$$\begin{vmatrix} 2 & 3 & 6 \\ 4 & 8 & 8 \\ 1 & 1 & 4 \end{vmatrix} = 4\begin{vmatrix} 2 & 3 & 6 \\ 1 & 2 & 2 \\ 1 & 1 & 4 \end{vmatrix} = (4)(2)\begin{vmatrix} 2 & 3 & 3 \\ 1 & 2 & 1 \\ 1 & 1 & 2 \end{vmatrix}.$$

We have simplified the given determinant by first taking the common factor 4 from the second row and then similarly taking the factor 2 from the third column.

***Theorem* IV.** *If two rows (or columns) are identical, then* $D = 0$. In order to prove this, note first that the theorem is true for the case in which $n = 2$:

$$\begin{vmatrix} a & b \\ a & b \end{vmatrix} = ab - ab = 0.$$

The student may complete the proof by induction by showing that, if the theorem is true for the case in which $n = k$, it is true for the case in which $n = k + 1$. HINT: Let the gth and ith rows of the determinant of order $k + 1$ be identical, and expand it using the minors of any other row. Observe that each of the $k + 1$ minors, which are of course determinants of order k, has two identical rows.

***Theorem* V.** *If two rows (or columns) are proportional, then* $D = 0$. For example,

$$\begin{vmatrix} 1 & 2 & -3 \\ 2 & 4 & -6 \\ 5 & -1 & 2 \end{vmatrix} = 2 \begin{vmatrix} 1 & 2 & -3 \\ 1 & 2 & -3 \\ 5 & -1 & 2 \end{vmatrix} = 0.$$

The method of proof is suggested in this example.

***Theorem* VI.** *If a determinant* D' *is constructed from a determinant* D *by interchanging any two rows (or any two columns) of* D*, then* $D' = -D$. For example,

$$\begin{vmatrix} 2 & 4 & 1 \\ 3 & 5 & 2 \\ 6 & 1 & 6 \end{vmatrix} = - \begin{vmatrix} 3 & 5 & 2 \\ 2 & 4 & 1 \\ 6 & 1 & 6 \end{vmatrix}.$$

The details of the proof by induction are left to the student. Observe first that the theorem is true for the case in which $n = 2$:

$$\begin{vmatrix} a_1 & b_1 \\ a_2 & b_2 \end{vmatrix} = a_1b_2 - a_2b_1 \quad \text{and} \quad \begin{vmatrix} a_2 & b_2 \\ a_1 & b_1 \end{vmatrix} = a_2b_1 - a_1b_2.$$

Now prove that if the theorem is true for the case $n = k$ it is true for the case $n = k + 1$. In order to do this, let the gth and ith rows of D be the ith and gth rows of D', respectively, all other rows of D and D' being identical. Expand D and D' using the minors of the elements of any other row.

We introduce the next theorem of this group by giving an example. Consider the two determinants D and D' where

$$D = \begin{vmatrix} a_1 & b_1 & c_1 \\ a_2 & b_2 & c_2 \\ a_3 & b_3 & c_3 \end{vmatrix} \qquad D' = \begin{vmatrix} a_1 + kb_1 & b_1 & c_1 \\ a_2 + kb_2 & b_2 & c_2 \\ a_3 + kb_3 & b_3 & c_3 \end{vmatrix}.$$

Here, D' is obtained from D by adding to the elements of the first column of D the corresponding elements of the second column, each multiplied by the same number k. We wish to prove that $D' = D$. For this purpose expand D' using the minors of the elements of the first column:

$$\begin{aligned} D' &= (a_1 + kb_1)A_1 - (a_2 + kb_2)A_2 + (a_3 + kb_3)A_3 \\ &= (a_1A_1 - a_2A_2 + a_3A_3) + k(b_1A_1 - b_2A_2 + b_3A_3) \\ &= D + k(0) = D. \end{aligned}$$

Observe that $b_1A_1 - b_2A_2 + b_3A_3 = 0$ because this is the expansion of the determinant

$$\begin{vmatrix} b_1 & b_1 & c_1 \\ b_2 & b_2 & c_2 \\ b_3 & b_3 & c_3 \end{vmatrix}$$

which has two identical columns.

The general theorem is as follows:

Theorem VII. *If a determinant **D'** is constructed from a determinant **D** by adding to each element of one row (or column) **k** times the corresponding element of any other row (or column), then **D'** = **D**.*

As indicated above, the proof is carried out by expanding D' using the minors of the elements of this row (or column). The result consists of the expansion of D plus k times the expansion of a determinant having two identical rows (or columns).

Theorem VIII. *If in the expansion of a determinant using the minors of the elements of a given row (or column) the elements of this row (or column) are replaced by the corresponding elements of any other row (or column), the resulting expression is equal to zero.*

The proof results from the observation that the expression in question is the expansion of a determinant having two identical rows or columns.

108. ***Evaluation of a determinant.*** There are various ways in which some of the theorems of the preceding section can be used to reduce the work involved in evaluating a given determinant. Consider, for example, the determinant

$$(1) \qquad \begin{vmatrix} 2 & 3 & 1 & 4 \\ 5 & 4 & 3 & 2 \\ -2 & -3 & -5 & -2 \\ 5 & 1 & 8 & -7 \end{vmatrix}.$$

If we expand it directly, we must evaluate four determinants of third order, but if we can first arrange things so that the first row consists of the numbers 1, 0, 0, 0, we shall need to evaluate only one such determinant.

In order to make $a_{11} = 1$, multiply each element of column three by -1 and add this to the corresponding element of column one. We then have

$$(2) \qquad \begin{vmatrix} 1 & 3 & 1 & 4 \\ 2 & 4 & 3 & 2 \\ 3 & -3 & -5 & 2 \\ -3 & 1 & 8 & -7 \end{vmatrix}$$

and this determinant is equal to the original one by virtue of Theorem VII. (We could have made $a_{11} = 1$ by interchanging columns one and three, and changing the sign, using Theorem VI.)

We now make the first row consist of the numbers 1, 0, 0, 0, as follows:

Multiply each element of column one by -3 *and add to column two.*

Multiply each element of column one by -1 *and add to column three.*

Multiply each element of column one by -4 *and add to column four.*

The result, which is still equal to the given determinant, is

$$(3) \qquad \begin{vmatrix} 1 & 0 & 0 & 0 \\ 2 & -2 & 1 & -6 \\ 3 & -12 & -8 & -10 \\ 3 & 10 & 11 & 5 \end{vmatrix}.$$

All the elements of the first column of (3) except a_{11} can now be replaced by zero without affecting the value of the determinant, for these elements do not enter into the expansion using the minors of the elements of the first row. We have then

$$(4)\qquad \begin{vmatrix} 1 & 3 & 1 & 4 \\ 2 & 4 & 3 & 2 \\ 3 & -3 & -5 & 2 \\ -3 & 1 & 8 & -7 \end{vmatrix} = \begin{vmatrix} 1 & 0 & 0 & 0 \\ 0 & -2 & 1 & -6 \\ 0 & -12 & -8 & -10 \\ 0 & 10 & 11 & 5 \end{vmatrix}.$$

The rule can be stated simply as follows:

STEP I. *Transform the given determinant into a determinant in which* $a_{11} = 1$.

STEP II. *In this determinant in which* $a_{11} = 1$, *replace all the other elements of row one and column one by zero, and at the same time subtract from each remaining element the product of the element at the left end of its row and the element at the top of its column.*

The right member of (4) is of course equal to

$$\begin{vmatrix} -2 & 1 & -6 \\ -12 & -8 & -10 \\ 10 & 11 & 5 \end{vmatrix} = -2 \begin{vmatrix} -2 & 1 & -6 \\ 6 & 4 & 5 \\ 10 & 11 & 5 \end{vmatrix}$$

$$= 4 \begin{vmatrix} 1 & 1 & -6 \\ -3 & 4 & 5 \\ -5 & 11 & 5 \end{vmatrix}$$

$$= 4 \begin{vmatrix} 1 & 0 & 0 \\ 0 & 7 & -13 \\ 0 & 16 & -25 \end{vmatrix}$$

$$= 4(-175 + 208) = 132.$$

PROBLEMS

Evaluate each of the following determinants:

1. $\begin{vmatrix} 2 & 1 & 8 & 2 \\ -2 & 3 & 1 & -5 \\ 6 & 0 & 1 & 3 \\ 4 & 4 & 2 & -1 \end{vmatrix}$

2. $\begin{vmatrix} 3 & 2 & -1 & 4 \\ 5 & 3 & 2 & -2 \\ 1 & 0 & 3 & 5 \\ 0 & 8 & -2 & 1 \end{vmatrix}$

3. $\begin{vmatrix} 6 & 8 & -4 & 4 \\ 3 & 2 & 0 & 6 \\ -12 & 10 & 3 & 1 \\ 3 & -6 & 1 & -2 \end{vmatrix}$

4. $\begin{vmatrix} -8 & -6 & 4 & 4 \\ 3 & 1 & 0 & 3 \\ 2 & 11 & 5 & -5 \\ 12 & 9 & -6 & -6 \end{vmatrix}$

5. $\begin{vmatrix} 2 & 1 & 1 & 0 \\ 6 & -1 & 0 & 1 \\ 0 & 3 & 4 & 2 \\ 6 & 2 & -1 & 5 \end{vmatrix}$

6. $\begin{vmatrix} 5 & -10 & 5 & 5 \\ 3 & 2 & -2 & 1 \\ 0 & -4 & -2 & 2 \\ 3 & 1 & 0 & -1 \end{vmatrix}$

7. $$\begin{vmatrix} 4 & -14 & 6 & 4 \\ 0 & 3 & 1 & -2 \\ -2 & 1 & -4 & 0 \\ 2 & 6 & 0 & 3 \end{vmatrix}$$

8. $$\begin{vmatrix} 5 & -2 & -1 & 3 \\ 3 & 7 & 2 & -2 \\ 4 & 8 & 4 & 4 \\ 2 & -3 & -3 & 1 \end{vmatrix}$$

9. $$\begin{vmatrix} 1 & 0 & 1 & 2 & -3 \\ 2 & 0 & 0 & 1 & 4 \\ -2 & 4 & 1 & 3 & 4 \\ 1 & 6 & 0 & 3 & 5 \\ 2 & 3 & -3 & 1 & 1 \end{vmatrix}$$

10. $$\begin{vmatrix} 2 & -2 & 1 & 6 & 1 \\ -4 & 4 & 2 & 8 & -2 \\ 3 & -5 & 0 & 1 & 1 \\ 2 & 0 & 6 & 2 & 2 \\ 8 & 4 & -8 & 4 & 4 \end{vmatrix}$$

Show that each of the following determinants is equal to zero:

11. $$\begin{vmatrix} 6 & 2 & 3 & -4 \\ -3 & 4 & 1 & 2 \\ 18 & -8 & 4 & -12 \\ 6 & 2 & 6 & -4 \end{vmatrix}$$

12. $$\begin{vmatrix} 4 & 2 & -1 & 4 \\ 4 & -4 & 6 & -4 \\ 2 & 3 & 3 & -2 \\ -6 & 6 & -9 & 6 \end{vmatrix}$$

13. $$\begin{vmatrix} 2 & 1 & 0 & -2 & 1 \\ 0 & 3 & 1 & 0 & 2 \\ -1 & 7 & -2 & 1 & 0 \\ 3 & 0 & 2 & -3 & 4 \\ 0 & 2 & -5 & 0 & 3 \end{vmatrix}$$

14. $$\begin{vmatrix} 5 & 0 & -3 & 6 & 0 \\ -3 & 2 & 2 & 1 & -3 \\ 2 & -4 & 0 & 1 & 6 \\ 0 & 4 & 1 & 2 & -6 \\ 7 & 8 & -1 & 5 & -12 \end{vmatrix}$$

15. Prove Theorem I (page 251).
16. Prove Theorem III (page 251), for the case in which $n = 3$.
17. Prove Theorem IV (page 252).
18. Prove Theorem VI (page 252).
19. Prove Theorem VII (page 253), for the case in which $n = 4$.
20. Write out a proof of the following special case of Theorem VIII (page 253): If in the expansion of a determinant of fourth order, using the minors of the elements of the first row, these elements are replaced by the corresponding elements of the second row, the resulting expression is equal to zero.

109. ***Solution of a system of n linear equations in n unknowns. Cramer's rule.*** We consider now the following system consisting of n linear equations in the n unknowns $x_1, x_2, x_3, \cdots, x_n$:

$$\begin{aligned} a_{11}x_1 + a_{12}x_2 + a_{13}x_3 + \cdots + a_{1n}x_n &= k_1, \\ a_{21}x_1 + a_{22}x_2 + a_{23}x_3 + \cdots + a_{2n}x_n &= k_2, \\ a_{31}x_1 + a_{32}x_2 + a_{33}x_3 + \cdots + a_{3n}x_n &= k_3, \\ \cdots\cdots\cdots\cdots\cdots\cdots\cdots & \\ a_{n1}x_1 + a_{n2}x_2 + a_{n3}x_3 + \cdots + a_{nn}x_n &= k_n. \end{aligned}$$

As in the special cases that we have studied previously, let D denote the determinant of the coefficients; let D_1 be the determinant obtained by replacing each coefficient of x_1 in D by the corresponding constant term k; let D_2 be the determinant obtained by replacing each coefficient of x_2 in D by the corresponding constant term; and so on. Then if $D \neq 0$, the system has the unique solution

$$x_1 = \frac{D_1}{D}; \qquad x_2 = \frac{D_2}{D}; \qquad x_3 = \frac{D_3}{D}; \qquad \cdots; \qquad x_n = \frac{D_n}{D}.$$

This theorem is usually called *Cramer's rule* (after G. Cramer, 1704–1752).

For the sake of simplicity, we shall use the case in which $n = 4$ in indicating the proof. The method is quite general and can easily be extended to the above case. If $n = 4$, we have the equations

$$(1) \qquad \begin{aligned} a_{11}x_1 + a_{12}x_2 + a_{13}x_3 + a_{14}x_4 &= k_1, \\ a_{21}x_1 + a_{22}x_2 + a_{23}x_3 + a_{24}x_4 &= k_2, \\ a_{31}x_1 + a_{32}x_2 + a_{33}x_3 + a_{34}x_4 &= k_3, \\ a_{41}x_1 + a_{42}x_2 + a_{43}x_3 + a_{44}x_4 &= k_4. \end{aligned}$$

Multiply both sides of the first equation by A_{11}, both sides of the second by $-A_{21}$, both sides of the third by A_{31}, both sides of the fourth by $-A_{41}$, and add. The coefficient of x_1 on the left will be

$$a_{11}A_{11} - a_{21}A_{21} + a_{31}A_{31} - a_{41}A_{41}$$

which is equal to D. The coefficient of x_2 will be

$$a_{12}A_{11} - a_{22}A_{21} + a_{32}A_{31} - a_{42}A_{41}$$

which is equal to zero by virtue of Theorem VIII (page 253). Similarly, the coefficients of x_3 and x_4 will be zero. The right side will be

$$k_1A_{11} - k_2A_{21} + k_3A_{31} - k_4A_{41}$$

which is the expansion of the determinant obtained by replacing each of the coefficients of x_1 in D by the corresponding k; *i.e.*, it is the expansion of the determinant that we have called D_1. The resulting equation is then

$$(2) \qquad x_1D = D_1.$$

In a similar fashion we find that if equations (1) are to be satisfied, then we must have

$$(3) \qquad x_2D = D_2; \qquad x_3D = D_3; \qquad x_4D = D_4.$$

It follows that if $D \neq 0$ the only possible solution of (1) is

$$(4) \qquad x_1 = \frac{D_1}{D}; \qquad x_2 = \frac{D_2}{D}; \qquad x_3 = \frac{D_3}{D}; \qquad x_4 = \frac{D_4}{D}.$$

It can be shown by direct substitution that the values of x_1, x_2, x_3, and x_4 given by (4) satisfy equations (1), so this is the solution for the case in which $D \neq 0$.

If $D = 0$, the system has no solution if any of the determinants D_1, D_2, D_3, D_4 is different from zero. In order to prove this, let us assume that $D = 0$; that one of the other determinants, say D_3, is different from zero; and that there is a solution $x_1 = p$, $x_2 = q$, $x_3 = r$, $x_4 = s$. Then from the second equation of (3) we must have

$$x_3 D = D_3 \qquad \text{or} \qquad r \cdot 0 = D_3.$$

As this is impossible, the assumption that there is a solution under these conditions is false.

If $D = 0$ and also $D_1 = D_2 = D_3 = D_4 = 0$, the system may or may not have solutions. We shall not discuss the various cases that may arise.

If the system has no solution, the equations are said to be *inconsistent*. If there are an unlimited number of solutions, they are said to be *dependent*.

Example

Solve the following system for y:

$$\begin{aligned} 2x + y - z + 2w &= 3, \\ 3x - 2y - 4z - 4w &= 2, \\ x + 2y + 6z + 4w &= 8, \\ 3x + y - 2w &= 0. \end{aligned}$$

Solution

We evaluate the following determinants:

$$D = \begin{vmatrix} 2 & 1 & -1 & 2 \\ 3 & -2 & -4 & -4 \\ 1 & 2 & 6 & 4 \\ 3 & 1 & 0 & -2 \end{vmatrix} = 152.$$

$$D_y = \begin{vmatrix} 2 & 3 & -1 & 2 \\ 3 & 2 & -4 & -4 \\ 1 & 8 & 6 & 4 \\ 3 & 0 & 0 & -2 \end{vmatrix} = -456.$$

Then we have $y = (D_y/D) = -\frac{456}{152} = -3$. The student should similarly solve for x, z, and w. The results are $x = 2$, $z = 1$, $w = \frac{3}{2}$.

An important special case of the above-considered system is that of a system of *homogeneous* linear equations—a homogeneous linear equation being one in which the constant term is zero:

$$\begin{aligned}
a_{11}x_1 + a_{12}x_2 + a_{13}x_3 + \cdots + a_{1n}x_n &= 0,\\
a_{21}x_1 + a_{22}x_2 + a_{23}x_3 + \cdots + a_{2n}x_n &= 0,\\
a_{31}x_1 + a_{32}x_2 + a_{33}x_3 + \cdots + a_{3n}x_n &= 0,\\
\cdots\cdots\cdots\cdots\cdots\cdots\cdots\cdots &\cdots\\
a_{n1}x_1 + a_{n2}x_2 + a_{n3}x_3 + \cdots + a_{nn}x_n &= 0.
\end{aligned}$$

In this case each of the determinants $D_1, D_2, \cdots, D_n$, has a column of zeros and consequently has the value zero. It follows immediately that if $D \neq 0$, the only solution of the system is the obvious "trivial" solution

$$x_1 = 0; \qquad x_2 = 0; \qquad x_3 = 0; \cdots ; \qquad x_n = 0.$$

We may therefore state the following theorem: *A necessary condition for the existence of a solution of a system of* **n** *homogeneous linear equations in* **n** *unknowns, other than the trivial solution, is that* $D = 0$.

It turns out that this condition is also sufficient; *i.e.*, if $D = 0$, then the system certainly has solutions other than the trivial one. We shall not go into the details but shall give an example to indicate the kind of solutions that may exist.

Example

The student may verify that $D = 0$ for the following system:

$$\begin{aligned}
3x - y - 4z - w &= 0,\\
x + 4y + 4z - w &= 0,\\
x - 7y - 3z + 2w &= 0,\\
5x + y + 2z - w &= 0.
\end{aligned}$$

If we tentatively let $x = a$, we get the following nonhomogeneous system in y, z, and w:

$$\begin{aligned}
y + 4z + w &= 3a,\\
4y + 4z - w &= -a,\\
7y + 3z - 2w &= a,\\
y + 2z - w &= -5a.
\end{aligned}$$

If we take three of these equations and solve them in the usual way for y, z, and w, we get $y = 2a$, $z = -a$, and $w = 5a$. These numbers satisfy the remaining equation. The given system then has the solution

$$x = a, \qquad y = 2a, \qquad z = -a, \qquad w = 5a,$$

where a is any number. The trivial solution is the one that corresponds to $a = 0$. A study of the steps used assures us that the system has no solution that is not contained in the one that we have found.

110. ***Systems in which the number of equations is not equal to the number of unknowns.*** We shall make no attempt to treat this case systematically or with any degree of completeness, but since one sometimes encounters such systems in the applications, the following brief remarks may be of some value.*

If there are more equations than unknowns, one would not expect any solution to exist except under special conditions. Consider, for example, the system

$$\begin{aligned} 4x - 3y &= 9, \\ 2x + y &= 12, \\ 6x - 7y &= 8. \end{aligned}$$

If we take the first two equations and solve them for x and y, we get $x = 4\frac{1}{2}$, $y = 3$. If we substitute these numbers for x and y in the third equation, we find that they do not satisfy it. We can conclude that the system has no solution, but if the third equation had been $6x - 7y = 6$, then $x = 4\frac{1}{2}$ and $y = 3$ would have satisfied it, and this would then have been the unique solution of the system.

In general, then, if one has n unknowns and more than n equations, he may select n of the equations and solve this system in the usual way. If these have a unique solution, he may determine whether or not these numbers satisfy the remaining equations by direct substitution. If they do not, the system has no solution. If it happens that the determinant D of the coefficients in the n

* The student may wish to study this topic in Dickson, "New First Course in the Theory of Equations," New York, John Wiley & Sons, Inc., 1939, or in Uspensky, "Theory of Equations," New York, McGraw-Hill Book Company, Inc., 1948.

equations that he has selected is zero, he should make another selection.

If there are more unknowns than equations, there are, in general, infinitely many solutions. Consider the system

$$\begin{aligned} 2x - 3y - 4z &= 6, \\ x + y - 2z &= -2. \end{aligned}$$

It would appear that we might be able to assign an arbitrary value to one of the unknowns and then solve the system for the other two. We may tentatively let $z = k$. The system becomes

$$\begin{aligned} 2x - 3y &= 4k + 6, \\ x + y &= 2k - 2. \end{aligned}$$

If we solve these for x and y in the usual way, we get $x = 2k$, $y = -2$. The system then has the solution

$$x = 2k, \qquad y = -2, \qquad z = k,$$

where k is any number.

An important special case is that of the following system of homogeneous equations:

$$\begin{aligned} a_1x + b_1y + c_1z &= 0, \\ a_2x + b_2y + c_2z &= 0. \end{aligned}$$

Our method of seeking the possible solutions is to divide through by z and then regard x/z and y/z as the unknowns:

$$\begin{aligned} a_1\left(\frac{x}{z}\right) + b_1\left(\frac{y}{z}\right) &= -c_1, \\ a_2\left(\frac{x}{z}\right) + b_2\left(\frac{y}{z}\right) &= -c_2. \end{aligned}$$

If we solve these in the usual way, we get

$$\frac{x}{z} = \frac{\begin{vmatrix} -c_1 & b_1 \\ -c_2 & b_2 \end{vmatrix}}{\begin{vmatrix} a_1 & b_1 \\ a_2 & b_2 \end{vmatrix}} = \frac{\begin{vmatrix} b_1 & c_1 \\ b_2 & c_2 \end{vmatrix}}{\begin{vmatrix} a_1 & b_1 \\ a_2 & b_2 \end{vmatrix}};$$

$$\frac{y}{z} = \frac{\begin{vmatrix} a_1 & -c_1 \\ a_2 & -c_2 \end{vmatrix}}{\begin{vmatrix} a_1 & b_1 \\ a_2 & b_2 \end{vmatrix}} = \frac{\begin{vmatrix} c_1 & a_1 \\ c_2 & a_2 \end{vmatrix}}{\begin{vmatrix} a_1 & b_1 \\ a_2 & b_2 \end{vmatrix}}.$$

The ratios of x to z and y to z will have these values if and only if we have

$$x = k\begin{vmatrix} b_1 & c_1 \\ b_2 & c_2 \end{vmatrix}, \qquad y = k\begin{vmatrix} c_1 & a_1 \\ c_2 & a_2 \end{vmatrix}, \qquad z = k\begin{vmatrix} a_1 & b_1 \\ a_2 & b_2 \end{vmatrix},$$

where k is any constant. One can show by direct substitution that these numbers satisfy the given equations.

PROBLEMS

Solve each of the following systems:

1. $x + 3y + 5z - w = 4,$
$3x - y - 2z + 4w = 3,$
$2x + 6y + 3z - w = -1,$
$5x + 7y + 2z + 2w = 1.$

2. $2x + 4y + 3z - w = -2,$
$3x - 5y - 3z + w = 7,$
$x + 5y + z - 2w = -1,$
$4x - 2y - z + w = 10.$

3. $3x + 3y + 3z + 2w = 3,$
$x - y - z + 2w = 7,$
$5x + 2y - z + 4w = 10,$
$3x - 2y + z + 2w = 1.$

4. $2x + y + z - 3w = 1,$
$3x + 4y + 2z + w = 3,$
$3x + 7y - 3z - 3w = 7,$
$x + y + 3z + 2w = 2.$

5. $3x - 2y + 6z = 11,$
$5y - 3z + 4w = 0,$
$x + 4z + 3w = 5,$
$2x + 2y + w = 7.$

6. $2x + y - z + w = 6,$
$x - 3y - 2z - w = 3,$
$5x - y - 2w = 8,$
$y + 5z + w = 0.$

Show that each of the following systems has no solution:

7. $3x + y - 2z + w = 6,$
$x - 7y + z + 3w = 5,$
$9x - 9y - 2z + 19w = 10,$
$2x + 2y - z + 6w = 12.$

8. $4x - 2y + 6z - w = 5,$
$x + 5y - z + 3w = 7,$
$3x + y + 12z - 4w = 8,$
$2x + 18y + 3z + 6w = 12.$

Show that each of the following systems has no solution other than the trivial one:

9. $x + 2y + z + w = 0,$
$2x - y - z + 3w = 0,$
$3x + y + 4z + w = 0,$
$x - 7y + z - 3w = 0.$

10. $x - 3y + 2z + w = 0,$
$2x + 6y - 5z - w = 0,$
$x - y + 4z + 3w = 0,$
$3x + 5y - 6z - 2w = 0.$

11. Show that $D = 0$ for the system

$$\begin{aligned} 3x - y + 3z - 4w &= 0, \\ x + 7y - z + 2w &= 0, \\ 5x + 4y + 4z - 3w &= 0, \\ 2x - 3y + 2z - 5w &= 0. \end{aligned}$$

By letting $w = k$ and solving the resulting system, show that the solution is $x = 3k$, $y = -k$, $z = -2k$, $w = k$, where k is any number.

12. Show that $D = 0$ for the system

$$\begin{aligned} 5x - 2y - 2z - 2w &= 0, \\ 3x + 4y + 3z - w &= 0, \\ x + 2y + \tfrac{1}{2}z - w &= 0, \\ 7x + 4y - z - 5w &= 0. \end{aligned}$$

By letting $x = k$ and solving the resulting system, show that the solution is $x = k$, $y = \frac{3}{2}k$, $z = -2k$, $w = 3k$, where k is any number.

13. Show that the following system has no solution:

$$\begin{aligned} 3x + 2y + 3z &= 4, \\ x - 4y - 6z &= 6, \\ 2x - y - 2z &= 7, \\ 4x + y + 2z &= 6. \end{aligned}$$

14. Find any solution that may exist for the following system:

$$\begin{aligned} x - 2y - 6z &= 7, \\ 2x - 3y + z &= 3, \\ 3x + 2y + 8z &= 11, \\ 2x + 5y + 12z &= 8. \end{aligned}$$

Find any solution that may exist for each of the following systems:

15. $2x - y - 2z = 7,$
$x + 3y - 8z = 0.$

16. $3x - 2y - 7z = 12,$
$5x + 2y - z = 36.$

17. $2x + y + 2z = 0,$
$5x - 2y - 10z = 0.$

18. $5x - 2y - 6z = 0,$
$7x + 5y + 2z = 0.$

19. $x - y + z + 4w = 2,$
$3x + y + 2z = 2,$
$5x - 2y + z - 2w = 7.$

20. $x + 2y + 4z + 5w = 0,$
$5x - 2y - 2z + 6w = 0,$
$6x + 4y + 2z - w = 0.$

CHAPTER XVI

COMPLEX NUMBERS

111. ***Introduction.*** In Chap. III we found that the system of real numbers was inadequate for our purposes because it contained no number whose square is a negative number. We then invented a new symbol i and adjoined it to our number system. It was to be regarded as a new number—a number that could be used along with the real numbers in the operations of addition, subtraction, multiplication, and division, subject to the rules that would apply if it represented a real number, and subject to the one additional rule that i^2 is equivalent to -1.

We called the "generalized" number $a + bi$, where a and b are any real numbers, a *complex number*. We called a the *real part* and b the *imaginary part*, and we saw that the real number can be regarded as the special case of this complex number that results if $b = 0$. We agreed to call the complex number an *imaginary number* if $b \neq 0$, and to call it a *pure imaginary number* if $a = 0$ and $b \neq 0$. The symbol i is often called the *imaginary unit*.

We shall say that a complex number is in *standard form* when it is written in the form $a + bi$ where a and b are real.

Example

$(6/i) - 2i^2$ is a complex number. It is equivalent to $2 - 6i$, and this is the standard form.

When we use the symbol $a + bi$ to denote a complex number, it is to be understood that a and b are real so that the number is in standard form. The standard form is often called the *rectangular form*, for reasons that will appear presently.

112. ***Graphical representation of complex numbers.*** We have seen that it is possible to set up a one-to-one correspondence between the real numbers and the points of a directed line. If we take as our line the x-axis, for example, then corresponding to any given real number there is precisely one point on the line, and conversely.

We seek now a corresponding graphical representation of the complex numbers, and for this purpose we proceed as follows:

Draw two mutually perpendicular lines that intersect at a point O which is called the origin. Call one of these lines, say the "horizontal" one, the *axis of real numbers*, and the other one the *axis of pure imaginary numbers*. Let the positive direction on the first be to the right and that on the second be upward. Choose a unit on each axis, and mark a scale along it, as shown in Fig. 30.

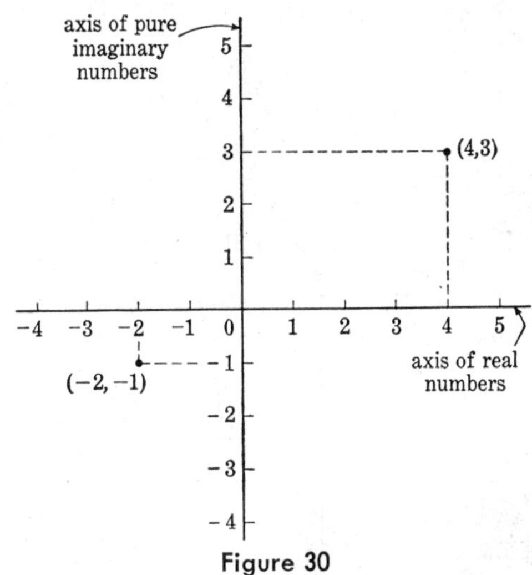

Figure 30

Now, corresponding to any given complex number $a + bi$, we have the point (a,b), and conversely. Thus we have a one-to-one correspondence between the complex numbers and the points of a plane.

Examples

Corresponding to the number $4 + 3i$ we have the point $(4,3)$; corresponding to the number $-2 - i$ we have the point $(-2,-1)$. The point that corresponds to any given real number lies on the axis of real numbers, and the point that corresponds to any given pure imaginary number lies on the axis of pure imaginary numbers. The origin corresponds to the number zero, or $0 + 0i$.

The plane determined by the two axes is usually called the *complex plane*.

In Fig. 31 let $A(a,b)$ represent the number $a + bi$, and $B(c,d)$ represent the number $c + di$. We shall show how to find by graphical construction the point that represents their sum: Through B draw a line parallel to OA and through A draw a line parallel to OB. These lines intersect at a point S which is the fourth vertex of the parallelogram having the origin and A and B as three vertices. It is easy to show that the coordinates of S

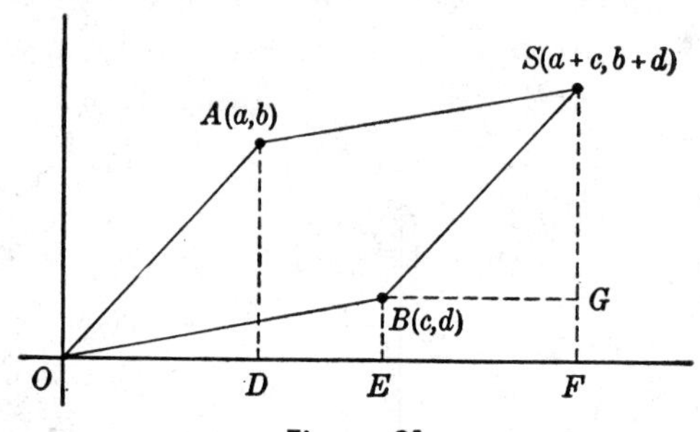

Figure 31

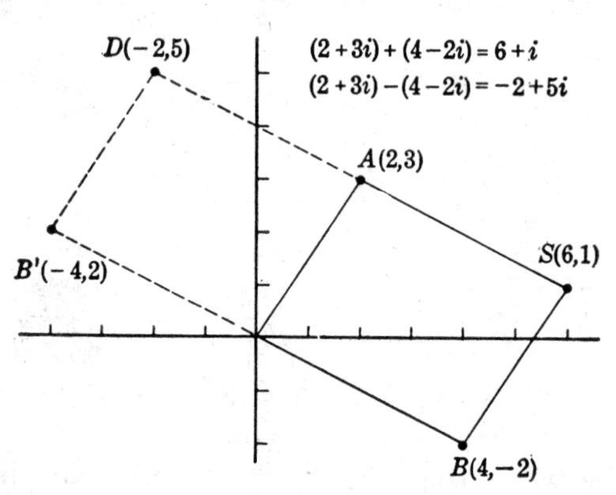

Figure 32

are $(a + c,\ b + d)$ so that S represents the sum of the given numbers. Thus, since the triangles ODA and BGS are congruent, $BG = OD = a$ so that $EF = a$. Then

$$OF = OE + EF = c + a.$$

Similarly, $FG = EB = d$ and $GS = DA = b$;

hence $FS = FG + GS = d + b.$

In order to subtract $c + di$ from $a + bi$, graphically, one may add $-c - di$ to $a + bi$ in the above manner.

Example

The point $A(2,3)$ in Fig. 32 represents the number $2 + 3i$ and $B(4,-2)$ represents $4 - 2i$. Their sum is represented by $S(6,1)$. In order to find the difference

$$(2 + 3i) - (4 - 2i)$$

we plot the point $B'(-4,2)$ which represents the number $-4 + 2i$ and then add as before. We thus get the point $D(-2,5)$; *i.e.*,

$$(2 + 3i) - (4 - 2i) = -2 + 5i.$$

It is often convenient to represent the complex number $a + bi$ by the directed line segment drawn from the origin to the point (a,b) as in Fig. 33, instead of by the point (a,b) itself. This directed segment is called a *vector*, and it is usually drawn with an arrowhead to indicate the positive direction.

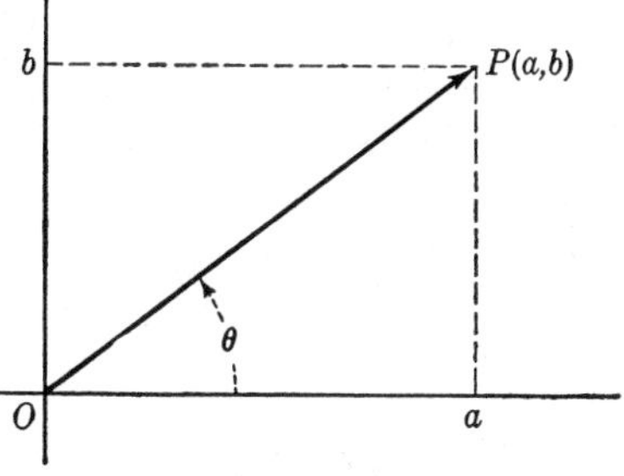

Figure 33

The *absolute value* or *modulus* of the complex number $a + bi$ is defined as the positive number $\sqrt{a^2 + b^2}$:

$$|a + bi| = \sqrt{a^2 + b^2}.$$

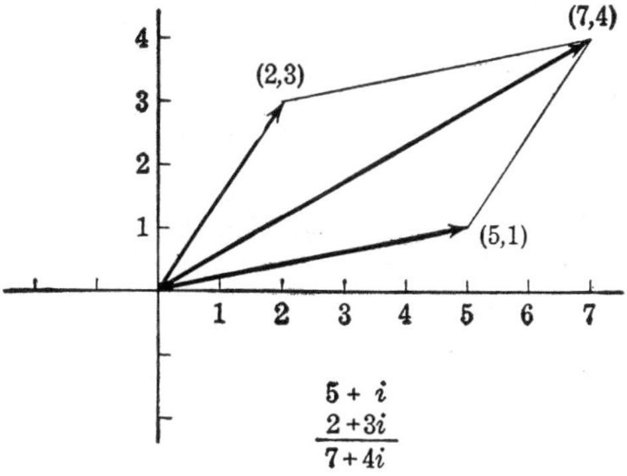

$$\begin{array}{r} 5 + i \\ 2 + 3i \\ \hline 7 + 4i \end{array}$$

Figure 34

This is equal to the length of the vector OP in Fig. 33. The angle θ that OP makes with the positive end of the axis of real numbers is called the *amplitude* or sometimes the *argument* of the complex number.

In Fig. 34 we have represented the numbers $5 + i$ and $2 + 3i$ by vectors drawn from the origin to (5,1) and (2,3), respectively. If we complete the parallelogram having these vectors as two adjacent sides, the vector drawn from the origin to the fourth vertex represents the sum.

The special case in which the vectors representing the two given numbers lie along the same straight line is left to the student.

PROBLEMS

In each of the following problems perform the indicated operations and express the result in the form $a + bi$:

1. $4i(2 + 5i)$.

2. $1 + 2i + i(i^3 - 1)$.

3. $3 + i(3 + i)(3 - i)$.

4. $(1 + i)(1 - i)(2i + 1)$.

5. $(i^2 + 2)(i^2 - 2) + 2i + 3$.

6. $i(i^3 + 1)(1 + i)$.

7. $\frac{4}{i} - \frac{3}{i^2} + i$.

8. $\frac{3}{i^3} - \frac{4}{i^2} + \frac{6}{i}$.

9. $\left(\frac{3}{i} + 1\right)\left(\frac{1 + i}{i}\right)$.

10. $\left(\frac{1}{i} + i^2\right)\left(\frac{i^3 - 1}{i}\right)$.

In each of the following problems represent each of the given complex numbers as a point in the complex plane and then find by graphical construction the point that represents their sum:

11. $4 + i;\ 1 + 3i$.

12. $5 - 2i;\ 1 + 3i$.

13. $-1 + 3i;\ 2 - i$.

14. $-8 + i;\ -3 - 3i$.

15. $-3 + 2i;\ -3 - 2i$.

16. $4 + 2i;\ -4 + 2i$.

17. $5 + 0i;\ 0 + 2i$.

18. $-3 + 0i;\ 0 - 4i$.

19. $2 + i;\ 4 + 2i$.

20. $-1 - i;\ -2 - 2i$.

21. $6 + 2i;\ -3 - i$.

22. $-4 + 6i;\ 2 - 3i$.

In each of the following cases represent each of the given complex numbers as a vector and then find by graphical construction the vector that represents their sum. When three or more numbers are given, find the sum of two of them, add this to the next one, and so on:

23. $5 + i;\ -1 + 3i$.

24. $-2 + 6i;\ -5 + i$.

25. $-3 - i;\ 2 - 3i$.

26. $-4 + 0i;\ 1 + 3i$.

27. $0 - 4i;\ -2 + 2i$.

28. $-4 + 0i;\ 0 + 3i$.

29. $2 + 6i;\ 1 + 3i$.

30. $4 - 2i;\ -2 + i$.

31. $5 + i;\ -2 + 3i;\ -1 + i$.

32. $-6 + 3i;\ -2 - 6i;\ 4 + 3i$.

33. $7 + 2i;\ 1 + 6i;\ -2 - 4i$.

34. $-5 - i;\ -2 + 3i;\ 7 + i$.

35. $6 + 0i;\ 0 + 5i;\ -3 - 3i$.

36. $5 + i;\ 1 + 3i;\ -4 + 2i;\ -2 - 4i$.

37. $8 + 2i;\ 2 - 4i;\ -3 - i;\ -2 + 3i$.

38. $-6 + i;\ 2 + 3i;\ -2 + 3i;\ 3 + i$.

39. Find the absolute value of each of the following complex numbers: $3 + 4i$; 6; $5 - 12i$; $-4 + 4i$; $-7i$.

40. Find the absolute value of each of the following complex numbers: $-3 + 2i$; $-6 - 8i$; -5; $2 - 2i$; $-6i$.

41. $\left|\frac{1}{i} + i^2\right| = ?$

42. $\left|(1 + 2i)\left(\frac{3}{i} - 1\right)\right| = ?$

43. $|(i^3 + 1)(i^3 - 1)| = ?$

44. $\left|\left(\frac{3 + i}{i}\right)\left(\frac{1}{i} - 1\right)\right| = ?$

In each of the following problems carry out the indicated operations and express the result in the form $a + bi$:

45. $\dfrac{3 + 5i}{1 + i}$.

46. $\dfrac{2 - 7i}{3 + i}$.

47. $i\left(\dfrac{4 + 7i}{2 + i}\right)$.

48. $\dfrac{2(19 - 4i)}{5i - 1}$.

49. $\dfrac{(3 - 5i)(1 + i)}{(2 + i)(-1 + 3i)}$.

50. $\dfrac{(5 - 3i)(7 + i)}{(-1 + i)(-1 - i)}$.

51. $\dfrac{(-5 - 7i)(-3 + 5i)}{(1 + i)^2}$.

52. $\dfrac{(1 + i)^3(1 - i)}{(2 + i)(2 - i)}$.

53. $\dfrac{1 + 3i}{50} + \dfrac{1 + i}{(1 - 3i)(3 + i)}$.

54. $\dfrac{2 + 6i}{(1 + i)^2} + \dfrac{25i}{3 + 4i}$.

113. ***Review of the principal trigonometric functions.*** In the remainder of this chapter we must assume that the student has studied trigonometry, but before proceeding, we shall review briefly the definitions of the sine, cosine, and tangent of a general angle.

Recall at first that the angle θ in Fig. 35 may be regarded as being generated by the rotation of OP about O from an initial position of coincidence with the positive end of the x-axis. The initial position of the rotating line is called the *initial side* of the angle, and the terminal position is called the *terminal side*. If the rotation is counterclockwise, the angle is regarded as positive; if it is clockwise, the angle is negative. We shall measure the angle in degrees, one right angle being 90 degrees (90°), one complete revolution being 360°, etc.

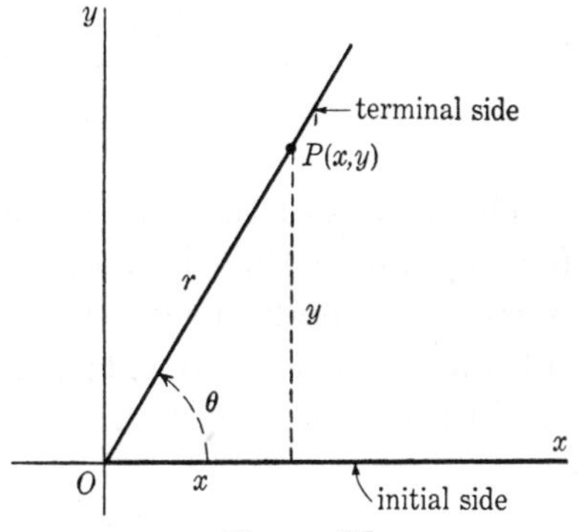

Figure 35

An angle is said to be in *standard position* with respect to the coordinate system when it is placed with its vertex at the origin and its initial side along the positive end of the x-axis, as shown in Fig. 35.

The definitions of the sine, cosine, and tangent of an angle θ (which may be positive or negative and of any number of degrees whatever) are as follows: With the angle in standard position, select any point P on the terminal side; this point has a pair of coordinates (x,y). Let r, where $r = \sqrt{x^2 + y^2}$, denote the distance of P from the origin. Then (Fig. 35),

$$\textbf{sine } \theta = \frac{y}{r};$$

$$\textbf{cosine } \theta = \frac{x}{r};$$

$$\textbf{tangent } \theta = \frac{y}{x}.$$

In the interest of brevity, we shall adopt the abbreviations sin θ, cos θ, and tan θ.

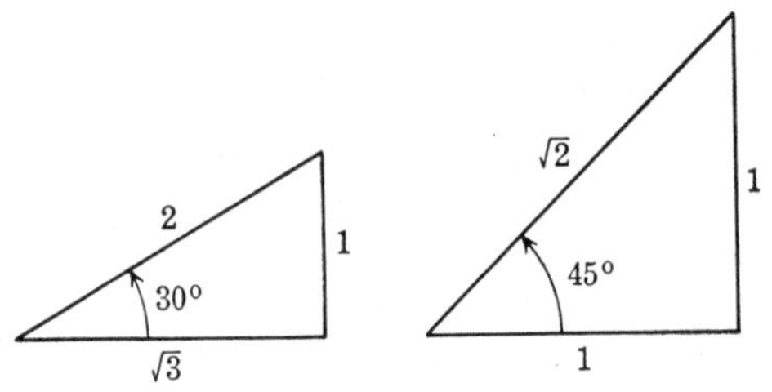

Figure 36

The values of these trigonometric functions can be found easily for any angle that is a multiple of either 30° or 45°. For this purpose we make use of the fact that in a 45° right triangle the sides are in the ratio $1{:}1{:}\sqrt{2}$, and in a 30-60° right triangle they are in the ratio $1{:}\sqrt{3}{:}2$. The situation is illustrated in Fig. 36.

Examples

Referring to Fig. 37 we can write down the sine, cosine, and tangent of 135° and 210° as follows:

$$\sin 135° = \frac{1}{\sqrt{2}} = 0.707; \qquad \sin 210° = \frac{-1}{2} = -0.500;$$

$$\cos 135° = \frac{-1}{\sqrt{2}} = -0.707; \qquad \cos 210° = \frac{-\sqrt{3}}{2} = -0.866;$$

$$\tan 135° = \frac{1}{-1} = -1.000. \qquad \tan 210° = \frac{-1}{-\sqrt{3}} = 0.577.$$

If θ is any of the quadrantal angles 0°, 90°, 180°, 270°, we have either $x = 0$ or $y = 0$, and the other is equal to r or $-r$. In order to write down the values of the trigonometric functions, we

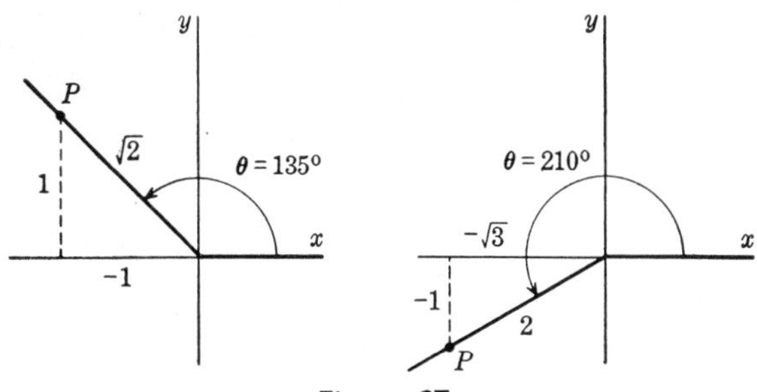

Figure 37

may therefore choose P so that its coordinates are (1,0), (0,1), (−1,0), or (0,−1), as the case may be, and we then have $r = 1$. Thus for 270° (Fig. 38) we have,

$$\sin 270° = \frac{-1}{1} = -1;$$

$$\cos 270° = \frac{0}{1} = 0;$$

$$\tan 270° = \frac{-1}{0} \text{ (undefined).}$$

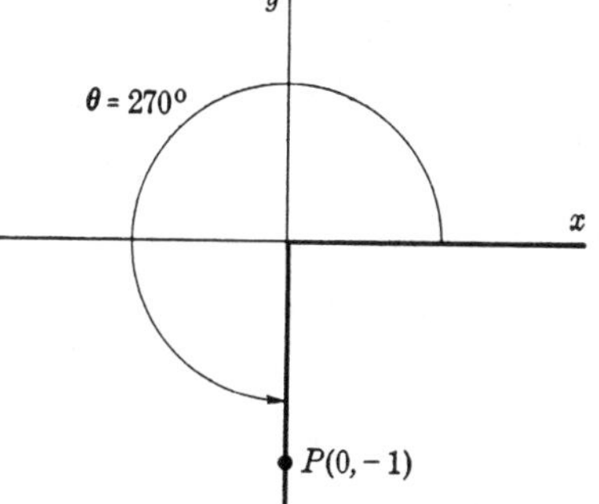

Figure 38

Our definition gives no value for tan 270° because we cannot divide by zero.

It should be observed that the values of the trigonometric functions of θ are equal to those of $(\theta + k \cdot 360°)$ where k is any integer. For example, the values of the functions of 390°, 750°, and −330°, are the same as those of 30°. When these angles are placed in standard position, their initial and terminal sides coincide.

Let r (a positive number) be the distance from the origin to the point $P(x,y)$, and let θ be the angle that OP makes with the positive direction on the x-axis (Fig. 39). It follows from our definitions of $\sin \theta$ and $\cos \theta$ that

$$x = r \cos \theta; \qquad y = r \sin \theta.$$

114. ***The polar or trigonometric form of a complex number.*** Consider the complex number $x + iy$, where x and y denote real numbers. Let $r = \sqrt{x^2 + y^2}$ be its absolute value or modulus, and let θ be its amplitude. The situation is then represented by Fig. 39 in which the point $P(x,y)$ represents the given number. Now since $x = r \cos \theta$ and $y = r \sin \theta$,

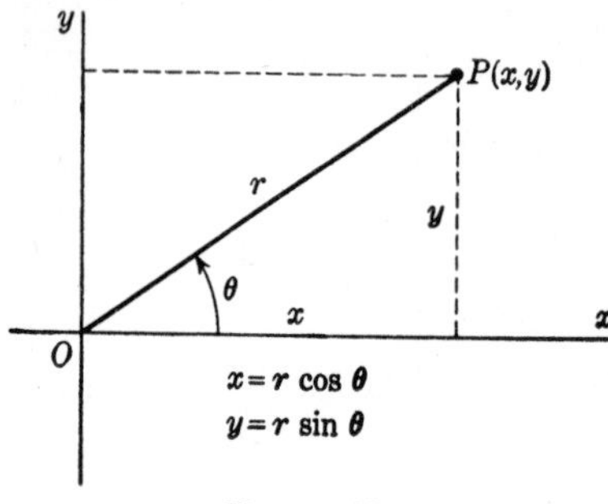

Figure 39

$$x + iy = r(\cos \theta + i \sin \theta).*$$

The form on the right is called the *polar* or *trigonometric* form of the number whose standard or rectangular form is $x + iy$.

We may replace θ by $\theta + k \cdot 360°$, where k is any integer, and thus obtain what is called the *complete polar form:*

$$x + iy = r[\cos (\theta + k \cdot 360°) + i \sin (\theta + k \cdot 360°)].$$

115. ***Multiplication and division of complex numbers in polar form.*** Let

$$r_1(\cos \alpha + i \sin \alpha) \qquad \text{and} \qquad r_2(\cos \beta + i \sin \beta)$$

be two complex numbers in polar form. By actual multiplication we find that their product is

$$\begin{aligned} &r_1r_2[\cos \alpha \cos \beta + i \cos \alpha \sin \beta + i \sin \alpha \cos \beta + i^2 \sin \alpha \sin \beta] \\ &= r_1r_2[(\cos \alpha \cos \beta - \sin \alpha \sin \beta) + i(\sin \alpha \cos \beta + \cos \alpha \sin \beta)] \\ &\qquad = r_1r_2[\cos (\alpha + \beta) + i \sin (\alpha + \beta)].\dagger \end{aligned}$$

Hence, *the product of two complex numbers is a complex number whose modulus is equal to the product of their moduli and whose amplitude is equal to the sum of their amplitudes.*

It can be proved in a similar fashion that the quotient

$$\frac{r_1 (\cos \alpha + i \sin \alpha)}{r_2 (\cos \beta + i \sin \beta)}$$

* The abbreviation r cis θ is often used to mean $r(\cos \theta + i \sin \theta)$.

† Recall from trigonometry that

$$\sin (\alpha + \beta) = \sin \alpha \cos \beta + \cos \alpha \sin \beta;$$
$$\cos (\alpha + \beta) = \cos \alpha \cos \beta - \sin \alpha \sin \beta.$$

is equal to

$$\frac{r_1}{r_2} [\cos (\alpha - \beta) + i \sin (\alpha - \beta)].$$

Thus the modulus of the quotient is equal to the modulus of the numerator divided by the modulus of the denominator; the amplitude of the quotient is equal to the amplitude of the numerator minus the amplitude of the denominator. The proof will be left as an exercise for the student.

Example

The numbers -2 and $1 + i$ have the following polar forms:

$$\begin{aligned} -2 &= 2(\cos 180° + i \sin 180°), \\ 1 + i &= \sqrt{2}(\cos 45° + i \sin 45°). \end{aligned}$$

In order to divide the first by the second, we have

$$\begin{aligned} \frac{-2}{1+i} &= \frac{2}{\sqrt{2}} [\cos (180° - 45°) + i \sin (180° - 45°)] \\ &= \sqrt{2} (\cos 135° + i \sin 135°) \\ &= \sqrt{2}\left(-\frac{1}{\sqrt{2}} + i\frac{1}{\sqrt{2}}\right) \\ &= -1 + i. \end{aligned}$$

PROBLEMS

Find, without using tables, the values of the sine, cosine, and tangent of each of the following angles:

1. 30°; 90°; 225°.
2. 45°; 150°; 180°.
3. 60°; 315°; 270°.
4. 120°; 390°; 135°.
5. 450°; 240°; 495°.
6. −60°; −360°; 690°.
7. 540°; −270°; −315°.
8. −630°; −225°; 210°.
9. 585°; −780°; 0°.
10. 840°; −675°; −810°.

Find, using tables, the values of the sine, cosine, and tangent of each of the following angles:

11. 20°; 35°; 70°.
12. 45°; 50°; 85°.
13. 64°; 205°; 380°.
14. 28°; 105°; 186°.
15. 82°; −130°; 124°.
16. 57°; −38°; 256°.
17. 76°; −306°; −20°.
18. 27°; 714°; −112°.
19. 86°; −138°; 460°.
20. 75°; −110°; 260°.

Plot the point representing each of the following numbers, and write each number in polar form using the value of θ that lies in the interval $0° \leqq \theta < 360°$:

21. $2 + 2i$; 8; $6i$.
22. $-1 - i$; -4; $-2i$.
23. $-4 + 4i$; -3; $3i$.
24. $5 - 5i$; 16; $-8i$.
25. $2\sqrt{2} - 2\sqrt{2}\,i$; $\sqrt{3} + i$; -1.
26. $2\sqrt{3} - 2i$; $1 + i$; $-3i$.
27. $3 - 3\sqrt{3}\,i$; $4 + 4i$; 32.
28. $-2 + 2\sqrt{3}\,i$; $-3 + 3i$; $-i$.

Plot the point representing each of the following numbers, and in each case write the number in the complete polar form:

29. $6i$; $1 + i$; -8.
30. $-2 + 2i$; 16; $-i$.
31. $-\sqrt{2} + \sqrt{2}\,i$; -1; $-4i$.
32. $\sqrt{3} + i$; $8i$; -32.
33. $3 + 3i\sqrt{3}$; -64; $12i$.
34. $-6 - 6i$; $-\sqrt{3} + i$; $4i$.

Write each of the following numbers in standard form and plot the corresponding point:

35. $4(\cos 45° + i \sin 45°)$; $3(\cos 90° + i \sin 90°)$.
36. $6(\cos 30° + i \sin 30°)$; $2(\cos 0° + i \sin 0°)$.
37. $8(\cos 150° + i \sin 150°)$; $2(\cos 180° + i \sin 180°)$.
38. $12(\cos 135° + i \sin 135°)$; $4(\cos 270° + i \sin 270°)$.
39. $\cos 225° + i \sin 225°$; $7(\cos 360° + i \sin 360°)$.
40. $2\sqrt{2}\,(\cos 315° + i \sin 315°)$; $8(\cos 120° + i \sin 120°)$.

In each of the following problems perform the indicated multiplication or division, using the polar form, and then reduce the answer to rectangular form, using tables if necessary. Where convenient, check by performing the same operation using the rectangular form:

41. $3(\cos 90° + i \sin 90°) \cdot 2(\cos 45° + i \sin 45°)$.
42. $12(\cos 45° + i \sin 45°) \cdot 2(\cos 135° + i \sin 135°)$.
43. $5(\cos 110° + i \sin 110°) \cdot 2(\cos 70° + i \sin 70°)$.
44. $4(\cos 165° + i \sin 165°) \cdot 3(\cos 105° + i \sin 105°)$.
45. $3(\cos 135° + i \sin 135°) \cdot 4(\cos 225° + i \sin 225°)$.
46. $5(\cos 55° + i \sin 55°) \cdot 2(\cos 70° + i \sin 70°)$.
47. $3(\cos 27° + i \sin 27°) \cdot 8(\cos 17° + i \sin 17°)$.
48. $8(\cos 90° + i \sin 90°) \div 2(\cos 0° + i \sin 0°)$.
49. $12(\cos 225° + i \sin 225°) \div 3(\cos 135° + i \sin 135°)$.
50. $10(\cos 60° + i \sin 60°) \div 2(\cos 40° + i \sin 40°)$.
51. $6(\cos 315° + i \sin 315°) \div (\cos 45° + i \sin 45°)$.
52. $\sqrt{2}\,(\cos 155° + i \sin 155°) \div 2(\cos 20° + i \sin 20°)$.
53. $\sqrt{2}\,(\cos 35° + i \sin 35°) \div 10(\cos 170° + i \sin 170°)$.
54. $4(\cos 127° + i \sin 127°) \div 2(\cos 54° + i \sin 54°)$.

55. Prove the theorem on page 272 concerning the quotient of two complex numbers.

56. Show that

$$\frac{1}{r(\cos\theta + i \sin\theta)} = \frac{1}{r}(\cos\theta - i \sin\theta).$$

116. ***Powers and roots of complex numbers. De Moivre's theorem.*** If we use the above method of multiplication to multiply the complex number $r(\cos\theta + i\sin\theta)$ by itself, we obtain the following result:

$$[r(\cos\theta + i\sin\theta)]^2 = r^2(\cos 2\theta + i\sin 2\theta).$$

If we again multiply by $r(\cos\theta + i\sin\theta)$, we get

$$[r(\cos\theta + i\sin\theta)]^3 = r^3(\cos 3\theta + i\sin 3\theta),$$

and so on. The student may readily prove, by means of mathematical induction, that if n is any positive integer, then

$$[r(\cos\theta + i\sin\theta)]^n = r^n(\cos n\theta + i\sin n\theta).$$

This is *De Moivre's theorem* (after Abraham De Moivre, 1667–1754).

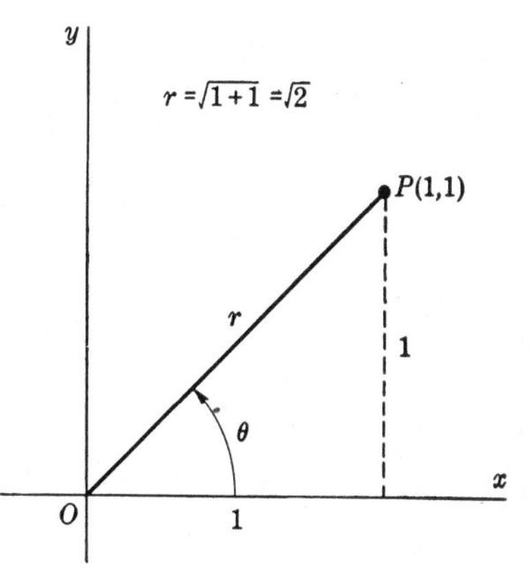

Figure 40

***Example* 1**

Evaluate $(1 + i)^{10}$.

Solution

As it would be tedious to carry out the multiplication in the usual way, we use De Moivre's theorem. We first plot the point representing $1 + i$ and observe (Fig. 40) that $r = \sqrt{2}$ and $\theta = 45°$. Thus

$$1 + i = \sqrt{2}\,(\cos 45° + i\sin 45°).$$

Then
$$\begin{aligned}(1 + i)^{10} &= (\sqrt{2})^{10}(\cos 450° + i\sin 450°)\\ &= 2^5(\cos 90° + i\sin 90°)\\ &= 32(0 + i) = 32i.\end{aligned}$$

***Example* 2**

Show that $-\frac{1}{2} + \frac{1}{2}\sqrt{3}\,i$ is a cube root of 1.

Solution

We must show that the cube of $-\frac{1}{2} + \frac{1}{2}\sqrt{3}\,i$ is 1, and this is most easily done by the use of De Moivre's theorem. In this case (Fig. 41) $r = 1$ and $\theta = 120°$, so that

$$-\tfrac{1}{2} + \tfrac{1}{2}\sqrt{3}\,i = 1(\cos 120^\circ + i \sin 120^\circ).$$

Then
$$\left(-\tfrac{1}{2} + \tfrac{1}{2}\sqrt{3}\,i\right)^3 = 1^3(\cos 360^\circ + i \sin 360^\circ)$$
$$= 1(1 + 0i) = 1.$$

The following examples show how De Moivre's theorem can be used to find the nth roots of a complex number.

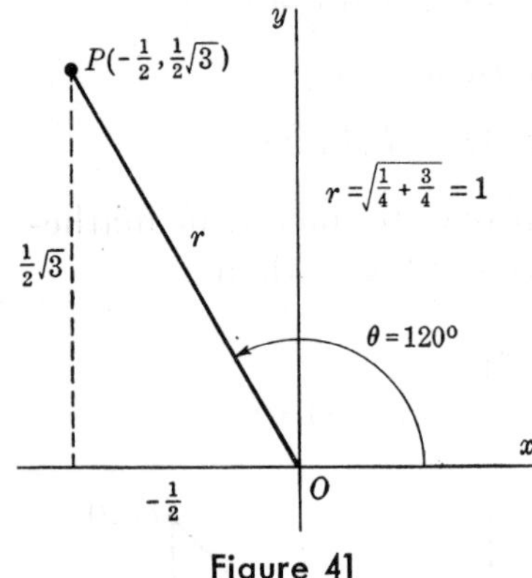

Figure 41

Example 1

Find the cube roots of $8i$.

Solution

We first write the given number in polar form (Fig. 42):

$$8i = 8(\cos 90^\circ + i \sin 90^\circ).$$

We now seek a number

$$r(\cos \theta + i \sin \theta)$$

whose cube is equal to this given number, and since the cube is $r^3(\cos 3\theta + i \sin 3\theta)$, we must determine the values of r and θ for which

$$(1) \quad r^3(\cos 3\theta + i \sin 3\theta) = 8(\cos 90^\circ + i \sin 90^\circ).$$

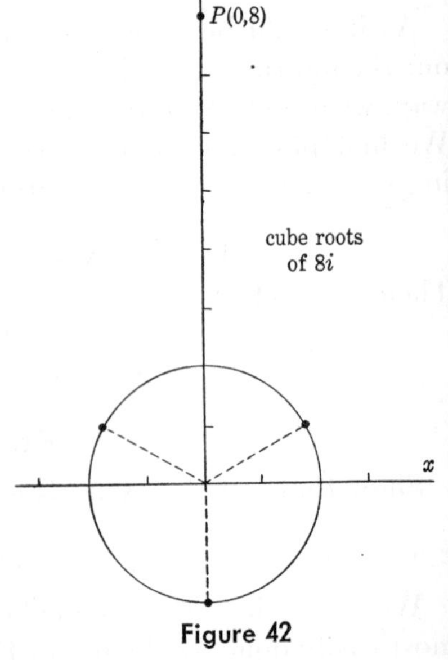

Figure 42

Each side of (1) is a complex number, and two complex numbers are equal if and only if their moduli are equal and their amplitudes are either equal or differ by an integral multiple of 360°. Hence (1) is true if and only if

$$r^3 = 8$$

and

$$3\theta = 90^\circ + k \cdot 360^\circ,$$

where k is an integer (positive, negative, or zero). Since r, in the number $r(\cos \theta + i \sin \theta)$ that we are seeking, must be a positive real number, we must have $r = 2$, and $\theta = 30^\circ + k \cdot 120^\circ$. The cube roots are then given by the expression

$$2 \cos [(30° + k \cdot 120°) + i \sin (30° + k \cdot 120°)].$$

We have the following cases:

$$\begin{aligned} &\textit{For } \mathbf{k} = 0: \quad 2(\cos 30° + i \sin 30°) = \sqrt{3} + i. \\ &\textit{For } \mathbf{k} = 1: \quad 2(\cos 150° + i \sin 150°) = -\sqrt{3} + i. \\ &\textit{For } \mathbf{k} = 2: \quad 2(\cos 270° + i \sin 270°) = 0 - 2i. \end{aligned}$$

For all other integral values of k, we get a repetition of one of these three numbers since increasing (or decreasing) k by a multiple of 3 simply adds (or subtracts) a multiple of 360° to the angle. Hence the number $8i$ has the three cube roots

$$\sqrt{3} + i, \qquad -\sqrt{3} + i, \qquad -2i,$$

and no others.

The geometrical interpretation of the result is shown in Fig. 42. The given number, $8i$, is represented by the point $P(0,8)$. All the cube roots have the modulus $\sqrt[3]{8}$ or 2, and consequently they all lie on the circle of radius 2. The amplitude of the one corresponding to $k = 0$ is equal to one-third of the amplitude of the given number $8i$; *i.e.*, its amplitude is 30°. The roots corresponding to $k = 1$ and $k = 2$ are formed by adding 120° and 2(120°), respectively, to the amplitude of this one. Consequently the three cube roots are represented by three points that are spaced at intervals of 120° on this circle.

Example 2

Find the fourth roots of $2 + 2\sqrt{3}\, i$.

Solution

Write the given number in polar form:

$$2 + 2\sqrt{3}\, i = 4(\cos 60° + i \sin 60°).$$

We then seek a number $r(\cos \theta + i \sin \theta)$ such that

$$r^4(\cos 4\theta + i \sin 4\theta) = 4(\cos 60° + i \sin 60°).$$

We must have

$$r^4 = 4 \quad \text{and} \quad 4\theta = 60° + k \cdot 360°,$$

or

$$r = \sqrt{2} \quad \text{and} \quad \theta = 15° + k \cdot 90°.$$

The required roots are then given by

$$\sqrt{2}\,[\cos(15° + k \cdot 90°) + i \sin(15° + k \cdot 90°)].$$

Corresponding to $k = 0, 1, 2, 3$, we have the following roots:

For $\mathbf{k} = 0$: $\sqrt{2}\,(\cos 15° + i \sin 15°)$.
For $\mathbf{k} = 1$: $\sqrt{2}\,(\cos 105° + i \sin 105°)$.
For $\mathbf{k} = 2$: $\sqrt{2}\,(\cos 195° + i \sin 195°)$.
For $\mathbf{k} = 3$: $\sqrt{2}\,(\cos 285° + i \sin 285°)$.

Other integral values of k yield only repetitions of these so that these are the four distinct fourth roots. By using tables, one may reduce each of them to the standard form $a + bi$. For example,

$$\sqrt{2}\,(\cos 15° + i \sin 15°) = 1.4142(0.96593 + 0.25882i) = 1.366 + 0.3660i,$$

correct to four significant figures.

The student should supply the geometrical interpretation.

The above procedure may be employed to find the roots of an equation of the form

$$x^n = k,$$

where n is a positive integer and k is any complex number.

Example 3

Find all solutions of the equation $x^5 + 32 = 0$.

Solution

The equation is equivalent to $x^5 = -32$, and the required solutions are the five fifth roots of -32. We therefore proceed as follows:

$$-32 = 32(\cos 180° + i \sin 180°).$$

If $r(\cos \theta + i \sin \theta)$ is a fifth root of -32, then

$$r^5(\cos 5\theta + i \sin 5\theta) = 32(\cos 180° + i \sin 180°),$$

from which we have

$$r^5 = 32 \quad \text{and} \quad 5\theta = 180° + k \cdot 360°,$$

or

$$r = 2 \quad \text{and} \quad \theta = 36° + k \cdot 72°.$$

The required solutions are then given by

$$2[\cos (36° + k \cdot 72°) + i \sin (36° + k \cdot 72°)],$$

where k is an integer. Corresponding to $k = 0, 1, 2, 3, 4$, we have the five distinct solutions:

For $k = 0$: $2(\cos 36° + i \sin 36°)$.
For $k = 1$: $2(\cos 108° + i \sin 108°)$.
For $k = 2$: $2(\cos 180° + i \sin 180°)$.
For $k = 3$: $2(\cos 252° + i \sin 252°)$.
For $k = 4$: $2(\cos 324° + i \sin 324°)$.

Observe that the one corresponding to $k = 2$ is equal to -2, and that this is the one real root of the given equation. The others form two pairs of conjugate imaginary numbers. Other values of k yield only repetitions of these.

PROBLEMS

In each of the following problems carry out the indicated operation using De Moivre's theorem. Express the result in the form $a + bi$:

1. $(-\frac{1}{2} - \frac{1}{2}\sqrt{3}\, i)^3$.
2. $(2 + 2i)^6$.
3. $(\sqrt{3} - i)^3$.
4. $(1 - \sqrt{3}\, i)^6$.
5. $(5 - 5i)^4$.
6. $(-\sqrt{2} + \sqrt{2}\, i)^{10}$.
7. $[2(\cos 5° + i \sin 5°)]^8$.
8. $(\cos 12° + i \sin 12°)^{10}$.

In each of the following problems find all the indicated roots of the given number and show the graphical interpretation:

9. The square roots of $-4i$.
10. The square roots of $4 + 4\sqrt{3}\, i$.
11. The cube roots of i.
12. The cube roots of -27.
13. The cube roots of 1.
14. The cube roots of $-3\sqrt{3} + 3i$.
15. The fourth roots of $16i$.
16. The fourth roots of $-8 + 8\sqrt{3}\, i$.
17. The fourth roots of -36.
18. The fifth roots of 32.
19. The sixth roots of 64.
20. The sixth roots of -1.

Find all solutions for each of the following equations:

21. $x^4 + 1 = 0$.
22. $x^4 - 16 = 0$.
23. $x^3 + 8i = 0$.
24. $x^3 + 1 = 0$.
25. $x^5 + 1 = 0$.
26. $x^6 + 64 = 0$.
27. $x^2 + 2 = 2\sqrt{3}\, i$.
28. $x^4 + 8 = 8\sqrt{3}\, i$.

29. Write out the proof of De Moivre's theorem, using mathematical induction.

CHAPTER XVII

PERMUTATIONS AND COMBINATIONS. PROBABILITY

117. ***Introduction.*** Suppose that there are five different highways connecting towns A and B. We may designate these as roads ①, ②, ③, ④, and ⑤, respectively. Suppose that there are three different highways connecting towns B and C. These may be designated as roads x, y, and z, respectively. By how many different routes could one drive from A to C via B?

The answer can be obtained in the following way: We may choose road ① for the trip from A to B, and with this choice we may select any one of the roads x, y, or z, for the trip from B to C. There are thus three possible routes employing road ①. Similarly there are three using road ②, three using road ③, etc. Altogether there are then

$$3 + 3 + 3 + 3 + 3 = 5 \cdot 3 = 15$$

different routes.

This is an illustration of the following principle which will play an important part in the work of this chapter.

Fundamental principle. *If a thing can be done in any one of* m *different ways and if, after it has been done, a second thing can be done in any one of* n *different ways, then the two things can be done in this order in* mn *different ways.*

The extension of this principle to the case of more than two things is obvious.

***Example* 1**

How many positive even integers, each having three digits, can be formed from the digits 1, 2, 3, 4, 5?

Solution

We can use any one of the five given digits in the hundreds place, so this place can be filled in any one of five different ways. Also, since it is not required that the digits be different, the tens place can be filled in any one of five ways. The digit in the units place must be either 2 or 4 since the number is to be even, so this place can be filled in only two ways. We have then

$$5 \cdot 5 \cdot 2 = 50$$

numbers of the specified kind.

***Example* 2**

How many positive integers less than 300 can be formed from the digits 1, 2, 3, 4, 5, if repetition of digits in a number is not allowed?

Solution

The three-digit numbers satisfying the given requirements must have either 1 or 2 in the hundreds place, hence this place can be filled in two ways. After the hundreds place has been filled in either of these ways, the tens place may be filled by any of the remaining four digits; and after this place has been filled, the units place may be filled by any one of the remaining three digits. We therefore have

$$2 \cdot 4 \cdot 3 = 24$$

numbers of three digits satisfying all the requirements. A two-digit number will satisfy the given conditions if it has one of the five given digits in the tens place, and any one of the remaining four digits in the units place. There are then

$$5 \cdot 4 = 20$$

such numbers. Finally, we must add the five one-digit numbers that satisfy the specified conditions. The total is then

$$24 + 20 + 5 = 49.$$

If repetition of digits is allowed we have

$$2 \cdot 5 \cdot 5 + 5 \cdot 5 + 5 = 80$$

numbers. Here the first, second, and third terms on the left represent the number of three-digit, two-digit, and one-digit numbers, respectively.

118. ***Permutations.*** Each arrangement of a given set of objects in some order along a straight line is called a *permutation* of these objects. Thus the permutations of the letters a, b, and c, are

$$abc,\ acb,\ bac,\ bca,\ cab,\ cba.$$

More generally, if we have a set consisting of n distinct objects, any arrangement of r of them ($r \leqq n$) in some order on a straight line is called *a permutation of the n objects taken r at a time.* Thus the permutations of the letters a, b, and c, taken two at a time are

$$ab,\ ac,\ ba,\ bc,\ ca,\ cb.$$

The number of permutations of n distinct objects taken r at a time may be denoted by the symbol ${}_nP_r$. This number is the same as the number of ways in which r places can be filled from n distinct objects, and can of course be computed by means of the fundamental principle of the preceding section. The result is embodied in the following:

Theorem. *The number of permutations of* **n** *distinct objects taken* **r** *at a time is*

$$\mathbf{{}_nP_r = n(n-1)(n-2)\cdots(n-r+1).}$$

Proof: The first place can be filled by any one of the n objects, hence in any one of n ways. After the first place has been filled in one of these ways, the second place can be filled by any one of the remaining $n - 1$ objects, and hence in $n - 1$ ways. Then the third place can be filled by any of the remaining $n - 2$ objects, and so on. The rth place can be filled by any one of the remaining $n - (r - 1)$ objects, and hence in any one of $n - r + 1$ ways. Consequently, by the fundamental principle,

$${}_nP_r = n(n-1)(n-2)\cdots(n-r+1).$$

Example

The number of three-digit integers that can be formed from the nine digits $1, 2, \cdots, 9$, if repetition of a digit in a number is not allowed, is

$${}_9P_3 = 9 \cdot 8 \cdot 7 = 504.$$

For the special case in which we have n distinct objects taken n at a time, we have the following:

Corollary. *The number of permutations of **n** distinct objects taken **n** at a time is*

$$_nP_n = n(n-1)(n-2)\cdots 3\cdot 2\cdot 1 = n!$$

The proof is left to the student.

Example

The number of permutations of the letters of the word HISTORY is $7! = 5{,}040$.

119. *Permutations of n objects some of which are alike.* We consider next the problem of finding the number of permutations of n objects taken n at a time if some of the objects are identical. As an example we shall first compute the number of permutations of the letters of the word ELEVEN.

If we assign subscripts to the E's so that they are distinguishable, we have the six distinct letters or objects.

$$E_1, L, E_2, V, E_3, N.$$

The number of permutations of these is $6!$, one of which is that just written down. Observe now that with the letters L, V, and N held in their respective positions, the E's could be permuted in $3!$ different ways, and that without the subscripts these $3!$ "different" arrangements would be undistinguishable. The same thing is true of any other arrangement that may be made. It follows that if P is the number of distinct permutations of the letters ELEVEN, without the subscripts, then

$$3!\cdot P = 6!,$$

or

$$P = \frac{6!}{3!}.$$

This is an example of the following:

Theorem. *The number **P** of distinct permutations of **n** objects taken **n** at a time, if **k** of the objects are identical and the others are distinct, is*

$$P = \frac{n!}{k!}.$$

The theorem can readily be extended to more general cases. Thus if k_1 of the objects are identical, k_2 others are identical, and the remaining ones are distinct, then the number of distinct permutations is

$$P = \frac{n!}{k_1!k_2!}.$$

Example

The number of distinct permutations of the letters of the word MISSISSIPPI is

$$P = \frac{11!}{4!4!2!} = 34{,}650.$$

120. ***Circular permutations.*** One sometimes wishes to consider the arrangements of objects on a circle, or other simple closed curve, instead of on a straight line. Consider, for example, the problem of computing the number of different ways in which 6 people can be seated at a round table. We agree in advance that two arrangements are regarded as the same, if one can be obtained from the other by rotation alone, for such arrangements would have all the people in the same positions relative to each other. In order to compute the number of permutations, we regard any one person as fixed in position and compute the number of arrangements of the remaining five relative to this one. The number is of course $5!$ and hence there are $5!$ different ways in which the six people can be seated. For the general case, we have the following:

Theorem. *The number of distinct permutations of* $\mathbf{n}$ *objects taken* $\mathbf{n}$ *at a time on a circle, or other simple closed curve, is* $(\mathbf{n} - 1)!$

In considering arrangements of keys on a ring or beads on a necklace, two arrangements should be regarded as identical if one can be obtained from the other by turning the ring or necklace over. There are then only $\frac{1}{2}(n - 1)!$ distinct arrangements of n keys on a ring ($n \geqq 3$). In particular, three keys can be arranged on a ring in only one way.

PROBLEMS

1. How many different signals can be made from 5 different flags if each signal consists of 3 flags hung in a vertical line?

2. In how many different ways can the four positions of president, vice president, secretary, and treasurer, of a club be filled from a membership of 20 if any one person can hold only one of the offices?

3. If a penny, a nickel, a dime, and a quarter are tossed together, in how many different ways may they fall?

4. How many four-digit numbers greater than 6,000 can be made from the digits 3, 5, 7, 9, if repetition of a digit in a number is not allowed? How many if repetitions are allowed?

5. In how many ways can the batting order of a baseball team of 9 men be arranged if it is specified that the pitcher is to bat last and that a certain group consisting of 4 men must occupy the first four positions in some (unspecified) order?

6. In how many ways can a set consisting of 5 different books on mathematics, 3 on physics, and 2 on chemistry, be arranged on a straight shelf that has space for 10 books if the books on each subject are to be kept together?

7. A telephone dial has 10 holes. How many different signals, each consisting of 6 impulses in succession, can be made if repetition of an impulse in a signal is not allowed? How many if repetitions are allowed?

8. How many odd integers, each consisting of four digits, can be made from the digits 1, 2, 3, 4, 5, 6, if repetition of a digit in a number is not allowed? How many if repetitions are allowed?

9. How many even integers, each consisting of four digits, can be formed from the digits 3, 4, 5, 6, 7, 9, if repetitions are not allowed?

10. How many integers less than 600 can be formed from the digits 1, 3, 5, 7, 9, if repetitions are not allowed?

11. Find the number of permutations of the letters of the word CALENDAR.

12. Find the number of permutations of the letters of the word TALLAHASSEE.

13. Find the number of permutations of the letters of the word COLORADO.

14. In how many ways can 10 beads of different colors be strung on a necklace?

15. In how many ways can 7 different keys be arranged on a key ring?

16. In how many ways can 8 persons be seated at a round table if a certain 2 must be seated together?

17. How many different signals can be made with 6 flags of different colors by displaying them one or more at a time in a vertical line?

18. A signal is made by displaying 12 flags in a vertical line on a flagpole. How many different signals are possible if 4 of the flags are red, 4 are blue, 2 are yellow, and 2 are white?

121. *Combinations.*

A set of r objects selected from a set of n distinct objects is called a *combination* of the n objects taken r at a time. The order in which they are chosen or arranged is immaterial.

Example

There are four different combinations of three letters that can be selected from the four letters a, b, c, d. They are

$$abc,\ abd,\ acd,\ bcd.$$

Taken two at a time there are six different combinations. They are

$$ab, ac, ad, bc, bd, cd.$$

We shall now prove that the number of combinations of n distinct objects taken r at a time, denoted by the symbol ${}_nC_r$, is given by the formula

$${}_nC_r = \frac{n!}{r!(n-r)!}.$$

Proof: The number of permutations of the n objects taken r at a time is

$${}_nP_r = n(n-1)(n-2)\cdots(n-r+1).$$

Each of the combinations that can be formed furnishes $r!$ of these, since the number of permutations of the r objects in each combination is $r!$. It follows that if ${}_nC_r$ is the number of combinations,

$${}_nC_r \cdot r! = n(n-1)(n-2)\cdots(n-r+1),$$

or

$${}_nC_r = \frac{n(n-1)(n-2)\cdots(n-r+1)}{r!}.$$

If we multiply numerator and denominator of the right member of this last equation by $(n-r)!$, we get

$${}_nC_r = \frac{n(n-1)(n-2)\cdots(n-r+1)}{r!}\cdot\frac{(n-r)!}{(n-r)!}.$$

The numerator is equivalent to

$$n(n-1)(n-2)\cdots(n-r+1)\cdot(n-r)(n-r-1)\cdots 3\cdot 2\cdot 1$$

and is thus equal to $n!$. Hence

$${}_nC_r = \frac{n!}{r!(n-r)!}.$$

The student may show, as a corollary, that *the number of combinations of* n *objects taken* r *at a time is equal to the number of combinations of* n *objects taken* $(n-r)$ *at a time.* For example, the number of combinations of 20 objects taken 3 at a time is the same as the number of combinations if they are taken 17 at a time:

$${}_{20}C_3 = {}_{20}C_{17} = \frac{20!}{17!3!} = \frac{20\cdot 19\cdot 18}{3\cdot 2} = 1{,}140.$$

It may be observed that selecting the 3 objects to take from the set of 20 objects is equivalent to selecting the 17 objects that are to be left behind, and obviously we should have ${}_{20}C_3 = {}_{20}C_{17}$.

Example

In how many ways can a set consisting of 3 white balls and 4 black balls be chosen from a box containing 5 white balls and 7 black ones?

Solution

The number of ways in which the three white balls can be chosen is $\dfrac{5!}{3!2!} = 10$. The number of ways in which the 4 black balls can be chosen is $\dfrac{7!}{4!3!} = 35$. The number of ways in which the selection can be made is then $(10)(35) = 350$.

122. *Division of n objects into two or more groups.* Suppose that we have $p + q$ distinct objects and wish to divide them into two groups, one containing p of the objects and the other containing q of them. For every selection of p of the objects for one group there remains q objects for the other. Consequently the number of ways in which the division can be made is equal to the number of ways in which p objects can be selected from a set of $p + q$ objects. This is

$$\frac{(p+q)!}{p!q!}.$$

Example

Five cards consisting of an ace, a king, a queen, a jack, and a 10, can be divided into two piles, one containing 2 cards and the other containing 3 cards, in

$$\frac{5!}{2!3!} = 10 \text{ ways.}$$

These are

$$\begin{cases} \text{A K} \\ \text{Q J 10} \end{cases} \quad \begin{cases} \text{A Q} \\ \text{K J 10} \end{cases} \quad \begin{cases} \text{A J} \\ \text{K Q 10} \end{cases} \quad \begin{cases} \text{A 10} \\ \text{K Q J} \end{cases} \quad \begin{cases} \text{K Q} \\ \text{A J 10} \end{cases}$$

$$\begin{cases} \text{K J} \\ \text{A Q 10} \end{cases} \quad \begin{cases} \text{K 10} \\ \text{A Q J} \end{cases} \quad \begin{cases} \text{Q J} \\ \text{A K 10} \end{cases} \quad \begin{cases} \text{Q 10} \\ \text{A K J} \end{cases} \quad \begin{cases} \text{J 10} \\ \text{A K Q} \end{cases}$$

If $p = q$, so that we have $2p$ objects to be divided into two equal groups, then there are

$$\frac{(2p)!}{p!p!}$$

ways in which the groups can be formed. However, if no distinction is made between the groups there are only

$$\frac{(2p)!}{p!p!2!}$$

distinct sets of two groups. Thus the four cards A, K, Q, J, can be dealt to two persons in

$$\frac{4!}{2!2!} = 6 \text{ ways.}$$

These are

A K	A Q	A J	K Q	K J	Q J
Q J	K J	K Q	A J	A Q	A K

If the cards are just being dealt into two piles and if no distinction is made between the piles, then the first of the above divisions (which has A K in the first pile and Q J in the second) is the same as the last (which has Q J in the first pile and A K in the second). Similarly, the second set of two piles is the same as the fifth and the third is the same as the fourth. There are only

$$\frac{4!}{2!2!2!} = 3$$

distinct sets of two piles.

The above results can be extended to the case of three or more groups in an obvious way. Thus if we have $p + q + r$ distinct objects, we can divide them into three groups of p objects, q objects, and r objects, respectively, in

$$\frac{(p + q + r)!}{p!q!r!} \text{ ways.}$$

Proof: We may first divide them into two groups consisting of $p + q$ and r of the objects, respectively. This can be done in

$$\frac{(p + q + r)!}{(p + q)!r!} \text{ ways.}$$

After this has been done in any one of these ways, we can divide the first group into two new groups of p and q objects, respectively. This can be done in

$$\frac{(p + q)!}{p!q!} \text{ ways.}$$

From the fundamental principle (page 280), we can conclude that the number of ways in which the $p + q + r$ objects can be divided into three groups containing p, q, and r of the objects, respectively, is

$$\frac{(p + q + r)!}{(p + q)!r!} \cdot \frac{(p + q)!}{p!q!} = \frac{(p + q + r)!}{p!q!r!}.$$

If $p = q = r$, so that we have $3p$ objects to divide into three equal groups, then there are

$$\frac{(3p)!}{(p!)^3}$$

ways in which the groups can be formed. Again, however, if no distinction is made among the groups, there are only

$$\frac{(3p)!}{(p!)^3 3!}$$

distinct sets of three groups.

Example

The 52 cards of a bridge deck can be dealt among 4 players in

$$\frac{52!}{(13!)^4} \text{ ways.}$$

If the cards are dealt into 4 piles, and no distinction is made among the piles, there are

$$\frac{52!}{(13!)^4 4!}$$

distinct sets of 4 piles. Observe that the first of these results can be obtained as follows: Deal the cards into 4 piles and then hand any one of the 4 piles to the first player, any one of the remaining 3 piles to the second, and so on. There would thus be

$$\frac{52!}{(13!)^4 4!} \cdot 4! = \frac{52!}{(13!)^4}$$

ways of dealing the cards.

123. ***Combinations and the binomial coefficients.*** The formula for ${}_nC_r$ can be written in the form

$${}_nC_r = \frac{n(n-1)(n-2) \cdots (n-r+1)}{r!}.$$

The expansion of $(a+b)^n$ by means of the binomial formula is

$$(a+b)^n = a^n + na^{n-1}b + \frac{n(n-1)}{2!}a^{n-2}b^2 + \cdots + \frac{n(n-1) \cdots (n-r+1)}{r!}a^{n-r}b^r + \cdots + b^n.$$

It follows that the coefficient of the term involving b^r is ${}_nC_r$. The above expansion could be written in the form:

$$(a+b)^n = a^n + {}_nC_1a^{n-1}b + {}_nC_2a^{n-2}b^2 + \cdots + {}_nC_ra^{n-r}b^r + \cdots + {}_nC_nb^n.$$

124. ***Total number of combinations.*** If we set $a = b = 1$ in the expansion of $(a+b)^n$, we have

$$(1+1)^n = 2^n = 1 + {}_nC_1 + {}_nC_2 + {}_nC_3 + \cdots + {}_nC_n.$$

It follows immediately that

$${}_nC_1 + {}_nC_2 + {}_nC_3 + \cdots + {}_nC_n = 2^n - 1.$$

The total number of combinations that can be made from **n** *distinct objects, by taking them one at a time, two at a time, three at a time, and on up to* **n** *at a time, is* $2^n - 1$.

Example

How many different sums of money can be formed from a penny, a nickel, a dime, a quarter, and a half dollar?

Solution

The total number of combinations that can be formed by taking the coins 1, 2, 3, 4, and 5 at a time, is

$$2^5 - 1 = 31.$$

PROBLEMS

1. How many straight lines are determined by 10 points, no 3 of which lie on the same straight line?

2. How many triangles are determined by 12 points, no 3 of which lie on the same straight line?

3. How many committees of 3 persons each can be chosen from among 12 persons?

4. In an examination a student is to choose any 10 questions from a set of 12. In how many ways can he make his selection? In how many ways can he do it if he is required to answer questions 1 and 2?

5. In how many ways can a group of 13 cards be selected from a deck containing 52 cards?

6. In how many ways can a set consisting of 4 black balls and 5 red ones be selected from a box containing 6 black and 6 red balls?

7. In how many ways can a set consisting of 8 red balls and 2 white ones be selected from a box containing 10 red and 5 white balls?

8. In how many ways can a hand consisting of 5 spades, 6 hearts, and 2 other cards be selected from a bridge deck containing 13 spades, 13 hearts, 13 diamonds and 13 clubs? HINT: We must select 5 of the 13 spades, 6 of the 13 hearts, and any 2 of the remaining 26 cards. This can be done in

$$\frac{13!}{5!8!} \cdot \frac{13!}{6!7!} \cdot \frac{26!}{2!24!} \text{ ways.}$$

9. In how many ways can a hand consisting of 11 spades and 2 diamonds be selected from a bridge deck? (See Prob. 8.)

10. In how many ways can a hand of 13 cards containing exactly 6 clubs be selected from a bridge deck? (See Prob. 8.)

11. In how many ways can a hand of 13 cards consisting of 4 aces, 4 kings, 2 queens, 2 jacks, and 1 other card be selected from a bridge deck? (See Prob. 8.)

12. In how many ways can 8 objects be divided into two equal groups?

13. In how many ways can 15 objects be divided into three equal groups?

14. In how many ways can a group of 9 boys be divided into two teams with 5 on one side and 4 on the other?

15. From 6 white balls, 8 black balls, and 3 red balls, how many selections consisting of 2 balls of each color can be made?

16. How many positive integers of 3 distinct digits can be formed from the digits 1, 2, 3, 4, 5?

17. How many different sums of money can be formed from a nickel, a dime, a quarter, and a dollar?

18. How many different sums of money can be formed from a penny, a nickel, a dime, a quarter, a half dollar, and a dollar, if each sum is to consist of 3 or more coins?

19. How many positive integers, each containing 5 distinct digits, can be formed from the digits 1, 2, 3, 4, 5, 6, 7, 8, 9, if each number is to contain 3 odd and 2 even digits?

20. A combination of 4 letters is picked at random from the letters $a, a, a, b, b, c, d, e, f$. In how many ways can the combination chosen consist of 4 distinct letters?

21. A combination of 4 balls is picked at random from a box containing 4 white, 3 red, 1 blue, 1 green, 1 yellow, and 1 black ball. In how many ways can the group chosen contain at least 2 red balls?

22. A combination of 4 balls is picked at random from a box containing 3 red, 4 blue, 2 green, 1 white, and 1 yellow ball. In how many ways can the set chosen contain at least 1 red ball? In how many ways can it fail to contain any red balls?

125. *Probability.* A bridge deck consists of 52 cards, 13 of which are spades. Suppose that a single card is drawn at random from the deck. It may be any one of the 13 spades, or it may be any one of the other 39 cards, and we assume that the drawing of any one specific card is just as likely as the drawing of any other. We say that the probability of drawing a spade is 13/52 or 1/4, and that the probability of failing to draw a spade is 39/52 or 3/4. This is in accordance with the following:

Definition. *If on any trial an event can happen in any one of* m *ways, and can fail to happen in any one of* n *ways, and if these* $m + n$ *ways are equally likely, then the probability of the event happening is*

$$p = \frac{m}{m+n},$$

and the probability of its failing to happen is

$$q = \frac{n}{m+n}.$$

It may be observed that the sum of the probability that an event will happen and the probability that it will fail, in any trial, is 1. Furthermore, if an event is certain to happen, the corresponding probability is 1; if it is certain to fail, the probability of its happening is zero.

As a further example, consider the ways in which a penny and a nickel may fall when they are tossed together. There are four possibilities: both heads; both tails; penny heads and nickel tails; penny tails and nickel heads. The probability of two heads is then 1/4; the probability of two tails is 1/4; the probability of a head and a tail is 2/4 or 1/2.

The ideas of the preceding sections on combinations are frequently useful in connection with problems on probability.

Example

From a box containing 4 red, 2 white, and 6 green balls, two balls are drawn at random. What is the probability that one is red and one is white?

Solution

The number of combinations of 12 balls taken 2 at a time is

$$\frac{12!}{2!10!} = 66.$$

There are 66 possible combinations, and when one reaches in and takes 2 balls he may get any one of these combinations. Now the number of combinations that would consist of 1 red ball and 1 white ball is

$$\frac{4!}{1!3!} \cdot \frac{2!}{1!1!} = 8.$$

This means that 8 of the 66 combinations consist of 1 red and 1 white ball. (The other combinations consist of a red one and a green one, 2 red ones, a green one and a white one, etc.).

The probability of drawing a red one and a white one in a single draw is then

$$\tfrac{8}{66} \quad \text{or} \quad \tfrac{4}{33}.$$

In the above example there are 66 possible combinations, all of which are assumed equally likely; 8 of these consist of 1 red and 1 white ball and 58 do not. We say that the *odds* are 58 to 8 *against* drawing this particular combination. In general, if an event can succeed in s ways and fail in f ways, then if $s > f$, we say that the odds are s to f *in favor* of the event. If $s < f$, we say that the odds are f to s *against* the event. The probabilities of its happening and failing are of course

$$p = \frac{s}{s+f} \qquad \text{and} \qquad q = \frac{f}{s+f}.$$

Suppose now, in connection with the above example, that the person who draws is to receive \$5 if he succeeds in drawing 1 red ball and 1 white one, and is to receive nothing if he fails to draw this combination. His chance of success is 4/33, so that if he paid

$$\tfrac{4}{33} \cdot \$5 = \$0.61$$

per chance, he could expect, in the long run, to come out even. This would be regarded as the mathematical value of a draw. If the operator of the game charged $1 per draw he would expect that, on the average, he would make a profit of $13 on every 33 customers.

These same ideas are involved in the following:

Example

A man holds 9 cards consisting of 3 aces, 2 kings, a ten, a nine, an eight, and a seven. He permits anyone to draw 3 of the cards for a fee of $6, and pays this person $14 if he succeeds in drawing both an ace and a king. What are the operator's long-term prospects for profit or loss on the game?

Solution

The player will win if he draws 2 aces and 1 king, 1 ace and 2 kings, or 1 ace, 1 king, and 1 other card. He can do these in 6, 3, and 24 ways, respectively, so that there are 33 winning combinations out of a total of

$$\frac{9!}{3!6!} = 84 \text{ combinations.}$$

The probability of drawing both an ace and a king in a single draw of 3 cards is then 33/84 or 11/28. This means that the odds are 17 to 11 against success for the player, and the mathematical value of a draw is

$$\tfrac{11}{28} \cdot \$14 = \$5.50.$$

It is of course possible that the first 3 or 4 or 10 customers may all win, but the operator of the game may expect that in the long run he will "pay out" only to 11 out of every 28 customers and that from 1,000 customers he will take in about $500 more than he will pay out.

126. ***Statistical probability.*** When the probability of an event is determined as in the preceding section—by an analysis of the number of equally likely ways in which the event can happen and fail to happen—it is called a *mathematical* or *a priori* probability.

Consider now the following example illustrating a somewhat different situation. A box contains black balls and white balls, the number of each being unknown. The probability of drawing a black ball in a single draw cannot be established by the methods employed above. Suppose, however, that 300 trial draws are made—each trial consisting of drawing a ball, noting its color, and returning it to the box—and assume that the balls are thoroughly mixed again before each new trial. If in the 300 trials one draws a black ball 40 times and a white one 260 times, he may conclude that the probability of drawing a black ball in a single draw is 40/300 or 2/15. A probability that is thus determined by observation and experiment is called a *statistical* or *empirical* probability.

The field of life insurance is one in which statistical probability plays an important part. Calculations in this field are based on the American Experience Table of Mortality from which the following figures are taken:

Mortality Table

Age	*Number living*	*Age*	*Number living*
10	100,000	60	57,917
20	92,637	70	38,569
30	85,441	80	14,474
40	78,106	90	847
50	69,804	95	3

The table tells us that out of 100,000 people alive at the age of ten, 38,569 were alive at the age of seventy. From this we would conclude that the probability that a child of ten will live to reach the age of seventy is

$$\frac{38{,}569}{100{,}000} = 0.386, \text{ approximately.}$$

Similarly, the probability that a person who is forty years old will die before reaching the age of fifty would be computed as follows: Out of 78,106 persons living at the age of forty, 69,804 were still alive at the age of fifty. This means that 8,302 out of the group died between the ages of forty and fifty, and the corresponding probability of dying is

$$\frac{8{,}302}{78{,}106} = 0.106, \text{ approximately.}$$

127. *Mutually exclusive events.* Two or more events are said to be *mutually exclusive* if not more than one of them can happen in a given trial.

Examples

The drawing of an ace and the drawing of a king, in a single draw of one card from a deck, are mutually exclusive events. Either event, but not both, could happen.

The drawing of an ace and the drawing of a heart are not mutually exclusive. It would be possible to draw the ace of hearts.

In dealing with mutually exclusive events we have the following:

Theorem. *The probability that one or another of several mutually exclusive events will occur on a single trial is equal to the sum of the probabilities of the single events.*

Example

The probability of drawing an ace in a single draw from a deck of cards is 4/52 or 1/13. The probability of drawing a king is also 1/13. The probability of drawing an ace *or* a king is then

$$\tfrac{1}{13} + \tfrac{1}{13} = \tfrac{2}{13}.$$

This example should lead the student to a method of proving the theorem.

128. *Independent and dependent events.* Two or more events are said to be *independent* if the happening of any one of them at a given trial does not affect the probability of occurrence of any of the others. They are *dependent* if the happening of one of them at a given trial does affect the probability of occurrence of another.

Example

Two balls are drawn in succession from a box containing 4 white and 4 black balls. If the ball that is drawn on the first trial is returned to the box before the second ball is drawn, then the two drawings are independent events. If the first ball is not returned to the box before the second ball is drawn, the two events are dependent—the probability of drawing a white ball on the second draw is in this case affected by the result of the first draw.

An important theorem concerning the probability of occurrence of all of a set of independent events, in a given trial, is as follows:

Theorem. *If the separate probabilities of* **n** *independent events are* $p_1, p_2, \cdots, p_n$, *respectively, then the probability that all of them will happen in a given single trial is the product*

$$p = p_1 \cdot p_2 \cdots p_n$$

of these separate probabilities.

Example 1

If 6 coins are tossed at the same time, the probability that all will be heads is

$$\tfrac{1}{2} \cdot \tfrac{1}{2} \cdot \tfrac{1}{2} \cdot \tfrac{1}{2} \cdot \tfrac{1}{2} \cdot \tfrac{1}{2} = \tfrac{1}{64}.$$

Example 2

From a box containing 6 black, 2 green, and 4 white balls, a ball is drawn and replaced, and then a second ball is drawn. What is the probability that one of the balls drawn is white and one is green?

Solution

The probability of drawing a white ball on the first draw is $4/12$; the probability of drawing a green one on the second draw is $2/12$, and these are independent events. The probability of getting a white one on the first draw *and* a green one on the second draw is then

$$\tfrac{4}{12} \cdot \tfrac{2}{12} = \tfrac{1}{18}.$$

The probability of drawing a green one first and then a white one is also $1/18$, and since the two events of drawing white then green and drawing green then white are mutually exclusive, the probability that either one or the other will occur is

$$\tfrac{1}{18} + \tfrac{1}{18} = \tfrac{1}{9}.$$

This is then the probability of drawing a white ball and a green ball (the order being immaterial) in two draws.

A corresponding theorem for the case of dependent events will be stated only for the case of two events. The extension to three or more events is obvious.

Theorem. *Let p_1 be the probability that a first event will occur. Let p_2 be the probability that a second event will occur, this being determined on the assumption that the first event has happened. Then the probability that both events will occur in succession is equal to the product p_1p_2.*

Example

Suppose that in the preceding example the first ball drawn is not returned to the box. What is then the probability of drawing a white one and a green one in the two draws?

Solution

The probability of drawing a white ball on the first draw is 4/12. Assuming that a white one is drawn, the probability of getting a green one on the second draw is 2/11. Hence the probability of getting a white one then a green one is

$$\tfrac{4}{12} \cdot \tfrac{2}{11} = \tfrac{2}{33}.$$

Similarly, the probability of getting a green one then a white one is

$$\tfrac{2}{12} \cdot \tfrac{4}{11} = \tfrac{2}{33}.$$

The probability of getting a white one and a green one (order immaterial) in two successive draws is then

$$\tfrac{2}{33} + \tfrac{2}{33} = \tfrac{4}{33}.$$

Observe now that this same result could be obtained by previous methods. Since the first ball is not returned to the box before the second one is drawn, the required probability is precisely the same as if both balls were drawn at a single draw. The total number of possible combinations of the 12 balls taken 2 at a time is

$$\frac{12!}{2!10!} = 66.$$

The number of these combinations that consist of 1 white ball and 1 green ball is

$$\frac{4!}{1!3!} \cdot \frac{2!}{1!1!} = 8.$$

The probability of getting one of these combinations is then 8/66 or 4/33.

PROBLEMS

1. What is the probability of throwing a 7 in a single throw with a pair of dice?

2. Two cards are drawn from a bridge deck in a single draw. What is the probability that both are spades?

3. Three cards are drawn from a bridge deck in a single draw. What is the probability that 2 are spades and 1 is a heart?

4. From a box containing 6 red, 4 white, and 4 green balls, 2 balls are drawn in a single draw. What is the probability that both are green?

5. From a box containing 12 white, 8 black, and 2 green balls, 4 balls are drawn in a single draw. What is the probability that 3 are white and 1 is black?

6. From a box containing 8 black, 8 white, and 8 red balls, 3 are drawn in a single draw. The person who makes the draw is to receive $5 if he does not draw a black ball. Find the mathematical value of a chance.

7. From a box containing 12 white, 8 red, 8 black, and 2 green balls, 3 balls are drawn at a single draw. The person who operates the game charges $1 per chance and pays $100 if the customer succeeds in drawing both of the green balls. How much profit per customer may he expect to make in the long run?

8. From a box containing 8 white, 6 red, and 10 green balls, 3 balls are drawn at a single draw. What are the odds in favor of or against drawing at least 2 green ones?

9. A person draws 4 balls in succession from a box containing 8 black, 8 red, and 4 white balls. The balls that are drawn are not returned to the box. What are the odds for or against drawing at least 2 black balls?

10. What is the probability of drawing either a white or a green ball in a single draw from a box containing 2 white, 6 green, and 12 black balls?

11. What is the probability of drawing either a spade or a heart in a single draw from a bridge deck?

12. Using the table on page 295, find the probability that a person who is twenty years old will still be alive at the age of seventy.

13. Using the table on page 295, find the probability that a person who is sixty years old will still be alive at the age of eighty.

14. A person is to draw a ball from a box containing 6 black, 6 green, and 4 white balls. Then, without returning this ball to the box, he is to draw a second ball. If he succeeds in drawing 2 balls of the same color, he is to receive a prize of $1. What is the mathematical value of a chance to draw?

15. A person is to draw 3 balls in succession from a box containing 12 black, 6 green, and 6 white balls. Each ball that is drawn is returned to the box before the next ball is drawn. He is to receive a prize of $4 if he succeeds in drawing 1 ball of each color. What is the mathematical value of a chance to draw?

16. One box contains 12 black and 6 white balls, and another contains 12 white and 6 black balls. A person draws 1 ball from each box. What is the probability that he will get one of each color?

17. A person draws 1 card from each of 3 bridge decks. Find the probability that he will get either 3 diamonds or 3 clubs.

18. What is the probability of throwing either a 7 or an 11 in a single throw with a pair of dice?

19. A person draws a ball from a box that contains 8 red balls and 12 white ones. If he gets a red one, he receives a prize of $5. If he does not get a red ball, he returns the ball to the box and draws again. If he gets a red ball on this second draw, he still receives the $5 prize. What is the mathematical value of a chance?

20. A person draws a ball from a box that contains 12 red, 10 black, 8 white, and 6 green balls. If he draws a green ball, he receives a prize of $18. If he does not draw a green ball, he returns the ball to the box and draws again. If he gets a green ball on this second draw, he still receives the $18 prize. What is the mathematical value of a chance?

TABLES

TABLE I.—COMMON LOGARITHMS

N	0	1	2	3	4	5	6	7	8	9
10	0000	0043	0086	0128	0170	0212	0253	0294	0334	0374
11	0414	0453	0492	0531	0569	0607	0645	0682	0719	0755
12	0792	0828	0864	0899	0934	0969	1004	1038	1072	1106
13	1139	1173	1206	1239	1271	1303	1335	1367	1399	1430
14	1461	1492	1523	1553	1584	1614	1644	1673	1703	1732
15	1761	1790	1818	1847	1875	1903	1931	1959	1987	2014
16	2041	2068	2095	2122	2148	2175	2201	2227	2253	2279
17	2304	2330	2355	2380	2405	2430	2455	2480	2504	2529
18	2553	2577	2601	2625	2648	2672	2695	2718	2742	2765
19	2788	2810	2833	2856	2878	2900	2923	2945	2967	2989
20	3010	3032	3054	3075	3096	3118	3139	3160	3181	3201
21	3222	3243	3263	3284	3304	3324	3345	3365	3385	3404
22	3424	3444	3464	3483	3502	3522	3541	3560	3579	3598
23	3617	3636	3655	3674	3692	3711	3729	3747	3766	3784
24	3802	3820	3838	3856	3874	3892	3909	3927	3945	3962
25	3979	3997	4014	4031	4048	4065	4082	4099	4116	4133
26	4150	4166	4183	4200	4216	4232	4249	4265	4281	4298
27	4314	4330	4346	4362	4378	4393	4409	4425	4440	4456
28	4472	4487	4502	4518	4533	4548	4564	4579	4594	4609
29	4624	4639	4654	4669	4683	4698	4713	4728	4742	4757
30	4771	4786	4800	4814	4829	4843	4857	4871	4886	4900
31	4914	4928	4942	4955	4969	4983	4997	5011	5024	5038
32	5051	5065	5079	5092	5105	5119	5132	5145	5159	5172
33	5185	5198	5211	5224	5237	5250	5263	5276	5289	5302
34	5315	5328	5340	5353	5366	5378	5391	5403	5416	5428
35	5441	5453	5465	5478	5490	5502	5514	5527	5539	5551
36	5563	5575	5587	5599	5611	5623	5635	5647	5658	5670
37	5682	5694	5705	5717	5729	5740	5752	5763	5775	5786
38	5798	5809	5821	5832	5843	5855	5866	5877	5888	5899
39	5911	5922	5933	5944	5955	5966	5977	5988	5999	6010
40	6021	6031	6042	6053	6064	6075	6085	6096	6107	6117
41	6128	6138	6149	6160	6170	6180	6191	6201	6212	6222
42	6232	6243	6253	6263	6274	6284	6294	6304	6314	6325
43	6335	6345	6355	6365	6375	6385	6395	6405	6415	6425
44	6435	6444	6454	6464	6474	6484	6493	6503	6513	6522
45	6532	6542	6551	6561	6571	6580	6590	6599	6609	6618
46	6628	6637	6646	6656	6665	6675	6684	6693	6702	6712
47	6721	6730	6739	6749	6758	6767	6776	6785	6794	6803
48	6812	6821	6830	6839	6848	6857	6866	6875	6884	6893
49	6902	6911	6920	6928	6937	6946	6955	6964	6972	6981
50	6990	6998	7007	7016	7024	7033	7042	7050	7059	7067
51	7076	7084	7093	7101	7110	7118	7126	7135	7143	7152
52	7160	7168	7177	7185	7193	7202	7210	7218	7226	7235
53	7243	7251	7259	7267	7275	7284	7292	7300	7308	7316
54	7324	7332	7340	7348	7356	7364	7372	7380	7388	7396
N	0	1	2	3	4	5	6	7	8	9

TABLE I.—COMMON LOGARITHMS.—(*Continued*)

N	0	1	2	3	4	5	6	7	8	9
55	7404	7412	7419	7427	7435	7443	7451	7459	7466	7474
56	7482	7490	7497	7505	7513	7520	7528	7536	7543	7551
57	7559	7566	7574	7582	7589	7597	7604	7612	7619	7627
58	7634	7642	7649	7657	7664	7672	7679	7686	7694	7701
59	7709	7716	7723	7731	7738	7745	7752	7760	7767	7774
60	7782	7789	7796	7803	7810	7818	7825	7832	7839	7846
61	7853	7860	7868	7875	7882	7889	7896	7903	7910	7917
62	7924	7931	7938	7945	7952	7959	7966	7973	7980	7987
63	7993	8000	8007	8014	8021	8028	8035	8041	8048	8055
64	8062	8069	8075	8082	8089	8096	8102	8109	8116	8122
65	8129	8136	8142	8149	8156	8162	8169	8176	8182	8189
66	8195	8202	8209	8215	8222	8228	8235	8241	8248	8254
67	8261	8267	8274	8280	8287	8293	8299	8306	8312	8319
68	8325	8331	8338	8344	8351	8357	8363	8370	8376	8382
69	8388	8395	8401	8407	8414	8420	8426	8432	8439	8445
70	8451	8457	8463	8470	8476	8482	8488	8494	8500	8506
71	8513	8519	8525	8531	8537	8543	8549	8555	8561	8567
72	8573	8579	8585	8591	8597	8603	8609	8615	8621	8627
73	8633	8639	8645	8651	8657	8663	8669	8675	8681	8686
74	8692	8698	8704	8710	8716	8722	8727	8733	8739	8745
75	8751	8756	8762	8768	8774	8779	8785	8791	8797	8802
76	8808	8814	8820	8825	8831	8837	8842	8848	8854	8859
77	8865	8871	8876	8882	8887	8893	8899	8904	8910	8915
78	8921	8927	8932	8938	8943	8949	8954	8960	8965	8971
79	8976	8982	8987	8993	8998	9004	9009	9015	9020	9025
80	9031	9036	9042	9047	9053	9058	9063	9069	9074	9079
81	9085	9090	9096	9101	9106	9112	9117	9122	9128	9133
82	9138	9143	9149	9154	9159	9165	9170	9175	9180	9186
83	9191	9196	9201	9206	9212	9217	9222	9227	9232	9238
84	9243	9248	9253	9258	9263	9269	9274	9279	9284	9289
85	9294	9299	9304	9309	9315	9320	9325	9330	9335	9340
86	9345	9350	9355	9360	9365	9370	9375	9380	9385	9390
87	9395	9400	9405	9410	9415	9420	9425	9430	9435	9440
88	9445	9450	9455	9460	9465	9469	9474	9479	9484	9489
89	9494	9499	9504	9509	9513	9518	9523	9528	9533	9538
90	9542	9547	9552	9557	9562	9566	9571	9576	9581	9586
91	9590	9595	9600	9605	9609	9614	9619	9624	9628	9633
92	9638	9643	9647	9652	9657	9661	9666	9671	9675	9680
93	9685	9689	9694	9699	9703	9708	9713	9717	9722	9727
94	9731	9736	9741	9745	9750	9754	9759	9763	9768	9773
95	9777	9782	9786	9791	9795	9800	9805	9809	9814	9818
96	9823	9827	9832	9836	9841	9845	9850	9854	9859	9863
97	9868	9872	9877	9881	9886	9890	9894	9899	9903	9908
98	9912	9917	9921	9926	9930	9934	9939	9943	9948	9952
99	9956	9961	9965	9969	9974	9978	9983	9987	9991	9996
N	0	1	2	3	4	5	6	7	8	9

TABLE II.—TRIGONOMETRIC FUNCTIONS

Angles	Sines		Cosines		Tangents		Cotangents		Angles
	Nat.	Log.	Nat.	Log.	Nat.	Log.	Nat.	Log.	
0° 00′	.0000	∞	1.0000	0.0000	.0000	∞	∞	∞	90° 00′
10	.0029	7.4637	1.0000	0000	.0029	7.4637	343.77	2.5363	50
20	.0058	7648	1.0000	0000	.0058	7648	171.89	2352	40
30	.0087	9408	1.0000	0000	.0087	9409	114.59	0591	30
40	.0116	8.0658	.9999	0000	.0116	8.0658	85.940	1.9342	20
50	.0145	1627	.9999	0000	.0145	1627	68.750	8373	10
1° 00′	.0175	8.2419	.9998	9.9999	.0175	8.2419	57.290	1.7581	89° 00′
10	.0204	3088	.9998	9999	.0204	3089	49.104	6911	50
20	.0233	3668	.9997	9999	.0233	3669	42.964	6331	40
30	.0262	4179	.9997	9999	.0262	4181	38.188	5819	30
40	.0291	4637	.9996	9998	.0291	4638	34.368	5362	20
50	.0320	5050	.9995	9998	.0320	5053	31.242	4947	10
2° 00′	.0349	8.5428	.9994	9.9997	.0349	8.5431	28.636	1.4569	88° 00′
10	.0378	5776	.9993	9997	.0378	5779	26.432	4221	50
20	.0407	6097	.9992	9996	.0407	6101	24.542	3899	40
30	.0436	6397	.9990	9996	.0437	6401	22.904	3599	30
40	.0465	6677	.9989	9995	.0466	6682	21.470	3318	20
50	.0494	6940	.9988	9995	.0495	6945	20.206	3055	10
3° 00′	.0523	8.7188	.9986	9.9994	.0524	8.7194	19.081	1.2806	87° 00′
10	.0552	7423	.9985	9993	.0553	7429	18.075	2571	50
20	.0581	7645	.9983	9993	.0582	7652	17.169	2348	40
30	.0610	7857	.9981	9992	.0612	7865	16.350	2135	30
40	.0640	8059	.9980	9991	.0641	8067	15.605	1933	20
50	.0669	8251	.9978	9990	.0670	8261	14.924	1739	10
4° 00′	.0698	8.8436	.9976	9.9989	.0699	8.8446	14.301	1.1554	86° 00′
10	.0727	8613	.9974	9989	.0729	8624	13.727	1376	50
20	.0756	8783	.9971	9988	.0758	8795	13.197	1205	40
30	.0785	8946	.9969	9987	.0787	8960	12.706	1040	30
40	.0814	9104	.9967	9986	.0816	9118	12.251	0882	20
50	.0843	9256	.9964	9985	.0846	9272	11.826	0728	10
5° 00′	.0872	8.9403	.9962	9.9983	.0875	8.9420	11.430	1.0580	85° 00′
10	.0901	9545	.9959	9982	.0904	9563	11.059	0437	50
20	.0929	9682	.9957	9981	.0934	9701	10.712	0299	40
30	.0958	9816	.9954	9980	.0963	9836	10.385	0164	30
40	.0987	9945	.9951	9979	.0992	9966	10.078	0034	20
50	.1016	9.0070	.9948	9977	.1022	9.0093	9.7882	0.9907	10
6° 00′	.1045	9.0192	.9945	9.9976	.1051	9.0216	9.5144	0.9784	84° 00′
10	.1074	0311	.9942	9975	.1080	0336	9.2553	9664	50
20	.1103	0426	.9939	9973	.1110	0453	9.0098	9547	40
30	.1132	0539	.9936	9972	.1139	0567	8.7769	9433	30
40	.1161	0648	.9932	9971	.1169	0678	8.5555	9322	20
50	.1190	0755	.9929	9969	.1198	0786	8.3450	9214	10
7° 00′	.1219	9.0859	.9925	9.9968	.1228	9.0891	8.1443	0.9109	83° 00′
10	.1248	0961	.9922	9966	.1257	0995	7.9530	9005	50
20	.1276	1060	.9918	9964	.1287	1096	7.7704	8904	40
30	.1305	1157	.9914	9963	.1317	1194	7.5958	8806	30
40	.1334	1252	.9911	9961	.1346	1291	7.4287	8709	20
50	.1363	1345	.9907	9959	.1376	1385	7.2687	8615	10
8° 00′	.1392	9.1436	.9903	9.9958	.1405	9.1478	7.1154	0.8522	82° 00′
10	.1421	1525	.9899	9956	.1435	1569	6.9682	8431	50
20	.1449	1612	.9894	9954	.1465	1658	6.8269	8342	40
30	.1478	1697	.9890	9952	.1495	1745	6.6912	8255	30
40	.1507	1781	.9886	9950	.1524	1831	6.5606	8169	20
50	.1536	1863	.9881	9948	.1554	1915	6.4348	8085	10
9° 00′	.1564	9.1943	.9877	9.9946	.1584	9.1997	6.3138	0.8003	81° 00′
	Nat.	Log.	Nat.	Log.	Nat.	Log.	Nat.	Log.	
Angles	Cosines		Sines		Cotangents		Tangents		Angles

TABLE II.—TRIGONOMETRIC FUNCTIONS.—(*Continued*)

Angles	Sines		Cosines		Tangents		Cotangents		Angles
	Nat.	Log.	Nat.	Log.	Nat.	Log.	Nat.	Log.	
9° 00′	.1564	9.1943	.9877	9.9946	.1584	9.1997	6.3138	0.8003	81° 00′
10	.1593	2022	.9872	9944	.1614	2078	6.1970	7922	50
20	.1622	2100	.9868	9942	.1644	2158	6.0844	7842	40
30	.1650	2176	.9863	9940	.1673	2236	5.9758	7764	30
40	.1679	2251	.9858	9938	.1703	2313	5.8708	7687	20
50	.1708	2324	.9853	9936	.1733	2389	5.7694	7611	10
10° 00′	.1736	9.2397	.9848	9.9934	.1763	9.2463	5.6713	0.7537	80° 00′
10	.1765	2468	.9843	9931	.1793	2536	5.5764	7464	50
20	.1794	2538	.9838	9929	.1823	2609	5.4845	7391	40
30	.1822	2606	.9833	9927	.1853	2680	5.3955	7320	30
40	.1851	2674	.9827	9924	.1883	2750	5.3093	7250	20
50	.1880	2740	.9822	9922	.1914	2819	5.2257	7181	10
11° 00′	.1908	9.2806	.9816	9.9919	.1944	9.2887	5.1446	0.7113	79° 00′
10	.1937	2870	.9811	9917	.1974	2953	5.0658	7047	50
20	.1965	2934	.9805	9914	.2004	3020	4.9894	6980	40
30	.1994	2997	.9799	9912	.2035	3085	4.9152	6915	30
40	.2022	3058	.9793	9909	.2065	3149	4.8430	6851	20
50	.2051	3119	.9787	9907	.2095	3212	4.7729	6788	10
12° 00′	.2079	9.3179	.9781	9.9904	.2126	9.3275	4.7046	0.6725	78° 00′
10	.2108	3238	.9775	9901	.2156	3336	4.6382	6664	50
20	.2136	3296	.9769	9899	.2186	3397	4.5736	6603	40
30	.2164	3353	.9763	9896	.2217	3458	4.5107	6542	30
40	.2193	3410	.9757	9893	.2247	3517	4.4494	6483	20
50	.2221	3466	.9750	9890	.2278	3576	4.3897	6424	10
13° 00′	.2250	9.3521	.9744	9.9887	.2309	9.3634	4.3315	0.6366	77° 00′
10	.2278	3575	.9737	9884	.2339	3691	4.2747	6309	50
20	.2306	3629	.9730	9881	.2370	3748	4.2193	6252	40
30	.2334	3682	.9724	9878	.2401	3804	4.1653	6196	30
40	.2363	3734	.9717	9875	.2432	3859	4.1126	6141	20
50	.2391	3786	.9710	9872	.2462	3914	4.0611	6086	10
14° 00′	.2419	9.3837	.9703	9.9869	.2493	9.3968	4.0108	0.6032	76° 00′
10	.2447	3887	.9696	9866	.2524	4021	3.9617	5979	50
20	.2476	3937	.9689	9863	.2555	4074	3.9136	5926	40
30	.2504	3986	.9681	9859	.2586	4127	3.8667	5873	30
40	.2532	4035	.9674	9856	.2617	4178	3.8208	5822	20
50	.2560	4083	.9667	9853	.2648	4230	3.7760	5770	10
15° 00′	.2588	9.4130	.9659	9.9849	.2679	9.4281	3.7321	0.5719	75° 00′
10	.2616	4177	.9652	9846	.2711	4331	3.6891	5669	50
20	.2644	4223	.9644	9843	.2742	4381	3.6470	5619	40
30	.2672	4269	.9636	9839	.2773	4430	3.6059	5570	30
40	.2700	4314	.9628	9836	.2805	4479	3.5656	5521	20
50	.2728	4359	.9621	9832	.2836	4527	3.5261	5473	10
16° 00′	.2756	9.4403	.9613	9.9828	.2867	9.4575	3.4874	0.5425	74° 00′
10	.2784	4447	.9605	9825	.2899	4622	3.4495	5378	50
20	.2812	4491	.9596	9821	.2931	4669	3.4124	5331	40
30	.2840	4533	.9588	9817	.2962	4716	3.3759	5284	30
40	.2868	4576	.9580	9814	.2994	4762	3.3402	5238	20
50	.2896	4618	.9572	9810	.3026	4808	3.3052	5192	10
17° 00′	.2924	9.4659	.9563	9.9806	.3057	9.4853	3.2709	0.5147	73° 00′
10	.2952	4700	.9555	9802	.3089	4898	3.2371	5102	50
20	.2979	4741	.9546	9798	.3121	4943	3.2041	5057	40
30	.3007	4781	.9537	9794	.3153	4987	3.1716	5013	30
40	.3035	4821	.9528	9790	.3185	5031	3.1397	4969	20
50	.3062	4861	.9520	9786	.3217	5075	3.1084	4925	10
18° 00′	.3090	9.4900	.9511	9.9782	.3249	9.5118	3.0777	0.4882	72° 00′
	Nat.	Log.	Nat.	Log.	Nat.	Log.	Nat.	Log.	
Angles	Cosines		Sines		Cotangents		Tangents		Angles

TABLE II.—TRIGONOMETRIC FUNCTIONS.—(*Continued*)

Angles	Sines		Cosines		Tangents		Cotangents		Angles
	Nat.	Log.	Nat.	Log.	Nat.	Log.	Nat.	Log.	
18° 00′	.3090	**9**.4900	.9511	**9**.9782	.3249	**9**.5118	3.0777	**0**.4882	72° 00′
10	.3118	4939	.9502	9778	.3281	5161	3.0475	4839	50
20	.3145	4977	.9492	9774	.3314	5203	3.0178	4797	40
30	.3173	5015	.9483	9770	.3346	5245	2.9887	4755	30
40	.3201	5052	.9474	9765	.3378	5287	2.9600	4713	20
50	.3228	5090	.9465	9761	.3411	5329	2.9319	4671	10
19° 00′	.3256	**9**.5126	.9455	**9**.9757	.3443	**9**.5370	2.9042	**0**.4630	71° 00′
10	.3283	5163	.9446	9752	.3476	5411	2.8770	4589	50
20	.3311	5199	.9436	9748	.3508	5451	2.8502	4549	40
30	.3338	5235	.9426	9743	.3541	5491	2.8239	4509	30
40	.3365	5270	.9417	9739	.3574	5531	2.7980	4469	20
50	.3393	5306	.9407	9734	.3607	5571	2.7725	4429	10
20° 00′	.3420	**9**.5341	.9397	**9**.9730	.3640	**9**.5611	2.7475	**0**.4389	70° 00′
10	.3448	5375	.9387	9725	.3673	5650	2.7228	4350	50
20	.3475	5409	.9377	9721	.3706	5689	2.6985	4311	40
30	.3502	5443	.9367	9716	.3739	5727	2.6746	4273	30
40	.3529	5477	.9356	9711	.3772	5766	2.6511	4234	20
50	.3557	5510	.9346	9706	.3805	5804	2.6279	4196	10
21° 00′	.3584	**9**.5543	.9336	**9**.9702	.3839	**9**.5842	2.6051	**0**.4158	69° 00′
10	.3611	5576	.9325	9697	.3872	5879	2.5826	4121	50
20	.3638	5609	.9315	9692	.3906	5917	2.5605	4083	40
30	.3665	5641	.9304	9687	.3939	5954	2.5386	4046	30
40	.3692	5673	.9293	9682	.3973	5991	2.5172	4009	20
50	.3719	5704	.9283	9677	.4006	6028	2.4960	3972	10
22° 00′	.3746	**9**.5736	.9272	**9**.9672	.4040	**9**.6064	2.4751	**0**.3936	68° 00′
10	.3773	5767	.9261	9667	.4074	6100	2.4545	3900	50
20	.3800	5798	.9250	9661	.4108	6136	2.4342	3864	40
30	.3827	5828	.9239	9656	.4142	6172	2.4142	3828	30
40	.3854	5859	.9228	9651	.4176	6208	2.3945	3792	20
50	.3881	5889	.9216	9646	.4210	6243	2.3750	3757	10
23° 00′	.3907	**9**.5919	.9205	**9**.9640	.4245	**9**.6279	2.3559	**0**.3721	67° 00′
10	.3934	5948	.9194	9635	.4279	6314	2.3369	3686	50
20	.3961	5978	.9182	9629	.4314	6348	2.3183	3652	40
30	.3987	6007	.9171	9624	.4348	6383	2.2998	3617	30
40	.4014	6036	.9159	9618	.4383	6417	2.2817	3583	20
50	.4041	6065	.9147	9613	.4417	6452	2.2637	3548	10
24° 00′	.4067	**9**.6093	.9135	**9**.9607	.4452	**9**.6486	2.2460	**0**.3514	66° 00′
10	.4094	6121	.9124	9602	.4487	6520	2.2286	3480	50
20	.4120	6149	.9112	9596	.4522	6553	2.2113	3447	40
30	.4147	6177	.9100	9590	.4557	6587	2.1943	3413	30
40	.4173	6205	.9088	9584	.4592	6620	2.1775	3380	20
50	.4200	6232	.9075	9579	.4628	6654	2.1609	3346	10
25° 00′	.4226	**9**.6259	.9063	**9**.9573	.4663	**9**.6687	2.1445	**0**.3313	65° 00′
10	.4253	6286	.9051	9567	.4699	6720	2.1283	3280	50
20	.4279	6313	.9038	9561	.4734	6752	2.1123	3248	40
30	.4305	6340	.9026	9555	.4770	6785	2.0965	3215	30
40	.4331	6366	.9013	9549	.4806	6817	2.0809	3183	20
50	.4358	6392	.9001	9543	.4841	6850	2.0655	3150	10
26° 00′	.4384	**9**.6418	.8988	**9**.9537	.4877	**9**.6882	2.0503	**0**.3118	64° 00′
10	.4410	6444	.8975	9530	.4913	6914	2.0353	3086	50
20	.4436	6470	.8962	9524	.4950	6946	2.0204	3054	40
30	.4462	6495	.8949	9518	.4986	6977	2.0057	3023	30
40	.4488	6521	.8936	9512	.5022	7009	1.9912	2991	20
50	.4514	6546	.8923	9505	.5059	7040	1.9768	2960	10
27° 00′	.4540	**9**.6570	.8910	**9**.9499	.5095	**9**.7072	1.9626	**0**.2928	63° 00′
	Nat.	Log.	Nat.	Log.	Nat.	Log.	Nat.	Log.	
Angles	Cosines		Sines		Cotangents		Tangents		Angles

TABLE II.—TRIGONOMETRIC FUNCTIONS.—(*Continued*)

Angles	Sines		Cosines		Tangents		Cotangents		Angles
	Nat.	Log.	Nat.	Log.	Nat.	Log.	Nat.	Log.	
27° 00′	.4540	9.6570	.8910	9.9499	.5095	9.7072	1.9626	0.2928	63° 00′
10	.4566	6595	.8897	9492	.5132	7103	1.9486	2897	50
20	.4592	6620	.8884	9486	.5169	7134	1.9347	2866	40
30	.4617	6644	.8870	9479	.5206	7165	1.9210	2835	30
40	.4643	6668	.8857	9473	.5243	7196	1.9074	2804	20
50	.4669	6692	.8843	9466	.5280	7226	1.8940	2774	10
28° 00′	.4695	9.6716	.8829	9.9459	.5317	9.7257	1.8807	0.2743	62° 00′
10	.4720	6740	.8816	9453	.5354	7287	1.8676	2713	50
20	.4746	6763	.8802	9446	.5392	7317	1.8546	2683	40
30	.4772	6787	.8788	9439	.5430	7348	1.8418	2652	30
40	.4797	6810	.8774	9432	.5467	7378	1.8291	2622	20
50	.4823	6833	.8760	9425	.5505	7408	1.8165	2592	10
29° 00′	.4848	9.6856	.8746	9.9418	.5543	9.7438	1.8040	0.2562	61° 00′
10	.4874	6878	.8732	9411	.5581	7467	1.7917	2533	50
20	.4899	6901	.8718	9404	.5619	7497	1.7796	2503	40
30	.4924	6923	.8704	9397	.5658	7526	1.7675	2474	30
40	.4950	6946	.8689	9390	.5696	7556	1.7556	2444	20
50	.4975	6968	.8675	9383	.5735	7585	1.7437	2415	10
30° 00′	.5000	9.6990	.8660	9.9375	.5774	9.7614	1.7321	0.2386	60° 00′
10	.5025	7012	.8646	9368	.5812	7644	1.7205	2356	50
20	.5050	7033	.8631	9361	.5851	7673	1.7090	2327	40
30	.5075	7055	.8616	9353	.5890	7701	1.6977	2299	30
40	.5100	7076	.8601	9346	.5930	7730	1.6864	2270	20
50	.5125	7097	.8587	9338	.5969	7759	1.6753	2241	10
31° 00′	.5150	9.7118	.8572	9.9331	.6009	9.7788	1.6643	0.2212	59° 00′
10	.5175	7139	.8557	9323	.6048	7816	1.6534	2184	50
20	.5200	7160	.8542	9315	.6088	7845	1.6426	2155	40
30	.5225	7181	.8526	9308	.6128	7873	1.6319	2127	30
40	.5250	7201	.8511	9300	.6168	7902	1.6212	2098	20
50	.5275	7222	.8496	9292	.6208	7930	1.6107	2070	10
32° 00′	.5299	9.7242	.8480	9.9284	.6249	9.7958	1.6003	0.2042	58° 00′
10	.5324	7262	.8465	9276	.6289	7986	1.5900	2014	50
20	.5348	7282	.8450	9268	.6330	8014	1.5798	1986	40
30	.5373	7302	.8434	9260	.6371	8042	1.5697	1958	30
40	.5398	7322	.8418	9252	.6412	8070	1.5597	1930	20
50	.5422	7342	.8403	9244	.6453	8097	1.5497	1903	10
33° 00′	.5446	9.7361	.8387	9.9236	.6494	9.8125	1.5399	0.1875	57° 00′
10	.5471	7380	.8371	9228	.6536	8153	1.5301	1847	50
20	.5495	7400	.8355	9219	.6577	8180	1.5204	1820	40
30	.5519	7419	.8339	9211	.6619	8208	1.5108	1792	30
40	.5544	7438	.8323	9203	.6661	8235	1.5013	1765	20
50	.5568	7457	.8307	9194	.6703	8263	1.4919	1737	10
34° 00′	.5592	9.7476	.8290	9.9186	.6745	9.8290	1.4826	0.1710	56° 00′
10	.5616	7494	.8274	9177	.6787	8317	1.4733	1683	50
20	.5640	7513	.8258	9169	.6830	8344	1.4641	1656	40
30	.5664	7531	.8241	9160	.6873	8371	1.4550	1629	30
40	.5688	7550	.8225	9151	.6916	8398	1.4460	1602	20
50	.5712	7568	.8208	9142	.6959	8425	1.4370	1575	10
35° 00′	.5736	9.7586	.8192	9.9134	.7002	9.8452	1.4281	0.1548	55° 00′
10	.5760	7604	.8175	9125	.7046	8479	1.4193	1521	50
20	.5783	7622	.8158	9116	.7089	8506	1.4106	1494	40
30	.5807	7640	.8141	9107	.7133	8533	1.4019	1467	30
40	.5831	7657	.8124	9098	.7177	8559	1.3934	1441	20
50	.5854	7675	.8107	9089	.7221	8586	1.3848	1414	10
36° 00′	.5878	9.7692	.8090	9.9080	.7265	9.8613	1.3764	0.1387	54° 00′
	Nat.	Log.	Nat.	Log.	Nat.	Log.	Nat.	Log.	
Angles	Cosines		Sines		Cotangents		Tangents		Angles

TABLE II.—TRIGONOMETRIC FUNCTIONS.—(*Continued*)

Angles	Sines		Cosines		Tangents		Cotangents		Angles
	Nat.	Log.	Nat.	Log.	Nat.	Log.	Nat.	Log.	
36° 00′	.5878	9.7692	.8090	9.9080	.7265	9.8613	1.3764	0.1387	54° 00′
10	.5901	7710	.8073	9070	.7310	8639	1.3680	1361	50
20	.5925	7727	.8056	9061	.7355	8666	1.3597	1334	40
30	.5948	7744	.8039	9052	.7400	8692	1.3514	1308	30
40	.5972	7761	.8021	9042	.7445	8718	1.3432	1282	20
50	.5995	7778	.8004	9033	.7490	8745	1.3351	1255	10
37° 00′	.6018	9.7795	.7986	9.9023	.7536	9.8771	1.3270	0.1229	53° 00′
10	.6041	7811	.7969	9014	.7581	8797	1.3190	1203	50
20	.6065	7828	.7951	9004	.7627	8824	1.3111	1176	40
30	.6088	7844	.7934	8995	.7673	8850	1.3032	1150	30
40	.6111	7861	.7916	8985	.7720	8876	1.2954	1124	20
50	.6134	7877	.7898	8975	.7766	8902	1.2876	1098	10
38° 00′	.6157	9.7893	.7880	9.8965	.7813	9.8928	1.2790	0.1072	52° 00′
10	.6180	7910	.7862	8955	.7860	8954	1.2723	1046	50
20	.6202	7926	.7844	8945	.7907	8980	1.2647	1020	40
30	.6225	7941	.7826	8935	.7954	9006	1.2572	0994	30
40	.6248	7957	.7808	8925	.8002	9032	1.2497	0968	20
50	.6271	7973	.7790	8915	.8050	9058	1.2423	0942	10
39° 00′	.6293	9.7989	.7771	9.8905	.8098	9.9084	1.2349	0.0916	51° 00′
10	.6316	8004	.7753	8895	.8146	9110	1.2276	0890	50
20	.6338	8020	.7735	8884	.8195	9135	1.2203	0865	40
30	.6361	8035	.7716	8874	.8243	9161	1.2131	0839	30
40	.6383	8050	.7698	8864	.8292	9187	1.2059	0813	20
50	.6406	8066	.7679	8853	.8342	9212	1.1988	0788	10
40° 00′	.6428	9.8081	.7660	9.8843	.8391	9.9238	1.1918	0.0762	50° 00′
10	.6450	8096	.7642	8832	.8441	9264	1.1847	0736	50
20	.6472	8111	.7623	8821	.8491	9289	1.1778	0711	40
30	.6494	8125	.7604	8810	.8541	9315	1.1708	0685	30
40	.6517	8140	.7585	8800	.8591	9341	1.1640	0659	20
50	.6539	8155	.7566	8789	.8642	9366	1.1571	0634	10
41° 00′	.6561	9.8169	.7547	9.8778	.8693	9.9392	1.1504	0.0608	49° 00′
10	.6583	8184	.7528	8767	.8744	9417	1.1436	0583	50
20	.6604	8198	.7509	8756	.8796	9443	1.1369	0557	40
30	.6626	8213	.7490	8745	.8847	9468	1.1303	0532	30
40	.6648	8227	.7470	8733	.8899	9494	1.1237	0506	20
50	.6670	8241	.7451	8722	.8952	9519	1.1171	0481	10
42° 00′	.6691	9.8255	.7431	9.8711	.9004	9.9544	1.1106	0.0456	48° 00′
10	.6713	8269	.7412	8699	.9057	9570	1.1041	0430	50
20	.6734	8283	.7392	8688	.9110	9595	1.0977	0405	40
30	.6756	8297	.7373	8676	.9163	9621	1.0913	0379	30
40	.6777	8311	.7353	8665	.9217	9646	1.0850	0354	20
50	.6799	8324	.7333	8653	.9271	9671	1.0786	0329	10
43° 00′	.6820	9.8338	.7314	9.8641	.9325	9.9697	1.0724	0.0303	47° 00′
10	.6841	8351	.7294	8629	.9380	9722	1.0661	0278	50
20	.6862	8365	.7274	8618	.9435	9747	1.0599	0253	40
30	.6884	8378	.7254	8606	.9490	9772	1.0538	0228	30
40	.6905	8391	.7234	8594	.9545	9798	1.0477	0202	20
50	.6926	8405	.7214	8582	.9601	9823	1.0416	0177	10
44° 00′	.6947	9.8418	.7193	9.8569	.9657	9.9848	1.0355	0.0152	46° 00′
10	.6967	8431	.7173	8557	.9713	9874	1.0295	0126	50
20	.6988	8444	.7153	8545	.9770	9899	1.0235	0101	40
30	.7009	8457	.7133	8532	.9827	9924	1.0176	0076	30
40	.7030	8469	.7112	8520	.9884	9949	1.0117	0051	20
50	.7050	8482	.7092	8507	.9942	9975	1.0058	0025	10
45° 00′	.7071	9.8495	.7071	9.8495	1.0000	0.0000	1.0000	0.0000	45° 00′
	Nat.	Log.	Nat.	Log.	Nat.	Log.	Nat.	Log.	
Angles	Cosines		Sines		Cotangents		Tangents		Angles

Table III.—Powers and Roots

No.	Sq.	Sq. Root	Cube	Cube Root	No.	Sq.	Sq. Root	Cube	Cube Root
1	1	1.000	1	1.000	51	2,601	7.141	132,651	3.708
2	4	1.414	8	1.260	52	2,704	7.211	140,608	3.733
3	9	1.732	27	1.442	53	2,809	7.280	148,877	3.756
4	16	2.000	64	1.587	54	2,916	7.348	157,464	3.780
5	25	2.236	125	1.710	55	3,025	7.416	166,375	3.803
6	36	2.449	216	1.817	56	3,136	7.483	175,616	3.826
7	49	2.646	343	1.913	57	3,249	7.550	185,193	3.849
8	64	2.828	512	2.000	58	3,364	7.616	195,112	3.871
9	81	3.000	729	2.080	59	3,481	7.681	205,379	3.893
10	100	3.162	1,000	2.154	60	3,600	7.746	216,000	3.915
11	121	3.317	1,331	2.224	61	3,721	7.810	226,981	3.936
12	144	3.464	1,728	2.289	62	3,844	7.874	238,328	3.958
13	169	3.606	2,197	2.351	63	3,969	7.937	250,047	3.979
14	196	3.742	2,744	2.410	64	4,096	8.000	262,144	4.000
15	225	3.873	3,375	2.466	65	4,225	8.062	274,625	4.021
16	256	4.000	4,096	2.520	66	4,356	8.124	287,496	4.041
17	289	4.123	4,913	2.571	67	4,489	8.185	300,763	4.062
18	324	4.243	5,832	2.621	68	4,624	8.246	314,432	4.082
19	361	4.359	6,859	2.668	69	4,761	8.307	328,509	4.102
20	400	4.472	8,000	2.714	70	4,900	8.367	343,000	4.121
21	441	4.583	9,261	2.759	71	5,041	8.426	357,911	4.141
22	484	4.690	10,648	2.802	72	5,184	8.485	373,248	4.160
23	529	4.796	12,167	2.844	73	5,329	8.544	389,017	4.179
24	576	4.899	13,824	2.884	74	5,476	8.602	405,224	4.198
25	625	5.000	15,625	2.924	75	5,625	8.660	421,875	4.217
26	676	5.099	17,576	2.962	76	5,776	8.718	438,976	4.236
27	729	5.196	19,683	3.000	77	5,929	8.775	456,533	4.254
28	784	5.291	21,952	3.037	78	6,084	8.832	474,552	4.273
29	841	5.385	24,389	3.072	79	6,241	8.888	493,039	4.291
30	900	5.477	27,000	3.107	80	6,400	8.944	512,000	4.309
31	961	5.568	29,791	3.141	81	6,561	9.000	531,441	4.327
32	1,024	5.657	32,768	3.175	82	6,724	9.055	551,368	4.344
33	1,089	5.745	35,937	3.208	83	6,889	9.110	571,787	4.362
34	1,156	5.831	39,304	3.240	84	7,056	9.165	592,704	4.380
35	1,225	5.916	42,875	3.271	85	7,225	9.220	614,125	4.397
36	1,296	6.000	46,656	3.302	86	7,396	9.274	636,056	4.414
37	1,369	6.083	50,653	3.332	87	7,569	9.327	658,503	4.431
38	1,444	6.164	54,872	3.362	88	7,744	9.381	681,472	4.448
39	1,521	6.245	59,319	3.391	89	7,921	9.434	704,969	4.465
40	1,600	6.325	64,000	3.420	90	8,100	9.487	729,000	4.481
41	1,681	6.403	68,921	3.448	91	8,281	9.539	753,571	4.498
42	1,764	6.481	74,088	3.476	92	8,464	9.592	778,688	4.514
43	1,849	6.557	79,507	3.503	93	8,649	9.644	804,357	4.531
44	1,936	6.633	85,184	3.530	94	8,836	9.695	830,584	4.547
45	2,025	6.708	91,125	3.557	95	9,025	9.747	857,375	4.563
46	2,116	6.782	97,336	3.583	96	9,216	9.798	884,736	4.579
47	2,209	6.856	103,823	3.609	97	9,409	9.849	912,673	4.595
48	2,304	6.928	110,592	3.634	98	9,604	9.899	941,192	4.610
49	2,401	7.000	117,649	3.659	99	9,801	9.950	970,299	4.626
50	2,500	7.071	125,000	3.684	100	10,000	10.000	1,000,000	4.642

TABLE IV.—NATURAL LOGARITHMS

N	0	1	2	3	4	5	6	7	8	9
1.0	0.0 000	100	198	296	392	488	583	677	770	862
1.1	953	*044	*133	*222	*310	*398	*484	*570	*655	*740
1.2	0.1 823	906	989	*070	*151	*231	*311	*390	*469	*546
1.3	0.2 624	700	776	852	927	*001	*075	*148	*221	*293
1.4	0.3 365	436	507	577	646	716	784	853	920	988
1.5	0.4 055	121	187	253	318	383	447	511	574	637
1.6	700	762	824	886	947	*008	*068	*128	*188	*247
1.7	0.5 306	365	423	481	539	596	653	710	766	822
1.8	878	933	988	*043	*098	*152	*206	*259	*313	*366
1.9	0.6 419	471	523	575	627	678	729	780	831	881
2.0	931	981	*031	*080	*129	*178	*227	*275	*324	*372
2.1	0.7 419	467	514	561	608	655	701	747	793	839
2.2	885	930	975	*020	*065	*109	*154	*198	*242	*286
2.3	0.8 329	372	416	459	502	544	587	629	671	713
2.4	755	796	838	879	920	961	*002	*042	*083	*123
2.5	0.9 163	203	243	282	322	361	400	439	478	517
2.6	555	594	632	670	708	746	783	821	858	895
2.7	933	969	*006	*043	*080	*116	*152	*188	*225	*260
2.8	1.0 296	332	367	403	438	473	508	543	578	613
2.9	647	682	716	750	784	818	852	886	919	953
3.0	986	*019	*053	*086	*119	*151	*184	*217	*249	*282
3.1	1.1 314	346	378	410	442	474	506	537	569	600
3.2	632	663	694	725	756	787	817	848	878	909
3.3	939	969	*000	*030	*060	*090	*119	*149	*179	*208
3.4	1.2 238	267	296	326	355	384	413	442	470	499
3.5	528	556	585	613	641	669	698	726	754	782
3.6	809	837	865	892	920	947	975	*002	*029	*056
3.7	1.3 083	110	137	164	191	218	244	271	297	324
3.8	350	376	402	429	455	481	507	533	558	584
3.9	610	635	661	686	712	737	762	788	813	838
4.0	863	888	913	938	962	987	*012	*036	*061	*085
4.1	1.4 110	134	159	183	207	231	255	279	303	327
4.2	351	375	398	422	446	469	493	516	540	563
4.3	586	609	633	656	679	702	725	748	770	793
4.4	816	839	861	884	907	929	951	974	996	*019
4.5	1.5 041	063	085	107	129	151	173	195	217	239
4.6	261	282	304	326	347	369	390	412	433	454
4.7	476	497	518	539	560	581	602	623	644	665
4.8	686	707	728	748	769	790	810	831	851	872
4.9	892	913	933	953	974	994	*014	*034	*054	*074
5.0	1.6 094	114	134	154	174	194	214	233	253	273

If given number $n = N \times 10^m$, then $\log_e n = \log_e N + m \log_e 10$. Find $m \log_e 10$ from the following table:

Multiples of $\log_e 10$

$\log_e 10 = 2.3026$	$-\log_e 10 = 7.6974 - 10$
$2 \log_e 10 = 4.6052$	$-2 \log_e 10 = 5.3948 - 10$
$3 \log_e 10 = 6.9078$	$-3 \log_e 10 = 3.0922 - 10$
$4 \log_e 10 = 9.2103$	$-4 \log_e 10 = 0.7897 - 10$
$5 \log_e 10 = 11.5129$	$-5 \log_e 10 = 9.4871 - 20$

Table IV.—Natural Logarithms.—(*Continued*)

N	0	1	2	3	4	5	6	7	8	9
5.0	1.6 094	114	134	154	174	194	214	233	253	273
5.1	292	312	332	351	371	390	409	429	448	467
5.2	487	506	525	544	563	582	601	620	639	658
5.3	677	696	715	734	752	771	790	808	827	845
5.4	864	882	901	919	938	956	974	993	*011	*029
5.5	1.7 047	066	084	102	120	138	156	174	192	210
5.6	228	246	263	281	299	317	334	352	370	387
5.7	405	422	440	457	475	492	509	527	544	561
5.8	579	596	613	630	647	664	681	699	716	733
5.9	750	766	783	800	817	834	851	867	884	901
6.0	918	934	951	967	984	*001	*017	*034	*050	*066
6.1	1.8 083	099	116	132	148	165	181	197	213	229
6.2	245	262	278	294	310	326	342	358	374	390
6.3	405	421	437	453	469	485	500	516	532	547
6.4	563	579	594	610	625	641	656	672	687	703
6.5	718	733	749	764	779	795	810	825	840	856
6.6	871	886	901	916	931	946	961	976	991	*006
6.7	1.9 021	036	051	066	081	095	110	125	140	155
6.8	169	184	199	213	228	242	257	272	286	301
6.9	315	330	344	359	373	387	402	416	430	445
7.0	459	473	488	502	516	530	544	559	573	587
7.1	601	615	629	643	657	671	685	699	713	727
7.2	741	755	769	782	796	810	824	838	851	865
7.3	879	892	906	920	933	947	961	974	988	*001
7.4	2.0 015	028	042	055	069	082	096	109	122	136
7.5	149	162	176	189	202	215	229	242	255	268
7.6	281	295	308	321	334	347	360	373	386	399
7.7	412	425	438	451	464	477	490	503	516	528
7.8	541	554	567	580	592	605	618	631	643	656
7.9	669	681	694	707	719	732	744	757	769	782
8.0	794	807	819	832	844	857	869	882	894	906
8.1	919	931	943	956	968	980	992	*005	*017	*029
8.2	2.1 041	054	066	080	090	102	114	126	138	150
8.3	163	175	187	199	211	223	235	247	258	270
8.4	282	294	306	318	330	342	353	365	377	389
8.5	401	412	424	436	448	460	471	483	494	506
8.6	518	529	541	552	564	576	587	599	610	622
8.7	633	645	656	668	679	691	702	713	725	736
8.8	748	759	770	782	793	804	815	827	838	849
8.9	861	872	883	894	905	917	928	939	950	961
9.0	972	983	994	*006	*017	*028	*039	*050	*061	*072
9.1	2.2 083	094	105	116	127	137	148	159	170	181
9.2	192	203	214	225	235	246	257	268	279	289
9.3	300	311	322	332	343	354	364	375	386	396
9.4	407	418	428	439	450	460	471	481	492	502
9.5	513	523	534	544	555	565	576	586	597	607
9.6	618	628	638	649	659	670	680	690	701	711
9.7	721	732	742	752	762	773	783	793	803	814
9.8	824	834	844	854	865	875	885	895	905	915
9.9	925	935	946	956	966	976	986	996	*006	*016
10.	2.3 026	036	046	056	066	076	086	096	106	115

TABLE V.—EXPONENTIAL AND HYPERBOLIC FUNCTIONS

x	e^x	e^{-x}	sinh x	cosh x	tanh x
.00	1.000	1.000	.000	1.000	.000
.01	1.010	.990	.010	1.000	.010
.02	1.020	.980	.020	1.000	.020
.03	1.030	.970	.030	1.000	.030
.04	1.041	.961	.040	1.001	.040
.05	1.051	.951	.050	1.001	.050
.06	1.062	.942	.060	1.002	.060
.07	1.073	.932	.070	1.002	.070
.08	1.083	.923	.080	1.003	.080
.09	1.094	.914	.090	1.004	.090
.1	1.105	.905	.100	1.005	.100
.2	1.221	.819	.201	1.020	.197
.3	1.350	.741	.305	1.045	.291
.4	1.492	.670	.411	1.081	.380
.5	1.649	.607	.521	1.128	.462
.6	1.822	.549	.637	1.185	.537
.7	2.014	.497	.759	1.255	.604
.8	2.226	.449	.888	1.337	.664
.9	2.460	.407	1.027	1.433	.716
1.0	2.718	.368	1.175	1.543	.762
1.1	3.004	.333	1.336	1.669	.800
1.2	3.320	.301	1.509	1.811	.834
1.3	3.669	.273	1.698	1.971	.862
1.4	4.055	.247	1.904	2.151	.885
1.5	4.482	.223	2.129	2.352	.905
1.6	4.953	.202	2.376	2.577	.922
1.7	5.474	.183	2.646	2.828	.935
1.8	6.050	.165	2.942	3.107	.947
1.9	6.686	.150	3.268	3.418	.956
2.0	7.389	.135	3.627	3.762	.964
2.1	8.166	.122	4.022	4.144	.970
2.2	9.025	.111	4.457	4.568	.976
2.3	9.974	.100	4.937	5.037	.980
2.4	11.023	.091	5.466	5.557	.984
2.5	12.182	.082	6.050	6.132	.987
2.6	13.464	.074	6.695	6.769	.989
2.7	14.880	.067	7.406	7.473	.991
2.8	16.445	.061	8.192	8.253	.993
2.9	18.174	.055	9.060	9.115	.994
3.0	20.086	.050	10.018	10.068	.995
3.1	22.20	.045	11.08	11.12	.996
3.2	24.53	.041	12.25	12.29	.997
3.3	27.11	.037	13.54	13.57	.997
3.4	29.96	.033	14.97	15.00	.998
3.5	33.12	.030	16.54	16.57	.998
3.6	36.60	.027	18.29	18.31	.999
3.7	40.45	.025	20.21	20.24	.999
3.8	44.70	.022	22.34	22.36	.999
3.9	49.40	.020	24.69	24.71	.999
4.0	54.60	.018	27.29	27.31	.999
4.1	60.34	.017	30.16	30.18	.999
4.2	66.69	.015	33.34	33.35	1.000
4.3	73.70	.014	36.84	36.86	1.000
4.4	81.45	.012	40.72	40.73	1.000
4.5	90.02	.011	45.00	45.01	1.000
4.6	99.48	.010	49.74	49.75	1.000
4.7	109.95	.0090	54.97	54.98	1.000
4.8	121.51	.0082	60.75	60.76	1.000
4.9	134.29	.0074	67.14	67.15	1.000
5.0	148.41	.0067	74.20	74.21	1.000
6.0	403.4	.0025	201.7		1.000
7.0	1096.6	.00091	548.3		1.000
8.0	2981.0	.00034	1490.5		1.000
9.0	8103.1	.00012	4051.5		1.000
10.0	22026.5	.000045	11013.2		1.000

Table VI.—Compound Amount of 1, $(1 + i)^n$

n	$\frac{1}{2}$ %	1 %	$1\frac{1}{4}$ %	$1\frac{1}{2}$ %	2 %	n
1	1.005 0000	1.010 0000	1.012 5000	1.015 0000	1.020 0000	1
2	1.010 0250	1.020 1000	1.025 1562	1.030 2250	1.040 4000	2
3	1.015 0751	1.030 3010	1.037 9707	1.045 6784	1.061 2080	3
4	1.020 1505	1.040 6040	1.050 9453	1.061 3636	1.082 4322	4
5	1.025 2512	1.051 0100	1.064 0822	1.077 2840	1.104 0808	5
6	1.030 3775	1.061 5202	1.077 3832	1.093 4433	1.126 1624	6
7	1.035 5294	1.072 1354	1.090 8505	1.109 8449	1.148 6857	7
8	1.040 7070	1.082 8567	1.104 4861	1.126 4926	1.171 6594	8
9	1.045 9106	1.093 6853	1.118 2922	1.143 3900	1.195 0926	9
10	1.051 1401	1.104 6221	1.132 2708	1.160 5408	1.218 9944	10
11	1.056 3958	1.115 6684	1.146 4242	1.177 9489	1.243 3743	11
12	1.061 6778	1.126 8250	1.160 7545	1.195 6182	1.268 2418	12
13	1.066 9862	1.138 0933	1.175 2640	1.213 5524	1.293 6066	13
14	1.072 3211	1.149 4742	1.189 9548	1.231 7557	1.319 4788	14
15	1.077 6827	1.160 9690	1.204 8292	1.250 2321	1.345 8683	15
16	1.083 0712	1.172 5786	1.219 8896	1.268 9856	1.372 7857	16
17	1.088 4865	1.184 3044	1.235 1382	1.288 0203	1.400 2414	17
18	1.093 9289	1.196 1475	1.250 5774	1.307 3406	1.428 2462	18
19	1.099 3986	1.208 1090	1.266 2096	1.326 9508	1.456 8112	19
20	1.104 8956	1.220 1900	1.282 0372	1.346 8550	1.485 9474	20
21	1.110 4201	1.232 3919	1.298 0627	1.367 0578	1.515 6663	21
22	1.115 9722	1.244 7159	1.314 2885	1.387 5637	1.545 9797	22
23	1.121 5520	1.257 1630	1.330 7171	1.408 3772	1.576 8993	23
24	1.127 1598	1.269 7346	1.347 3510	1.429 5028	1.608 4372	24
25	1.132 7956	1.282 4320	1.364 1929	1.450 9454	1.640 6060	25
26	1.138 4596	1.295 2563	1.381 2454	1.472 7095	1.673 4181	26
27	1.144 1518	1.308 2089	1.398 5109	1.494 8002	1.706 8865	27
28	1.149 8726	1.321 2910	1.415 9923	1.517 2222	1.741 0242	28
29	1.155 6220	1.334 5039	1.433 6922	1.539 9805	1.775 8447	29
30	1.161 4001	1.347 8489	1.451 6134	1.563 0802	1.811 3616	30
31	1.167 2071	1.361 3274	1.469 7585	1.586 5264	1.847 5888	31
32	1.173 0431	1.374 9407	1.488 1305	1.610 3243	1.884 5406	32
33	1.178 9083	1.388 6901	1.506 7321	1.634 4792	1.922 2314	33
34	1.184 8029	1.402 5770	1.525 5663	1.658 9964	1.960 6760	34
35	1.190 7269	1.416 6028	1.544 6359	1.683 8813	1.999 8896	35
36	1.196 6805	1.430 7688	1.563 9438	1.709 1395	2.039 8873	36
37	1.202 6639	1.445 0765	1.583 4931	1.734 7766	2.080 6851	37
38	1.208 6772	1.459 5272	1.603 2868	1.760 7983	2.122 2988	38
39	1.214 7206	1.474 1225	1.623 3279	1.787 2102	2.164 7448	39
40	1.220 7942	1.488 8637	1.643 6195	1.814 0184	2.208 0397	40
41	1.226 8982	1.503 7524	1.664 1647	1.841 2287	2.252 2005	41
42	1.233 0327	1.518 7899	1.684 9668	1.868 8471	2.297 2445	42
43	1.239 1979	1.533 9778	1.706 0288	1.896 8798	2.343 1894	43
44	1.245 3938	1.549 3176	1.727 3542	1.925 3330	2.390 0531	44
45	1.251 6208	1.564 8108	1.748 9461	1.954 2130	2.437 8542	45
46	1.257 8789	1.580 4588	1.770 8080	1.983 5262	2.486 6113	46
47	1.264 1683	1.596 2634	1.792 9431	2.013 2791	2.536 3435	47
48	1.270 4892	1.612 2261	1.815 3548	2.043 4783	2.587 0704	48
49	1.276 8416	1.628 3483	1.838 0468	2.074 1305	2.638 8118	49
50	1.283 2258	1.644 6318	1.861 0224	2.105 2424	2.691 5880	50
60	1.348 8502	1.816 6967	2.107 1814	2.443 2198	3.281 0308	60
70	1.417 8305	2.006 7634	2.385 9000	2.835 4563	3.999 5582	70
80	1.490 3386	2.216 7152	2.701 4849	3.290 6628	4.875 4392	80
90	1.566 5547	2.448 6327	3.058 8126	3.818 9485	5.943 1331	90
100	1.646 6685	2.704 8138	3.463 4043	4.432 0456	7.244 6461	100

TABLE VI.—COMPOUND AMOUNT OF 1, $(1 + i)^n$.—(*Continued*)

n	$2\frac{1}{2}$ %	3 %	$3\frac{1}{2}$ %	4 %	$4\frac{1}{2}$ %	n
1	1.025 0000	1.030 0000	1.035 0000	1.040 0000	1.045 0000	1
2	1.050 6250	1.060 9000	1.071 2250	1.081 6000	1.092 0250	2
3	1.076 8906	1.092 7270	1.108 7179	1.124 8640	1.141 1661	3
4	1.103 8129	1.125 5088	1.147 5230	1.169 8586	1.192 5186	4
5	1.131 4082	1.159 2741	1.187 6863	1.216 6529	1.246 1819	5
6	1.159 6934	1.194 0523	1.229 2553	1.265 3190	1.302 2601	6
7	1.188 6858	1.229 8739	1.272 2793	1.315 9318	1.360 8618	7
8	1.218 4029	1.266 7701	1.316 8090	1.368 5690	1.422 1006	8
9	1 248 8630	1.304 7732	1.362 8974	1.423 3118	1.486 0951	9
10	1.280 0845	1.343 9164	1.410 5988	1.480 2443	1.552 9694	10
11	1.312 0867	1.384 2339	1.459 9697	1.539 4541	1.622 8530	11
12	1.344 8888	1.425 7609	1.511 0687	1.601 0322	1.695 8814	12
13	1.378 5110	1.468 5337	1.563 9561	1.665 0735	1.772 1961	13
14	1.412 9738	1.512 5897	1.618 6945	1.731 6764	1.851 9449	14
15	1.448 2982	1.557 9674	1.675 3488	1.800 9435	1.935 2824	15
16	1.484 5056	1.604 7064	1.733 9860	1.872 9812	2.022 3702	16
17	1.521 6183	1.652 8476	1.794 6756	1.947 9005	2.113 3768	17
18	1.559 6587	1.702 4331	1.857 4892	2.025 8165	2.208 4788	18
19	1.598 6502	1.753 5060	1.922 5013	2.106 8492	2.307 8603	19
20	1.638 6164	1.806 1112	1.989 7889	2.191 1231	2.411 7140	20
21	1.679 5818	1.860 2946	2.059 4315	2.278 7681	2.520 2412	21
22	1.721 5714	1.916 1034	2.131 5116	2.369 9188	2.633 6520	22
23	1.764 6107	1.973 5865	2.206 1145	2.464 7155	2.752 1664	23
24	1.808 7260	2.032 7941	2.283 3285	2.563 3042	2.876 0138	24
25	1.853 9441	2.093 7779	2.363 2450	2.665 8363	3.005 4345	25
26	1.900 2927	2.156 5913	2.445 9586	2.772 4698	3.140 6790	26
27	1.947 8000	2.221 2890	2.531 5671	2.883 3686	3.282 0096	27
28	1.996 4950	2.287 9277	2.620 1720	2.998 7033	3.429 7000	28
29	2.046 4074	2.356 5655	2.711 8780	3.118 6514	3.584 0365	29
30	2.097 5676	2.427 2625	2.806 7937	3.243 3975	3.745 3181	30
31	2.150 0068	2.500 0804	2.905 0315	3.373 1334	3.913 8574	31
32	2.203 7569	2.575 0828	3.006 7076	3.508 0588	4.089 9810	32
33	2.258 8509	2.652 3352	3.111 9424	3.648 3811	4.274 0302	33
34	2.315 3221	2.731 9053	3.220 8603	3.794 3163	4.466 3615	34
35	2.373 2052	2.813 8624	3.333 5904	3.946 0890	4.667 3478	35
36	2.432 5353	2.898 2783	3.450 2661	4.103 9326	4.877 3785	36
37	2.493 3487	2.985 2267	3.571 0254	4.268 0899	5.096 8605	37
38	2.555 6824	3.074 7835	3.696 0113	4.438 8134	5.326 2192	38
39	2.619 5745	3.167 0270	3.825 3717	4.616 3660	5.565 8991	39
40	2.685 0638	3.262 0378	3.959 2597	4.801 0206	5.816 3645	40
41	2.752 1904	3.359 8989	4.097 8338	4.993 0614	6.078 1009	41
42	2.820 9952	3.460 6959	4.241 2580	5.192 7839	6.351 6155	42
43	2.891 5201	3.564 5168	4.389 7020	5.400 4953	6.637 4382	43
44	2.963 8081	3.671 4523	4.543 3416	5.616 5151	6.936 1229	44
45	3.037 9033	3.781 5958	4.702 3586	5.841 1757	7.248 2484	45
46	3.113 8509	3.895 0437	4.866 9411	6.074 8227	7.574 4196	46
47	3.191 6971	4.011 8950	5.037 2840	6.317 8156	7.915 2685	47
48	3.271 4896	4.132 2519	5.213 5890	6.570 5282	8.271 4556	48
49	3.353 2768	4.256 2194	5.396 0646	6.833 3494	8.643 6711	49
50	3.437 1087	4.383 9060	5.584 9269	7.106 6834	9.032 6363	50
60	4.399 7898	5.891 6031	7.878 0909	10.519 6274	14.027 4079	60
70	5.632 1029	7.917 8219	11.112 8253	15.571 6184	21.784 1356	70
80	7.209 5678	10.640 8906	15.675 7375	23.049 7991	33.830 0964	80
90	9.228 8563	14.300 4671	22.112 1760	34.119 3333	52.537 1053	90
100	11.813 7164	19.218 6320	31.191 4080	50.504 9482	81.588 5180	100

TABLE VI.—COMPOUND AMOUNT OF 1, $(1 + i)^n$.—(*Continued*)

n	5 %	5½ %	6 %	7 %	8 %	n
1	1.050 0000	1.055 0000	1.060 0000	1.070 0000	1.080 0000	1
2	1.102 5000	1.113 0250	1.123 6000	1.144 9000	1.166 4000	2
3	1.157 6250	1.174 2414	1.191 0160	1.225 0430	1.259 7120	3
4	1.215 5062	1.238 8247	1.262 4770	1.310 7960	1.360 4890	4
5	1.276 2816	1.306 9600	1.338 2256	1.402 5517	1.469 3281	5
6	1.340 0956	1.378 8428	1.418 5191	1.500 7304	1.586 8743	6
7	1.407 1004	1.454 6792	1.503 6303	1.605 7815	1.713 8243	7
8	1.477 4554	1.534 6865	1.593 8481	1.718 1862	1.850 9302	8
9	1.551 3282	1.619 0943	1.689 4790	1.838 4592	1.999 0046	9
10	1.628 8946	1.708 1445	1.790 8477	1.967 1514	2.158 9250	10
11	1.710 3394	1.802 0924	1.898 2986	2.104 8520	2.331 6390	11
12	1.795 8563	1.901 2075	2.012 1965	2.252 1916	2.518 1701	12
13	1.885 6491	2.005 7739	2.132 9283	2.409 8450	2.719 6237	13
14	1.979 9316	2.116 0915	2.260 9040	2.578 5342	2.937 1936	14
15	2.078 9282	2.232 4765	2.396 5582	2.759 0315	3.172 1691	15
16	2.182 8746	2.355 2627	2.540 3517	2.952 1638	3.425 9426	16
17	2.292 0183	2.484 8021	2.692 7728	3.158 8152	3.700 0180	17
18	2.406 6192	2.621 4663	2.854 3392	3.379 9323	3.996 0195	18
19	2.526 9502	2.765 6469	3.025 5995	3.616 5275	4.315 7011	19
20	2.653 2977	2.917 7575	3.207 1355	3.869 6845	4.660 9571	20
21	2.785 9626	3.078 2342	3.399 5636	4.140 5624	5.033 8337	21
22	2.925 2607	3.247 5370	3.603 5374	4.430 4017	5.436 5404	22
23	3.071 5238	3.426 1516	3.819 7497	4.740 5299	5.871 4636	23
24	3.225 0999	3.614 5899	4.048 9346	5.072 3670	6.341 1807	24
25	3.386 3549	3.813 3923	4.291 8707	5.427 4326	6.848 4752	25
26	3.555 6727	4.023 1289	4.549 3830	5.807 3529	7.396 3532	26
27	3.733 4563	4.244 4010	4.822 3459	6.213 8676	7.988 0615	27
28	3.920 1291	4.477 8431	5.111 6867	6.648 8384	8.627 1064	28
29	4.116 1356	4.724 1244	5.418 3879	7.114 2570	9.317 2749	29
30	4.321 9424	4.983 9513	5.743 4912	7.612 2550	10.062 6569	30
31	4.538 0395	5.258 0686	6.088 1006	8.145 1129	10.867 6694	31
32	4.764 9415	5.547 2624	6.453 3867	8.715 2708	11.737 0830	32
33	5.003 1885	5.852 3618	6.840 5899	9.325 3398	12.676 0496	33
34	5.253 3480	6.174 2417	7.251 0253	9.978 1135	13.690 1336	34
35	5.516 0154	6.513 8250	7.686 0868	10.676 5815	14.785 3443	35
36	5.791 8161	6.872 0854	8.147 2520	11.423 9422	15.968 1718	36
37	6.081 4069	7.250 0501	8.636 0871	12.223 6181	17.245 6256	37
38	6.385 4773	7.648 8028	9.154 2524	13.079 2714	18.625 2756	38
39	6.704 7512	8.069 4870	9.703 5075	13.994 8204	20.115 2977	39
40	7.039 9887	8.513 3088	10.285 7179	14.974 4578	21.724 5215	40
41	7.391 9882	8.981 5408	10.902 8610	16.022 6699	23.462 4832	41
42	7.761 5876	9.475 5255	11.557 0327	17.144 2568	25.339 4819	42
43	8.149 6669	9.996 6794	12.250 4546	18.344 3548	27.366 6404	43
44	8.557 1503	10.546 4968	12.985 4819	19.628 4596	29.555 9717	44
45	8.985 0078	11.126 5541	13.764 6108	21.002 4518	31.920 4494	45
46	9.434 2582	11.738 5146	14.590 4875	22.472 6234	34.474 0853	46
47	9.905 9711	12.384 1329	15.465 9167	24.045 7070	37.232 0122	47
48	10.401 2696	13.065 2602	16.393 8717	25.728 9065	40.210 5731	48
49	10.921 3331	13.783 8495	17.377 5040	27.529 9300	43.427 4190	49
50	11.467 3998	14.541 9612	18.420 1543	29.457 0251	46.901 6125	50
60	18.679 1859	24.839 7704	32.987 6908	57.946 4268	101.257 0637	60
70	30.426 4255	42.429 9162	59.075 9302	113.989 3922	218.606 4059	70
80	49.561 4411	72.476 4263	105.795 9935	224.234 3876	471.954 8343	80
90	80.730 3650	123.800 2059	189.464 5112	441.102 9799	1018.915 0893	90
100	131.501 2578	211.468 6357	339.302 0835	867.716 3256	2199.761 2563	100

TABLE VII.—PRESENT VALUE OF 1, $(1 + i)^{-n}$

n	$\frac{1}{2}$ %	1 %	$1\frac{1}{4}$ %	$1\frac{1}{2}$ %	2 %	n
1	0.995 0249	0.990 0990	0.987 6543	0.985 2217	0.980 3922	**1**
2	0.990 0745	0.980 2960	0.975 4611	0.970 6618	0.961 1688	**2**
3	0.985 1488	0.970 5902	0.963 4183	0.956 3170	0.942 3223	**3**
4	0.980 2475	0.960 9803	0.951 5243	0.942 1842	0.923 8454	**4**
5	0.975 3707	0.951 4657	0.939 7771	0.928 2603	0.905 7308	**5**
6	0.970 5181	0.942 0452	0.928 1749	0.914 5422	0.887 9714	**6**
7	0.965 6896	0.932 7180	0.916 7159	0.901 0268	0.870 5602	**7**
8	0.960 8852	0.923 4832	0.905 3984	0.887 7111	0.853 4904	**8**
9	0.956 1047	0.914 3398	0.894 2207	0.874 5922	0.836 7553	**9**
10	0.951 3479	0.905 2870	0.883 1809	0.861 6672	0.820 3483	**10**
11	0.946 6149	0.896 3237	0.872 2775	0.848 9332	0.804 2630	**11**
12	0.941 9053	0.887 4492	0.861 5086	0.836 3874	0.788 4932	**12**
13	0.937 2192	0.878 6626	0.850 8727	0.824 0270	0.773 0325	**13**
14	0.932 5565	0.869 9630	0.840 3681	0.811 8493	0.757 8750	**14**
15	0.927 9169	0.861 3495	0.829 9932	0.799 8515	0.743 0147	**15**
16	0.923 3004	0.852 8213	0.819 7464	0.788 0310	0.728 4458	**16**
17	0.918 7068	0.844 3775	0.809 6260	0.776 3853	0.714 1626	**17**
18	0.914 1362	0.836 0173	0.799 6306	0.764 9116	0.700 1594	**18**
19	0.909 5882	0.827 7399	0.789 7587	0.753 6075	0.686 4308	**19**
20	0.905 0629	0.819 5445	0.780 0086	0.742 4704	0.672 9713	**20**
21	0.900 5601	0.811 4302	0.770 3788	0.731 4980	0.659 7758	**21**
22	0.896 0797	0.803 3962	0.760 8680	0.720 6876	0.646 8390	**22**
23	0.891 6216	0.795 4418	0.751 4745	0.710 0371	0.634 1559	**23**
24	0.887 1857	0.787 5661	0.742 1971	0.699 5439	0.621 7215	**24**
25	0.882 7718	0.779 7684	0.733 0341	0.689 2058	0.609 5309	**25**
26	0.878 3799	0.772 0480	0.723 9843	0.679 0205	0.597 5793	**26**
27	0.874 0099	0.764 4039	0.715 0463	0.668 9857	0.585 8620	**27**
28	0.869 6616	0.756 8356	0.706 2185	0.659 0992	0.574 3746	**28**
29	0.865 3349	0.749 3422	0.697 4998	0.649 3589	0.563 1123	**29**
30	0.861 0297	0.741 9229	0.688 8887	0.639 7624	0.552 0709	**30**
31	0.856 7460	0.734 5772	0.680 3839	0.630 3078	0.541 2460	**31**
32	0.852 4836	0.727 3041	0.671 9841	0.620 9929	0.530 6333	**32**
33	0.848 2424	0.720 1031	0.663 6880	0.611 8157	0.520 2287	**33**
34	0.844 0223	0.712 9733	0.655 4943	0.602 7741	0.510 0282	**34**
35	0.839 8231	0.705 9142	0.647 4018	0.593 8661	0.500 0276	**35**
36	0.835 6449	0.698 9250	0.639 4092	0.585 0897	0.490 2232	**36**
37	0.831 4875	0.692 0049	0.631 5152	0.576 4431	0.480 6109	**37**
38	0.827 3507	0.685 1534	0.623 7187	0.567 9242	0.471 1872	**38**
39	0.823 2346	0.678 3697	0.616 0185	0.559 5313	0.461 9482	**39**
40	0.819 1389	0.671 6531	0.608 4133	0.551 2623	0.452 8904	**40**
41	0.815 0635	0.665 0031	0.600 9021	0.543 1156	0.444 0102	**41**
42	0.811 0085	0.658 4189	0.593 4835	0.535 0892	0.435 3041	**42**
43	0.806 9736	0.651 8999	0.586 1566	0.527 1815	0.426 7688	**43**
44	0.802 9588	0.645 4455	0.578 9201	0.519 3907	0.418 4007	**44**
45	0.798 9640	0.639 0549	0.571 7729	0.511 7149	0.410 1968	**45**
46	0.794 9891	0.632 7276	0.564 7140	0.504 1526	0.402 1537	**46**
47	0.791 0339	0.626 4630	0.557 7422	0.496 7021	0.394 2684	**47**
48	0.787 0984	0.620 2604	0.550 8565	0.489 3617	0.386 5376	**48**
49	0.783 1825	0.614 1192	0.544 0558	0.482 1298	0.378 9584	**49**
50	0.779 2861	0.608 0388	0.537 3390	0.475 0047	0.371 5279	**50**
60	0.741 3722	0.550 4496	0.474 5676	0.409 2960	0.304 7823	**60**
70	0.705 3029	0.498 3149	0.419 1290	0.352 6769	0.250 0276	**70**
80	0.670 9885	0.451 1179	0.370 1668	0.303 8902	0.205 1097	**80**
90	0.638 3435	0.408 3912	0.326 9242	0.261 8522	0.168 2614	**90**
100	0.607 2868	0.369 7112	0.288 7333	0.225 6294	0.138 0330	**100**

TABLE VII.—PRESENT VALUE OF 1, $(1+i)^{-n}$.—(*Continued*)

n	$2\frac{1}{2}$ %	3 %	$3\frac{1}{2}$ %	4 %	$4\frac{1}{2}$ %	n
1	0.975 6098	0.970 8738	0.966 1836	0.961 5385	0.956 9378	1
2	0.951 8144	0.942 5959	0.933 5107	0.924 5562	0.915 7300	2
3	0.928 5994	0.915 1417	0.901 9427	0.888 9964	0.876 2966	3
4	0.905 9506	0.888 4870	0.871 4422	0.854 8042	0.838 5613	4
5	0.883 8543	0.862 6088	0.841 9732	0.821 9271	0.802 4510	5
6	0.862 2969	0.837 4843	0.813 5006	0.790 3145	0.767 8957	6
7	0.841 2652	0.813 0915	0.785 9910	0.759 9178	0.734 8285	7
8	0.820 7466	0.789 4092	0.759 4116	0.730 6902	0.703 1851	8
9	0.800 7284	0.766 4167	0.733 7310	0.702 5867	0.672 9044	9
10	0.781 1984	0.744 0939	0.708 9188	0.675 5642	0.643 9277	10
11	0.762 1448	0.722 4213	0.684 9457	0.649 5809	0.616 1987	11
12	0.743 5559	0.701 3799	0.661 7833	0.624 5970	0.589 6639	12
13	0.725 4204	0.680 9513	0.639 4042	0.600 5741	0.564 2716	13
14	0.707 7272	0.661 1178	0.617 7818	0.577 4751	0.539 9729	14
15	0.690 4656	0.641 8620	0.596 8906	0.555 2645	0.516 7204	15
16	0.673 6249	0.623 1669	0.576 7059	0.533 9082	0.494 4693	16
17	0.657 1951	0.605 0164	0.557 2038	0.513 3732	0.473 1764	17
18	0.641 1659	0.587 3946	0.538 3611	0.493 6281	0.452 8004	18
19	0.625 5277	0.570 2860	0.520 1557	0.474 6424	0.433 3018	19
20	0.610 2709	0.553 6758	0.502 5659	0.456 3870	0.414 6429	20
21	0.595 3863	0.537 5493	0.485 5709	0.438 8336	0.396 7874	21
22	0.580 8647	0.521 8925	0.469 1506	0.421 9554	0.379 7009	22
23	0.566 6972	0.506 6918	0.453 2856	0.405 7263	0.363 3501	23
24	0.552 8754	0.491 9337	0.437 9571	0.390 1215	0.347 7035	24
25	0.539 3906	0.477 6056	0.423 1470	0.375 1168	0.332 7306	25
26	0.526 2347	0.463 6947	0.408 8377	0.360 6892	0.318 4025	26
27	0.513 3997	0.450 1891	0.395 0122	0.346 8166	0.304 6914	27
28	0.500 8778	0.437 0768	0.381 6543	0.333 4775	0.291 5707	28
29	0.488 6612	0.424 3464	0.368 7482	0.320 6514	0.279 0150	29
30	0.476 7427	0.411 9868	0.356 2784	0.308 3187	0.267 0000	30
31	0.465 1148	0.399 9872	0.344 2304	0.296 4603	0.255 5024	31
32	0.453 7706	0.388 3370	0.332 5897	0.285 0579	0.244 4999	32
33	0.442 7030	0.377 0262	0.321 3427	0.274 0942	0.233 9712	33
34	0.431 9053	0.366 0449	0.310 4760	0.263 5521	0.223 8959	34
35	0.421 3711	0.355 3834	0.299 9769	0.253 4155	0.214 2544	35
36	0.411 0937	0.345 0324	0.289 8327	0.243 6687	0.205 0282	36
37	0.401 0670	0.334 9829	0.280 0316	0.234 2968	0.196 1992	37
38	0.391 2849	0.325 2262	0.270 5619	0.225 2854	0.187 7504	38
39	0.381 7414	0.315 7536	0.261 4125	0.216 6206	0.179 6655	39
40	0.372 4306	0.306 5568	0.252 5725	0.208 2890	0.171 9287	40
41	0.363 3470	0.297 6280	0.244 0314	0.200 2779	0.164 5251	41
42	0.354 4848	0.288 9592	0.235 7791	0.192 5749	0.157 4403	42
43	0.345 8389	0.280 5429	0.227 8059	0.185 1682	0.150 6605	43
44	0.337 4038	0.272 3718	0.220 1023	0.178 0464	0.144 1728	44
45	0.329 1744	0.264 4386	0.212 6592	0.171 1984	0.137 9644	45
46	0.321 1458	0.256 7365	0.205 4679	0.164 6139	0.132 0233	46
47	0.313 3129	0.249 2588	0.198 5197	0.158 2826	0.126 3381	47
48	0.305 6712	0.241 9988	0.191 8064	0.152 1948	0.120 8977	48
49	0.298 2158	0.234 9503	0.185 3202	0.146 3411	0.115 6916	49
50	0.290 9422	0.228 1071	0.179 0534	0.140 7126	0.110 7096	50
60	0.227 2836	0.169 7331	0.126 9343	0.095 0604	0.071 2890	60
70	0.177 5536	0.126 2974	0.089 9861	0.064 2194	0.045 9050	70
80	0.138 7046	0.093 9771	0.063 7928	0.043 3843	0.029 5595	80
90	0.108 3558	0.069 9278	0.045 2240	0.029 3089	0.019 0342	90
100	0.084 6474	0.052 0328	0.032 0601	0.019 8000	0.012 2566	100

TABLE VII.—PRESENT VALUE OF 1, $(1 + i)^{-n}$.—(*Continued*)

n	5 %	5½ %	6 %	7 %	8 %	n
1	0.952 3810	0.947 8673	0.943 3962	0.934 5794	0.925 9259	**1**
2	0.907 0295	0.898 4524	0.889 9964	0.873 4387	0.857 3388	**2**
3	0.863 8376	0.851 6137	0.839 6193	0.816 2979	0.793 8322	**3**
4	0.822 7025	0.807 2167	0.792 0937	0.762 8952	0.735 0298	**4**
5	0.783 5262	0.765 1344	0.747 2582	0.712 9862	0.680 5832	**5**
6	0.746 2154	0.725 2458	0.704 9605	0.666 3422	0.630 1696	**6**
7	0.710 6813	0.687 4368	0.665 0571	0.622 7497	0.583 4904	**7**
8	0.676 8394	0.651 5989	0.627 4124	0.582 0091	0.540 2689	**8**
9	0.644 6089	0.617 6293	0.591 8985	0.543 9337	0.500 2490	**9**
10	0.613 9132	0.585 4306	0.558 3948	0.508 3493	0.463 1935	**10**
11	0.584 6793	0.554 9105	0.526 7875	0.475 0928	0.428 8829	**11**
12	0.556 8374	0.525 9815	0.496 9694	0.444 0120	0.397 1138	**12**
13	0.530 3214	0.498 5607	0.468 8390	0.414 9644	0.367 6979	**13**
14	0.505 0680	0.472 5694	0.442 3010	0.387 8172	0.340 4610	**14**
15	0.481 0171	0.447 9331	0.417 2651	0.362 4460	0.315 2417	**15**
16	0.458 1115	0.424 5811	0.393 6463	0.338 7346	0.291 8905	**16**
17	0.436 2967	0.402 4465	0.371 3644	0.316 5744	0.270 2690	**17**
18	0.415 5206	0.381 4659	0.350 3438	0.295 8639	0.250 2490	**18**
19	0.395 7340	0.361 5791	0.330 5130	0.276 5083	0.231 7121	**19**
20	0.376 8895	0.342 7290	0.311 8047	0.258 4190	0.214 5482	**20**
21	0.358 9424	0.324 8616	0.294 1554	0.241 5131	0.198 6558	**21**
22	0.341 8499	0.307 9257	0.277 5051	0.225 7132	0.183 9405	**22**
23	0.325 5713	0.291 8727	0.261 7973	0.210 9469	0.170 3153	**23**
24	0.310 0679	0.276 6566	0.246 9786	0.197 1466	0.157 6993	**24**
25	0.295 3028	0.262 2337	0.232 9986	0.184 2492	0.146 0179	**25**
26	0.281 2407	0.248 5627	0.219 8100	0.172 1955	0.135 2018	**26**
27	0.267 8483	0.235 6045	0.207 3680	0.160 9304	0.125 1868	**27**
28	0.255 0936	0.223 3218	0.195 6301	0.150 4022	0.115 9137	**28**
29	0.242 9463	0.211 6794	0.184 5567	0.140 5628	0.107 3275	**29**
30	0.231 3774	0.200 6440	0.174 1101	0.131 3671	0.099 3773	**30**
31	0.220 3595	0.190 1839	0.164 2548	0.122 7730	0.092 0160	**31**
32	0.209 8662	0.180 2691	0.154 9574	0.114 7411	0.085 2000	**32**
33	0.199 8725	0.170 8712	0.146 1862	0.107 2347	0.078 8889	**33**
34	0.190 3548	0.161 9632	0.137 9115	0.100 2193	0.073 0453	**34**
35	0.181 2903	0.153 5196	0.130 1052	0.093 6629	0.067 6345	**35**
36	0.172 6574	0.145 5162	0.122 7408	0.087 5355	0.062 6246	**36**
37	0.164 4356	0.137 9301	0.115 7932	0.081 8088	0.057 9857	**37**
38	0.156 6054	0.130 7394	0.109 2388	0.076 4569	0.053 6905	**38**
39	0.149 1480	0.123 9236	0.103 0555	0.071 4550	0.049 7134	**39**
40	0.142 0457	0.117 4631	0.097 2222	0.066 7804	0.046 0309	**40**
41	0.135 2816	0.111 3395	0.091 7190	0.062 4116	0.042 6212	**41**
42	0.128 8396	0.105 5350	0.086 5274	0.058 3286	0.039 4641	**42**
43	0.122 7044	0.100 0332	0.081 6296	0.054 5127	0.036 5408	**43**
44	0.116 8613	0.094 8182	0.077 0091	0.050 9464	0.033 8341	**44**
45	0.111 2965	0.089 8751	0.072 6501	0.047 6135	0.031 3279	**45**
46	0.105 9967	0.085 1897	0.068 5378	0.044 4986	0.029 0073	**46**
47	0.100 9492	0.080 7485	0.064 6583	0.041 5875	0.026 8586	**47**
48	0.096 1421	0.076 5389	0.060 9984	0.038 8668	0.024 8691	**48**
49	0.091 5639	0.072 5487	0.057 5457	0.036 3241	0.023 0269	**49**
50	0.087 2037	0.068 7665	0.054 2884	0.033 9478	0.021 3212	**50**
60	0.053 5355	0.040 2580	0.030 3143	0.017 2573	0.009 8758	**60**
70	0.032 8662	0.023 5683	0.016 9274	0.008 7728	0.004 5744	**70**
80	0.020 1770	0.013 7976	0.009 4522	0.004 4596	0.002 1188	**80**
90	0.012 3869	0.008 0775	0.005 2780	0.002 2670	0.000 9814	**90**
100	0.007 6045	0.004 7288	0.002 9472	0.001 1524	0.000 4546	**100**

TABLE VIII.—AMOUNT OF ANNUITY OF 1 PER PERIOD, $S_{\overline{n}|i}$

n	$\frac{1}{2}$ %	1 %	$1\frac{1}{4}$ %	$1\frac{1}{2}$ %	2 %	n
1	1.000 0000	1.000 0000	1.000 0000	1.000 0000	1.000 0000	**1**
2	2.005 0000	2.010 0000	2.012 5000	2.015 0000	2.020 0000	**2**
3	3.015 0250	3.030 1000	3.037 6562	3.045 2250	3.060 4000	**3**
4	4.030 1001	4.060 4010	4.075 6270	4.090 9034	4.121 6080	**4**
5	5.050 2506	5.101 0050	5.126 5723	5.152 2669	5.204 0402	**5**
6	6.075 5019	6.152 0151	6.190 6544	6.229 5509	6.308 1210	**6**
7	7.105 8794	7.213 5352	7.268 0376	7.322 9942	7.434 2834	**7**
8	8.141 4088	8.285 6706	8.358 8881	8.432 8391	8.582 9690	**8**
9	9.182 1158	9.368 5273	9.463 3742	9.559 3317	9.754 6284	**9**
10	10.228 0264	10.462 2125	10.581 6664	10.702 7217	10.949 7210	**10**
11	11.279 1665	11.566 8347	11.713 9372	11.863 2625	12.168 7154	**11**
12	12.335 5624	12.682 5030	12.860 3614	13.041 2114	13.412 0897	**12**
13	13.397 2402	13.809 3280	14.021 1159	14.236 8296	14.680 3315	**13**
14	14.464 2264	14.947 4213	15.196 3799	15.450 3820	15.973 9382	**14**
15	15.536 5475	16.096 8955	16.386 3346	16.682 1378	17.293 4169	**15**
16	16.614 2303	17.257 8645	17.591 1638	17.932 3698	18.639 2852	**16**
17	17.697 3014	18.430 4431	18.811 0534	19.201 3554	20.012 0710	**17**
18	18.785 7879	19.614 7476	20.046 1915	20.489 3757	21.412 3124	**18**
19	19.879 7168	20.810 8950	21.296 7689	21.796 7164	22.840 5586	**19**
20	20.979 1154	22.019 0040	22.562 9785	23.123 6671	24.297 3698	**20**
21	22.084 0110	23.239 1940	23.845 0158	24.470 5221	25.783 3172	**21**
22	23.194 4311	24.471 5860	25.143 0785	25.837 5799	27.298 9835	**22**
23	24.310 4032	25.716 3018	26.457 3670	27.225 1436	28.844 9632	**23**
24	25.431 9552	26.973 4648	27.788 0840	28.633 5208	30.421 8625	**24**
25	26.559 1150	28.243 1995	29.135 4351	30.063 0236	32.030 2997	**25**
26	27.691 9106	29.525 6315	30.499 6280	31.513 9690	33.670 9057	**26**
27	28.830 3702	30.820 8878	31.880 8734	32.986 6785	35.344 3238	**27**
28	29.974 5220	32.129 0967	33.279 3843	34.481 4787	37.051 2103	**28**
29	31.124 3946	33.450 3877	34.695 3766	35.998 7008	38.792 2345	**29**
30	32.280 0166	34.784 8915	36.129 0688	37.538 6814	40.568 0792	**30**
31	33.441 4167	36.132 7404	37.580 6822	39.101 7616	42.379 4408	**31**
32	34.608 6238	37.494 0678	39.050 4407	40.688 2880	44.227 0296	**32**
33	35.781 6669	38.869 0085	40.538 5712	42.298 6123	46.111 5702	**33**
34	36.960 5752	40.257 6986	42.045 3033	43.933 0915	48.033 8016	**34**
35	38.145 3781	41.660 2756	43.570 8696	45.592 0879	49.994 4776	**35**
36	39.336 1050	43.076 8784	45.115 5055	47.275 9692	51.994 3672	**36**
37	40.532 7855	44.507 6471	46.679 4493	48.985 1087	54.034 2545	**37**
38	41.735 4494	45.952 7236	48.262 9424	50.719 8854	56.114 9396	**38**
39	42.944 1267	47.412 2508	49.866 2292	52.480 6837	58.237 2384	**39**
40	44.158 8473	48.886 3734	51.489 5571	54.267 8939	60.401 9832	**40**
41	45.379 6415	50.375 2371	53.133 1765	56.081 9123	62.610 0228	**41**
42	46.606 5397	51.878 9895	54.797 3412	57.923 1410	64.862 2233	**42**
43	47.839 5724	53.397 7794	56.482 3080	59.791 9881	67.159 4678	**43**
44	49.078 7703	54.931 7572	58.188 3369	61.688 8679	69.502 6571	**44**
45	50.324 1642	56.481 0747	59.915 6911	63.614 2010	71.892 7103	**45**
46	51.575 7850	58.045 8855	61.664 6372	65.568 4140	74.330 5645	**46**
47	52.833 6639	59.626 3443	63.435 4452	67.551 9402	76.817 1758	**47**
48	54.097 8322	61.222 6078	65.228 3882	69.565 2193	79.353 5193	**48**
49	55.368 3214	62.834 8338	67.043 7431	71.608 6976	81.940 5897	**49**
50	56.645 1630	64.463 1822	68.881 7899	73.682 8280	84.579 4014	**50**
60	69.770 0305	81.669 6699	88.574 5078	96.214 6517	114.051 5394	**60**
70	83.566 1055	100.676 3368	110.871 9978	122.363 7530	149.977 9111	**70**
80	98.067 7136	121.671 5217	136.118 7953	152.710 8525	193.771 9578	**80**
90	113.310 9358	144.863 2675	164.705 0076	187.929 9004	247.156 6563	**90**
100	129.333 6984	170.481 3829	197.072 3420	228.803 0433	312.232 3059	**100**

TABLE VIII.—AMOUNT OF ANNUITY OF 1 PER PERIOD, $S_{\overline{n}|i}$.—(*Continued*)

n	$2\frac{1}{2}$ %	3 %	$3\frac{1}{2}$ %	4 %	$4\frac{1}{2}$ %	n
1	1.000 0000	1.000 0000	1.000 0000	1.000 0000	1.000 0000	**1**
2	2.025 0000	2.030 0000	2.035 0000	2.040 0000	2.045 0000	**2**
3	3.075 6250	3.090 9000	3.106 2250	3.121 6000	3.137 0250	**3**
4	4.152 5156	4.183 6270	4.214 9429	4.246 4640	4.278 1911	**4**
5	5.256 3285	5.309 1358	5.362 4659	5.416 3226	5.470 7097	**5**
6	6.387 7367	6.468 4099	6.550 1522	6.632 9755	6.716 8917	**6**
7	7.547 4302	7.662 4622	7.779 4075	7.898 2945	8.019 1518	**7**
8	8.736 1159	8.892 3360	9.051 6868	9.214 2263	9.380 0136	**8**
9	9.954 5188	10.159 1061	10.368 4958	10.582 7953	10.802 1142	**9**
10	11.203 3818	11.463 8793	11.731 3932	12.006 1071	12.288 2094	**10**
11	12.483 4663	12.807 7957	13.141 0919	13.486 3514	13.841 1788	**11**
12	13.795 5530	14.192 0296	14.601 9616	15.025 8055	15.464 0318	**12**
13	15.140 4418	15.617 7904	16.113 0303	16.626 8377	17.159 9133	**13**
14	16.518 9528	17.086 3242	17.676 9864	18.291 9112	18.932 1094	**14**
15	17.931 9267	18.598 9139	19.295 6809	20.023 5876	20.784 0543	**15**
16	19.380 2248	20.156 8813	20.971 0297	21.824 5311	22.719 3367	**16**
17	20.864 7304	21.761 5877	22.705 0158	23.697 5124	24.741 7069	**17**
18	22.386 3487	23.414 4354	24.499 6913	25.645 4129	26.855 0837	**18**
19	23.946 0074	25.116 8684	26.357 1805	27.671 2294	29.063 5625	**19**
20	25.544 6576	26.870 3745	28.279 6818	29.778 0786	31.371 4228	**20**
21	27.183 2740	28.676 4857	30.269 4707	31.969 2017	33.783 1368	**21**
22	28.862 8559	30.536 7803	32.328 9022	34.247 9698	36.303 3780	**22**
23	30.584 4273	32.452 8837	34.460 4137	36.617 8886	38.937 0300	**23**
24	32.349 0380	34.426 4702	36.666 5282	39.082 6041	41.689 1963	**24**
25	34.157 7639	36.459 2643	38.949 8567	41.645 9083	44.565 2102	**25**
26	36.011 7080	38.553 0422	41.313 1017	44.311 7446	47.570 6446	**26**
27	37.912 0007	40.709 6335	43.759 0602	47.084 2144	50.711 3236	**27**
28	39.859 8008	42.930 9225	46.290 6273	49.967 5830	53.993 3332	**28**
29	41.856 2958	45.218 8502	48.910 7993	52.966 2863	57.423 0332	**29**
30	43.902 7032	47.575 4157	51.622 6773	56.084 9378	61.007 0697	**30**
31	46.000 2707	50.002 6782	54.429 4710	59.328 3353	64.752 3878	**31**
32	48.150 2775	52.502 7585	57.334 5025	62.701 4687	68.666 2452	**32**
33	50.354 0344	55.077 8413	60.341 2100	66.209 5274	72.756 2263	**33**
34	52.612 8853	57.730 1765	63.453 1524	69.857 9085	77.030 2565	**34**
35	54.928 2074	60.462 0818	66.674 0127	73.652 2249	81.496 6180	**35**
36	57.301 4126	63.275 9443	70.007 6032	77.598 3138	86.163 9658	**36**
37	59.733 9479	66.174 2226	73.457 8693	81.702 2464	91.041 3443	**37**
38	62.227 2966	69.159 4493	77.028 8947	85.970 3363	96.138 2048	**38**
39	64.782 9791	72.234 2328	80.724 9060	90.409 1497	101.464 4240	**39**
40	67.402 5535	75.401 2597	84.550 2778	95.025 5157	107.030 3231	**40**
41	70.087 6174	78.663 2975	88.509 5375	99.826 5363	112.846 6876	**41**
42	72.839 8078	82.023 1964	92.607 3713	104.819 5978	118.924 7885	**42**
43	75.660 8030	85.483 8923	96.848 6293	110.012 3817	125.276 4040	**43**
44	78.552 3231	89.048 4091	101.238 3313	115.412 8770	131.913 8422	**44**
45	81.516 1312	92.719 8614	105.781 6729	121.029 3920	138.849 9651	**45**
46	84.554 0344	96.501 4572	110.484 0314	126.870 5677	146.098 2135	**46**
47	87.667 8853	100.396 5010	115.350 9726	132.945 3904	153.672 6331	**47**
48	90.859 5824	104.408 3960	120.388 2566	139.263 2060	161.587 9016	**48**
49	94.131 0720	108.540 6478	125.601 8456	145.833 7343	169.859 3572	**49**
50	97.484 3488	112.796 8673	130.997 9102	152.667 0837	178.503 0283	**50**
60	135.991 5900	163.053 4368	196.516 8829	237.990 6852	289.497 9540	**60**
70	185.284 1142	230.594 0637	288 937 8646	364.290 4588	461.869 6796	**70**
80	248.382 7126	321.363 0186	419.306 7868	551.244 9768	729.557 6985	**80**
90	329.154 2533	443.348 9036	603.205 0270	827.983 3335	1145.269 0066	**90**
100	432.548 6540	607.287 7327	862.611 6567	1237.623 7046	1790.855 9563	**100**

Table VIII.—Amount of Annuity of 1 per Period, $S_{\overline{n}|i}$.—(*Continued*)

n	5 %	5½ %	6 %	7 %	8 %	n
1	1.000 0000	1.000 0000	1.000 0000	1.000 0000	1.000 0000	1
2	2.050 0000	2.055 0000	2.060 0000	2.070 0000	2.080 0000	2
3	3.152 5000	3.168 0250	3.183 6000	3.214 9000	3.246 4000	3
4	4.310 1250	4.342 2664	4.374 6160	4.439 9430	4.506 1120	4
5	5.525 6312	5.581 0910	5.637 0930	5.750 7390	5.866 6010	5
6	6.801 9128	6.888 0510	6.975 3185	7.153 2907	7.335 9290	6
7	8.142 0084	8.266 8938	8.393 8376	8.654 0211	8.922 8034	7
8	9.549 1089	9.721 5730	9.897 4679	10.259 8026	10.636 6276	8
9	11.026 5643	11.256 2595	11.491 3160	11.977 9888	12.487 5578	9
10	12.577 8925	12.875 3538	13.180 7949	13.816 4480	14.486 5625	10
11	14.206 7872	14.583 4983	14.971 6426	15.783 5993	16.645 4875	11
12	15.917 1265	16.385 5907	16.869 9412	17.888 4513	18.977 1265	12
13	17.712 9828	18.286 7981	18.882 1377	20.140 6429	21.495 2966	13
14	19.598 6320	20.292 5720	21.015 0659	22.550 4879	24.214 9203	14
15	21.578 5636	22.408 6635	23.275 9699	25.129 0220	27.152 1139	15
16	23.657 4918	24.641 1400	25.672 5281	27.888 0536	30.324 2830	16
17	25.840 3664	26.996 4027	28.212 8798	30.840 2173	33.750 2257	17
18	28.132 3847	29.481 2048	30.905 6526	33.999 0325	37.450 2437	18
19	30.539 0039	32.102 6711	33.759 9917	37.378 9648	41.446 2632	19
20	33.065 9541	34.868 3180	36.785 5912	40.995 4923	45.761 9643	20
21	35.719 2518	37.786 0755	39.992 7267	44.865 1768	50.422 9214	21
22	38.505 2144	40.864 3097	43.392 2903	49.005 7392	55.456 7552	22
23	41.430 4751	44.111 8467	46.995 8277	53.436 1409	60.893 2956	23
24	44.501 9989	47.537 9983	50.815 5774	58.176 6708	66.764 7592	24
25	47.727 0988	51.152 5882	54.864 5120	63.249 0377	73.105 9400	25
26	51.113 4538	54.965 9805	59.156 3827	68.676 4704	79.954 4152	26
27	54.669 1264	58.989 1094	63.705 7657	74.483 8233	87.350 7684	27
28	58.402 5828	63.233 5105	68.528 1116	80.697 6909	95.338 8298	28
29	62.322 7119	67.711 3535	73.639 7983	87.346 5293	103.965 9362	29
30	66.438 8475	72.435 4780	79.058 1862	94.460 7863	113.283 2111	30
31	70.760 7899	77.419 4293	84.801 6774	102.073 0414	123.345 8680	31
32	75.298 8294	82.677 4979	90.889 7780	110.218 1543	134.213 5374	32
33	80.063 7708	88.224 7603	97.343 1647	118.933 4251	145.950 6204	33
34	85.066 9594	94.077 1221	104.183 7546	128.258 7648	158.626 6701	34
35	90.320 3074	100.251 3638	111.434 7799	138.236 8784	172.316 8037	35
36	95.836 3227	106.765 1888	119.120 8667	148.913 4598	187.102 1480	36
37	101.628 1389	113.637 2742	127.268 1187	160.337 4020	203.070 3198	37
38	107.709 5458	120.887 3243	135.904 2058	172.561 0202	220.315 9454	38
39	114.095 0231	128.536 1271	145.058 4581	185.640 2916	238.941 2210	39
40	120.799 7742	136.605 6141	154.761 9656	199.635 1120	259.056 5187	40
41	127.839 7630	145.118 9229	165.047 6836	214.609 5698	280.781 0402	41
42	135.231 7511	154.100 4636	175.950 5446	230.632 2397	304.243 5234	42
43	142.993 3387	163.575 9891	187.507 5772	247.776 4965	329.583 0053	43
44	151.143 0056	173.572 6685	199.758 0319	266.120 8512	356.949 6457	44
45	159.700 1559	184.119 1653	212.743 5138	285.749 3108	386.505 6174	45
46	168.685 1637	195.245 7194	226.508 1246	306.751 7626	418.426 0668	46
47	178.119 4218	206.984 2339	241.098 6121	329.224 3860	452.900 1521	47
48	188.025 3929	219.368 3668	256.564 5288	353.270 0930	490.132 1643	48
49	198.426 6626	232.433 6270	272.958 4006	378.998 9995	530.342 7374	49
50	209.347 9957	246.217 4765	290.335 9046	406.528 9295	573.770 1564	50
60	353.583 7179	433.450 3717	533.128 1809	813.520 3834	1253.213 2958	60
70	588.528 5107	753.271 2042	967.932 1696	1614.134 1742	2720.080 0738	70
80	971.228 8213	1299.571 3869	1746.599 8914	3189.062 6797	5886.935 4283	80
90	1594.607 3010	2232.731 0166	3141.075 1872	6287.185 4268	12723.938 6160	90
100	2610.025 1569	3826.702 4668	5638.368 0586	12381.661 7938	27484.515 7043	100

TABLE IX.—PRESENT VALUE OF AN ANNUITY OF 1 PER PERIOD, $a_{\overline{n}|i}$

n	$\frac{1}{2}$ %	1 %	$1\frac{1}{4}$ %	$1\frac{1}{2}$ %	2 %	n
1	0.995 0249	0.990 0990	0.987 6543	0.985 2217	0.980 3922	1
2	1.985 0994	1.970 3951	1.963 1154	1.955 8834	1.941 5609	2
3	2.970 2481	2.940 9852	2.926 5337	2.912 2004	2.883 8833	3
4	3.950 4957	3.901 9656	3.878 0580	3.854 3846	3.807 7287	4
5	4.925 8663	4.853 4312	4.817 8350	4.782 6450	4.713 4595	5
6	5.896 3844	5.795 4765	5.746 0099	5.697 1872	5.601 4309	6
7	6.862 0740	6.728 1945	6.662 7258	6.598 2140	6.471 9911	7
8	7.822 9592	7.651 6778	7.568 1243	7.485 9251	7.325 4814	8
9	8.779 0639	8.566 0176	8.462 3450	8.360 5173	8.162 2367	9
10	9.730 4119	9.471 3045	9.345 5259	9.222 1846	8.982 5850	10
11	10.677 0267	10.367 6282	10.217 8034	10.071 1178	9.786 8480	11
12	11.618 9321	11.255 0775	11.079 3120	10.907 5052	10.575 3412	12
13	12.556 1513	12.133 7401	11.930 1847	11.731 5322	11.348 3738	13
14	13.488 7078	13.003 7030	12.770 5528	12.543 3815	12.106 2488	14
15	14.416 6246	13.865 0525	13.600 5459	13.343 2330	12.849 2635	15
16	15.339 9250	14.717 8738	14.420 2923	14.131 2640	13.577 7093	16
17	16.258 6319	15.562 2513	15.229 9183	14.907 6493	14.291 8719	17
18	17.172 7680	16.398 2686	16.029 5489	15.672 5609	14.992 0312	18
19	18.082 3562	17.226 0085	16.819 3076	16.426 1684	15.678 4620	19
20	18.987 4192	18.045 5530	17.599 3161	17.168 6388	16.351 4333	20
21	19.887 9792	18.856 9831	18.369 6950	17.900 1367	17.011 2092	21
22	20.784 0590	19.660 3793	19.130 5629	18.620 8244	17.658 0482	22
23	21.675 6806	20.455 8211	19.882 0374	19.330 8614	18.292 2041	23
24	22.562 8662	21.243 3873	20.624 2345	20.030 4054	18.913 9256	24
25	23.445 6380	22.023 1557	21.357 2686	20.719 6112	19.523 4565	25
26	24.324 0179	22.795 2037	22.081 2530	21.398 6317	20.121 0358	26
27	25.198 0278	23.559 6076	22.796 2992	22.067 6175	20.706 8978	27
28	26.067 6894	24.316 4432	23.502 5178	22.726 7167	21.281 2724	28
29	26.933 0242	25.065 7853	24.200 0176	23.376 0756	21.844 3847	29
30	27.794 0540	25.807 7082	24.888 9062	24.015 8380	22.396 4556	30
31	28.650 8000	26.542 2854	25.569 2901	24.646 1458	22.937 7015	31
32	29.503 2836	27.269 5895	26.241 2742	25.267 1387	23.468 3348	32
33	30.351 5259	27.989 6926	26.904 9622	25.878 9544	23.988 5636	33
34	31.195 5482	28.702 6659	27.560 4564	26.481 7285	24.498 5917	34
35	32.035 3713	29.408 5801	28.207 8582	27.075 5946	24.998 6193	35
36	32.871 0162	30.107 5050	28.847 2674	27.660 6843	25.488 8425	36
37	33.702 5037	30.799 5099	29.478 7826	28.237 1274	25.969 4534	37
38	34.529 8544	31.484 6633	30.102 5013	28.805 0516	26.440 6406	38
39	35.353 0890	32.163 0330	30.718 5198	29.364 5829	26.902 5888	39
40	36.172 2279	32.834 6861	31.326 9332	29.915 8452	27.355 4792	40
41	36.987 2914	33.499 6892	31.927 8352	30.458 9608	27.799 4894	41
42	37.798 2999	34.158 1081	32.521 3187	30.994 0500	28.234 7936	42
43	38.605 2735	34.810 0081	33.107 4753	31.521 2316	28.661 5623	43
44	39.408 2324	35.455 4535	33.686 3954	32.040 6222	29.079 9631	44
45	40.207 1964	36.094 5084	34.258 1682	32.552 3372	29.490 1599	45
46	41.002 1855	36.727 2361	34.822 8822	33.056 4898	29.892 3136	46
47	41.793 2194	37.353 6991	35.380 6244	33.553 1920	30.286 5820	47
48	42.580 3178	37.973 9595	35.931 4809	34.042 5536	30.673 1196	48
49	43.363 5003	38.588 0787	36.475 5367	34.524 6834	31.052 0780	49
50	44.142 7864	39.196 1175	37.012 8757	34.999 6881	31.423 6059	50
60	51.725 5608	44.955 0384	42.034 5918	39.380 2689	34.760 8867	60
70	58.939 4176	50.168 5144	46.469 6756	43.154 8718	37.498 6193	70
80	65.802 3054	54.888 2061	50.386 6571	46.407 3235	39.744 5136	80
90	72.331 2996	59.160 8815	53.846 0604	49.209 8545	41.586 9292	90
100	78.542 6448	63.028 8788	56.901 3394	51.624 7037	43.098 3516	100

TABLE IX.—PRESENT VALUE OF AN ANNUITY OF 1 PER PERIOD, $a_{\overline{n}|i}$.—(*Continued*)

n	2½ %	3 %	3½ %	4 %	4½ %	n
1	0.975 6098	0.970 8738	0.966 1836	0.961 5385	0.956 9378	1
2	1.927 4242	1.913 4697	1.899 6943	1.886 0947	1.872 6678	2
3	2.856 0236	2.828 6114	2.801 6370	2.775 0910	2.748 9644	3
4	3.761 9742	3.717 0984	3.673 0792	3.629 8952	3.587 5257	4
5	4.645 8285	4.579 7072	4.515 0524	4.451 8223	4.389 9767	5
6	5.508 1254	5.417 1914	5.328 5530	5.242 1369	5.157 8725	6
7	6.349 3906	6.230 2830	6.114 5440	6.002 0547	5.892 7009	7
8	7.170 1372	7.019 6922	6.873 9555	6.732 7449	6.595 8861	8
9	7.970 8655	7.786 1089	7.607 6865	7.435 3316	7.268 7905	9
10	8.752 0639	8.530 2028	8.316 6053	8.110 8958	7.912 7182	10
11	9.514 2087	9.252 6241	9.001 5510	8.760 4767	8.528 9169	11
12	10.257 7646	9.954 0040	9.663 3343	9.385 0738	9.118 5808	12
13	10.983 1850	10.634 9553	10.302 7385	9.985 6478	9.682 8524	13
14	11.690 9122	11.296 0731	10.920 5203	10.563 1229	10.222 8253	14
15	12.381 3777	11.937 9351	11.517 4109	11.118 3874	10.739 5457	15
16	13.055 0027	12.561 1020	12.094 1168	11.652 2956	11.234 0150	16
17	13.712 1977	13.166 1185	12.651 3206	12.165 6688	11.707 1914	17
18	14.353 3636	13.753 5131	13.189 6817	12.659 2970	12.159 9918	18
19	14.978 8913	14.323 7991	13.709 8374	13.133 9394	12.593 2936	19
20	15.589 1623	14.877 4749	14.212 4033	13.590 3263	13.007 9364	20
21	16.184 5486	15.415 0241	14.697 9742	14.029 1600	13.404 7239	21
22	16.765 4132	15.936 9166	15.167 1248	14.451 1153	13.784 4248	22
23	17.332 1105	16.443 6084	15.620 4105	14.856 8417	14.147 7749	23
24	17.884 9858	16.935 5421	16.058 3676	15.246 9631	14.495 4784	24
25	18.424 3764	17.413 1477	16.481 5146	15.622 0799	14.828 2090	25
26	18.950 6111	17.876 8424	16.890 3523	15.982 7692	15.146 6114	26
27	19.464 0109	18.327 0315	17.285 3645	16.329 5858	15.451 3028	27
28	19.964 8887	18.764 1082	17.667 0188	16.663 0632	15.742 8735	28
29	20.453 5499	19.188 4546	18.035 7670	16.983 7146	16.021 8885	29
30	20.930 2926	19.600 4414	18.392 0454	17.292 0333	16.288 8885	30
31	21.395 4074	20.000 4285	18.736 2758	17.588 4936	16.544 3910	31
32	21.849 1780	20.388 7655	19.068 8655	17.873 5515	16.788 8909	32
33	22.291 8809	20.765 7918	19.390 2082	18.147 6457	17.022 8621	33
34	22.723 7863	21.131 8367	19.700 6842	18.411 1978	17.246 7580	34
35	23.145 1573	21.487 2201	20.000 6611	18.664 6132	17.461 0124	35
36	23.556 2511	21.832 2525	20.290 4938	18.908 2820	17.666 0406	36
37	23.957 3181	22.167 2354	20.570 5254	19.142 5788	17.862 2398	37
38	24.348 6030	22.492 4616	20.841 0874	19.367 8642	18.049 9902	38
39	24.730 3444	22.808 2151	21.102 4999	19.584 4848	18.229 6557	39
40	25.102 7750	23.114 7720	21.355 0723	19.792 7739	18.401 5844	40
41	25.466 1220	23.412 4000	21.599 1037	19.993 0518	18.566 1095	41
42	25.820 6068	23.701 3592	21.834 8828	20.185 6267	18.723 5498	42
43	26.166 4457	23.981 9021	22.062 6887	20.370 7949	18.874 2103	43
44	26.503 8494	24.254 2739	22.282 7910	20.548 8413	19.018 3830	44
45	26.833 0239	24.518 7125	22.495 4503	20.720 0397	19.156 3474	45
46	27.154 1696	24.775 4491	22.700 9181	20.884 6536	19.288 3707	46
47	27.467 4826	25.024 7078	22.899 4378	21.042 9361	19.414 7088	47
48	27.773 1537	25.266 7066	23.091 2442	21.195 1309	19.535 6065	48
49	28.071 3695	25.501 6569	23.276 5645	21.341 4720	19.651 2981	49
50	28.362 3117	25.729 7640	23.455 6179	21.482 1846	19.762 0078	50
60	30.908 6565	27.675 5637	24.944 7341	22.623 4900	20.638 0220	60
70	32.897 8570	29.123 4214	26.000 3966	23.394 5150	21.202 1119	70
80	34.451 8172	30.200 7634	26.748 7757	23.915 3918	21.565 3449	80
90	35.665 7685	31.002 4071	27.279 3156	24.267 2776	21.799 2408	90
100	36.614 1053	31.598 9053	27.655 4254	24.504 9990	21.949 8527	100

Table IX.—Present Value of an Annuity of 1 per Period $a_{\overline{n}|i}$.—(*Continued*)

n	5%	$5\frac{1}{2}$%	6%	7%	8%	n
1	0.952 3810	0.947 8673	0.943 3962	0.934 5794	0.925 9259	**1**
2	1.859 4104	1.846 3197	1.833 3927	1.808 0182	1.783 2648	**2**
3	2.723 2480	2.697 9334	2.673 0120	2.624 3160	2.577 0970	**3**
4	3.545 9505	3.505 1501	3.465 1056	3.387 2113	3.312 1268	**4**
5	4.329 4767	4.270 2845	4.212 3638	4.100 1974	3.992 7100	**5**
6	5.075 6921	4.995 5303	4.917 3243	4.766 5397	4.622 8797	**6**
7	5.786 3734	5.682 9671	5.582 3814	5.389 2894	5.206 3701	**7**
8	6.463 2128	6.334 5660	6.209 7938	5.971 2985	5.746 6389	**8**
9	7.107 8217	6.952 1953	6.801 6923	6.515 2322	6.246 8879	**9**
10	7.721 7349	7.537 6258	7.360 0870	7.023 5816	6.710 0814	**10**
11	8.306 4142	8.092 5363	7.886 8746	7.498 6744	7.138 9643	**11**
12	8.863 2516	8.618 5179	8.383 8439	7.942 6863	7.536 0780	**12**
13	9.393 5703	9.117 0785	8.852 6830	8.357 6508	7.903 7759	**13**
14	9.898 6489	9.589 6479	9.294 9839	8.745 4680	8.244 2370	**14**
15	10.379 6500	10.037 5809	9.712 2490	9.107 9140	8.559 4787	**15**
16	10.837 7696	10.462 1620	10.105 8953	9.446 6486	8.851 3692	**16**
17	11.274 0662	10.864 6086	10.477 2597	9.763 2230	9.121 6381	**17**
18	11.689 5869	11.246 0745	10.827 6035	10.059 0869	9.371 8871	**18**
19	12.085 3209	11.607 6535	11.158 1165	10.335 5952	9.603 5992	**19**
20	12.462 2103	11.950 3825	11.469 9212	10.594 0143	9.818 1474	**20**
21	12.821 1527	12.275 2441	11.764 0766	10.835 5273	10.016 8032	**21**
22	13.163 0026	12.583 1697	12.041 5817	11.061 2405	10.200 7437	**22**
23	13.488 5739	12.875 0424	12.303 3790	11.272 1874	10.371 0590	**23**
24	13.798 6418	13.151 6990	12.550 3575	11.469 3340	10.528 7583	**24**
25	14.093 9446	13.413 9327	12.783 3562	11.653 5832	10.674 7762	**25**
26	14.375 1853	13.662 4954	13.003 1662	11.825 7787	10.809 9780	**26**
27	14.643 0336	13.898 0999	13.210 5341	11.986 7090	10.935 1648	**27**
28	14.898 1273	14.121 4217	13.406 1643	12.137 1113	11.051 0785	**28**
29	15.141 0736	14.333 1012	13.590 7210	12.277 6741	11.158 4060	**29**
30	15.372 4510	14.533 7452	13.764 8312	12.409 0412	11.257 7833	**30**
31	15.592 8105	14.723 9291	13.929 0860	12.531 8142	11.349 7994	**31**
32	15.802 6767	14.904 1982	14.084 0434	12.646 5553	11.434 9994	**32**
33	16.002 5492	15.075 0694	14.230 2296	12.753 7900	11.513 8884	**33**
34	16.192 9040	15.237 0326	14.368 1411	12.854 0094	11.586 9337	**34**
35	16.374 1943	15.390 5522	14.498 2464	12.947 6723	11.654 5682	**35**
36	16.546 8517	15.536 0684	14.620 9871	13.035 2078	11.717 1928	**36**
37	16.711 2873	15.673 9985	14.736 7803	13.117 0166	11.775 1785	**37**
38	16.867 8927	15.804 7379	14.846 0192	13.193 4735	11.828 8690	**38**
39	17.017 0407	15.928 6615	14.949 0747	13.264 9285	11.878 5824	**39**
40	17.159 0864	16.046 1247	15.046 2969	13.331 7088	11.924 6133	**40**
41	17.294 3680	16.157 4642	15.138 0159	13.394 1204	11.967 2346	**41**
42	17.423 2076	16.262 9992	15.224 5433	13.452 4490	12.006 6987	**42**
43	17.545 9120	16.363 0324	15.306 1729	13.506 9617	12.043 2395	**43**
44	17.662 7733	16.457 8506	15.383 1820	13.557 9081	12.077 0736	**44**
45	17.774 0698	16.547 7257	15.455 8321	13.605 5216	12.108 4015	**45**
46	17.880 0665	16.632 9154	15.524 3699	13.650 0202	12.137 4088	**46**
47	17.981 0157	16.713 6639	15.589 0282	13.691 6076	12.164 2674	**47**
48	18.077 1578	16.790 2027	15.650 0266	13.730 4744	12.189 1365	**48**
49	18.168 7217	16.862 7514	15.707 5723	13.766 7986	12.212 1634	**49**
50	18.255 9255	16.931 5179	15.761 8606	13.800 7463	12.233 4846	**50**
60	18.929 2895	17.449 8542	16.161 4277	14.039 1812	12.376 5518	**60**
70	19.342 6766	17.753 3041	16.384 5439	14.160 3893	12.442 8196	**70**
80	19.596 4605	17.930 9529	16.509 1308	14.222 0054	12.473 5144	**80**
90	19.752 2617	18.034 9540	16.578 6994	14.253 3279	12.487 7320	**90**
100	19.847 9102	18.095 8394	16.617 5462	14.269 2507	12.494 3176	**100**

ANSWERS

NOTE: Answers have been omitted in those cases in which the student can quickly check the correctness of his results in a simple way, and in a few additional cases in which it appeared obvious for other reasons that the answer should be omitted. With these exceptions, answers are given for all problems.

CHAPTER I

Pages 6–7; Lesson 1

19. 72; 2. **20.** 210; 7. **21.** 48; 6. **22.** 1,729; 1.
23. 8,400; 30. **24.** 2,730; 5. **25.** 2,304; 8. **26.** 400,980; 20.

Pages 12–13; Lesson 2

1. $\frac{5}{13}$; $\frac{z}{y}$. **2.** $\frac{14}{15}$; $\frac{2x}{3b}$. **3.** $\frac{3}{8}$; $\frac{a}{bc}$. **4.** $\frac{15}{4}$; $\frac{ac}{b}$.
5. $\frac{6}{7}$; $\frac{2x}{y}$. **6.** $\frac{29}{21}$; $\frac{ad + bc}{bd}$. **7.** $\frac{47}{30}$; $\frac{6a + 5c}{6b}$.
8. $\frac{41}{105}$; $\frac{4x + 3y}{3xy}$. **9.** $\frac{9}{20}$; $\frac{3x}{10a}$. **10.** 9; $\frac{ab + a}{c}$.
11. $\frac{5}{24}$; $\frac{b + a}{abc}$. **12.** $\frac{27}{20}$; $\frac{c(b + a)}{ab}$.
13. $\frac{319}{189}$; $\frac{f(ad + bc)}{bde}$. **14.** $\frac{8}{15}$; $\frac{2x}{3y}$. **15.** $\frac{13}{5}$; $\frac{a + 2b}{b}$.
16. $\frac{75}{14}$; $\frac{6(2a + 2b - 1)}{ac}$. **19.** $\frac{bc + ac + ab}{abc}$.
22. $\frac{25}{6}$. **23.** $\frac{11}{25}$. **24.** $\frac{13}{44}$. **25.** $\frac{2}{3}$. **26.** 2. **27.** $\frac{31}{6}$.

Pages 19–21; Lesson 3

1. 2. **2.** -14. **3.** $\frac{41}{48}$. **4.** $\frac{20}{9}$.
5. 9. **6.** 0. **7.** $\frac{3}{2}$. **8.** $-\frac{4}{7}$.
9. $\frac{9}{2}$. **10.** -4. **11.** $-\frac{4}{25}$. **12.** $\frac{16}{9}$.
15. 6; $\frac{9}{2}$. **16.** 10; $\frac{3}{2}$. **17.** $-\frac{17}{3}$; $\frac{3}{2}$. **18.** -6; $-\frac{17}{3}$.
19. $\frac{29}{8}$; $\frac{7}{16}$. **20.** $-\frac{45}{4}$; $-\frac{15}{8}$. **21.** $\frac{89}{9}$. **23.** 3.
24. $\frac{4}{5}$. **25.** $\frac{7}{5}$. **26.** 3. **27.** $-\frac{2}{3}$.
28. -1. **29.** -8. **30.** $\frac{3}{4}$.

Pages 27–29; Lesson 4

6. $-\frac{18}{5}$. **7.** 9. **8.** -190. **9.** $\frac{11}{6}$.
10. $\frac{5}{18}$. **11.** 1. **12.** x; $2y - x$.
15. $0.\dot{7}$. **16.** 0.82. **17.** $0.\dot{1}5384\dot{6}$. **18.** $3.\dot{5}7142\dot{8}$.
19. $3.\dot{2}3809\dot{5}$. **20.** $2.\dot{0}\dot{9}$. **21.** $0.\dot{7}5\dot{0}$. **22.** $0.23\dot{1}\dot{6}$.

25. $\frac{b^4}{2a}$. **26.** $2ab$. **27.** $2xy^3$. **28.** xy^3z^5.

29. $\frac{2a}{b}$. **30.** ax. **31.** $\frac{1}{a}$. **32.** $\frac{65xy(x+y)}{64}$.

33. $\frac{a^2}{4}$. **34.** x. **35.** $3x$; $5x - 4y$; 12; 23.

36. $x + 3y$; $x > 2y$; -2; 6; 7. **37.** $5x$; $17x - 6y$; 10; 13.

CHAPTER II

Pages 36–37; Lesson 5

8. $x^2 + x - 12$; $6x^4 + 11x^2 - 35$.
9. $3x^2 + 5xy - 2y^2$; $4x^4 - 23x^2y - 35y^2$.
10. $6x^2y^2 + 11xy^2 - 10y^2$; $x^4y^4 - 1$.
11. $x^6 + 9x^4y + 27x^2y^2 + 27y^3$; $27a^3 - 108a^2b + 144ab^2 - 64b^3$.
12. $27x^3 - 54x^2y^2 + 36xy^4 - 8y^6$; $x^6 + 6x^4y^2 + 12x^2y^4 + 8y^6$.
13. $25a^4b^4 - 16k^2$; $k^2 + a^2bk - 6a^4b^2$.
14. $x^2 + 2xy + y^2 + 2x + 2y + 1$; $4u^4 - 4u^2v + 4u^2y + v^2 - 2vy + y^2$.
15. $a^2 - b^2 + 2bc - c^2$; $9 - y^2 + 8yz - 16z^2$.
16. $a^2 + 2ab + b^2 - c^2$; $9x^2 + 24xy + 16y^2 - 4$.
17. $a^3 + 3a^2b + 3ab^2 + b^3 - 6a^2 - 12ab - 6b^2 + 12a + 12b - 8$;
$8x^3 + 12x^2y + 6xy^2 + y^3 + 12x^2 + 12xy + 3y^2 + 6x + 3y + 1$.
18. $u^3 + 3u^2v + 3uv^2 + v^3 - 3u^2 - 6uv - 3v^2 + 3u + 3v - 1$;
$8u^3 - 12u^2v + 6uv^2 - v^3 + 12u^2 - 12uv + 3v^2 + 6u - 3v + 1$.

Pages 40–41; Lesson 6

1. $\frac{27}{41}$; $3(x - y)$ if $y \neq -x$; $\frac{5x}{4}$ if $y \neq -1$.

2. $\frac{5}{7}$; $\frac{-6}{x+y}$ if $x^2 \neq y^2$; $\frac{x-2}{2}$ if $x \neq -4$.

3. $\frac{11}{19}$; $\frac{x^2 - 3x}{2}$ if $x \neq -\frac{5}{2}$; $\frac{x^2 + xy + y^2}{x+y}$ if $x^2 \neq y^2$.

4. $\frac{24}{29}$; $\frac{x-2y}{2}$ if $x \neq -2y$; $\frac{x+2y}{3(x-3y)}$ if $x^2 \neq 9y^2$.

5. $\frac{x-3y}{y-x}$. **6.** $\frac{x}{2(x+2)}$. **7.** $\frac{(2x-5)(3x-1)}{2x}$.

8. $\frac{12x^3(x-1)}{x^2+4}$. **9.** $\frac{x-y}{x+y}$. **10.** $x(x+2y)$.

11. $-\frac{(x-2y)^2}{y^2}$. **12.** $\frac{-4xy}{3x+y}$. **13.** $\frac{-(2x+5)(3x+2)}{2(7+x)}$.

14. $\frac{2}{x-y}$. **15.** $\frac{x^2}{x^2-y^2}$. **16.** $\frac{z^2-2y^2}{2xyz}$. **17.** x.

18. $\frac{1}{1+x}$. **19.** $\frac{10(x+1)}{4x-5}$. **20.** $\frac{1}{x-y}$. **21.** $\frac{x(x+1)}{x-1}$.

22. $\frac{y-x}{y+x}$. **23.** x. **24.** $\frac{x-4}{x+4}$. **25.** $-\frac{x}{y}$.

26. $\dfrac{-2}{x+2}$. **27.** $y - x$. **28.** -2. **29.** $\dfrac{3}{(x+2)^2(x-1)}$.
30. $\dfrac{x+1}{x-3y}$.

CHAPTER III

Pages 48–49; Lesson 7

19. 6. **20.** $\frac{9}{7}$. **21.** -1. **22.** $\frac{4}{3}$.
23. 0.1. **24.** -1.9. **25.** $\frac{5}{3}$. **26.** $-\frac{11}{2}$.
27. $\frac{15}{2}$. **28.** 3. **29.** No solution. **30.** -8.
31. No solution. **32.** 0. **33.** No solution. **34.** $\frac{5}{2}$.
35. $\frac{19}{3}$. **36.** 64. **37.** 8. **38.** $\frac{1}{8}$.
39. No solution. **40.** 5.

Pages 55–57; Lesson 8

3. $2, -3$. **4.** $\frac{3}{2}, -5$. **5.** $4, \frac{3}{7}$. **6.** $-1, -\frac{9}{4}$.
7. $-\frac{1}{2}, -\frac{15}{4}$. **8.** $\frac{7}{2}, 11$. **9.** $4, -\frac{3}{2}$. **10.** $\frac{1}{5}, -5$.
11. $-1, -\frac{5}{3}$. **12.** $\frac{1}{2}, -1$. **13.** $\frac{3}{2}, -4$. **14.** $\frac{7}{2}, -2$.
15. $3, -12$. **16.** $3, -2$. **17.** $3, 7$. **18.** $-\frac{3}{2} \pm \frac{1}{2}\sqrt{7}$.
19. $2 \pm \sqrt{5}$. **20.** $-3 \pm \sqrt{2}$. **21.** $-\frac{3}{2} \pm \frac{1}{2}\sqrt{6}$. **22.** $\frac{2}{3} \pm \frac{1}{3}\sqrt{6}$.
23. $\frac{1}{4} \pm \frac{1}{2}\sqrt{5}$. **24.** $\frac{3}{5} \pm \frac{1}{5}\sqrt{3}$. **25.** $2, 12$. **26.** $1 \pm \sqrt{6}$.
27. $-1, -5$. **28.** $-6, -16$. **29.** $\frac{3}{4} \pm \frac{1}{4}\sqrt{37}$. **30.** $-3, \frac{5}{3}$.
31. $\frac{1}{3} \pm \frac{1}{3}\sqrt{5}$. **32.** $\frac{7}{2} \pm \frac{1}{2}\sqrt{137}$. **33.** $3 \pm \sqrt{15}$. **34.** $\frac{4}{3} \pm \frac{2}{3}\sqrt{2}$.
35. $-\frac{5}{2} \pm \frac{1}{2}\sqrt{5}$. **36.** $\frac{7}{2} \pm \sqrt{2}$. **37.** $2, 4$. **38.** $6, \frac{5}{9}$.
39. $\frac{11}{2}, 7$. **40.** $-3, \frac{16}{3}$. **41.** $1.9, 14$. **42.** $3.5, -7.5$.
43. $-3, -9$. **44.** 13. **45.** 4.5. **46.** 17.
47. No solution. **48.** $-6, \frac{26}{9}$. **49.** 2.
50. 5. **51.** 8, 9. **52.** 18.

Pages 61–62; Lesson 9

1. $4 + 4i$. **2.** $4 - 9i$. **3.** $10 - 6i$. **4.** $25 - i$.
5. $13 + 10i$. **6.** $-12 + 6i$. **7.** $10 - i$. **8.** 0.
9. $-23 + 24i$. **10.** $5 + 4i$. **19.** $-1 \pm i$. **20.** $-3 \pm 2i$.
21. $4 \pm 3i$. **22.** $7 \pm 4i$. **23.** $\frac{1}{2} \pm 2i$. **24.** $-\frac{2}{3} \pm \frac{1}{3}i$.
25. $-3 \pm \sqrt{5}$. **26.** $-1 \pm \sqrt{10}$. **27.** $\frac{1}{2}, -\frac{19}{2}$. **28.** $-\frac{1}{2} \pm \frac{1}{2}\sqrt{3}\,i$.
29. $\frac{1}{2} \pm \frac{1}{2}\sqrt{6}\,i$. **30.** $\frac{1}{2} \pm \frac{1}{6}\sqrt{15}\,i$. **32.** $-10i$. **33.** $-4 - 3i$.
34. $-9 - 8i$. **35.** -2. **36.** $4 - 6i$. **37.** $-4i$.
41. $-2 + 5i$. **42.** $1.2 + 1.6i$. **43.** $0.38 - 0.14i$. **44.** $-\frac{23}{20} + \frac{21}{20}i$.
45. $0.7 - 0.9i$. **46.** $\frac{8}{17} - \frac{32}{17}i$. **47.** $2.16 + 2.12i$. **48.** $-6 - 2i$.
49. $\frac{10}{13} + \frac{24}{13}i$. **50.** $7.16 - 0.88i$.

Pages 68–70; Lesson 10

9. $2 \pm 4i$. **10.** $1 \pm \sqrt{5}$. **11.** $\frac{5}{3}, -\frac{4}{3}$. **12.** $-\frac{3}{2} \pm \sqrt{5}$.
13. $2 \pm \frac{1}{2}\sqrt{3}$. **14.** $5 \pm 3i$. **15.** $\frac{1}{2} \pm \frac{2}{3}i$. **16.** $4, -\frac{2}{3}$.

17. $2, -\frac{4}{3}$. **18.** $\frac{-1 \pm \sqrt{86}}{5}$. **19.** $\frac{-2 \pm \sqrt{4 + 3y^2}}{y}$.

20. $\frac{y \pm \sqrt{y^2 - y}}{y}$. **21.** $-y \pm 2$. **22.** $-2y \pm \sqrt{y^2 + 2}$.

23. $-y \pm \sqrt{4y - y^2}$. **24.** $-4 \pm \sqrt{y^2 + 4y}$.
25. $x \pm \sqrt{16 - x^2}$. **26.** $x + 1, -4x - 1$.
27. $x - 1 \pm \sqrt{x + 4}$. **28.** $\frac{1}{2}(-x - 6 \pm \sqrt{9x^2 + 24})$.
29. 4. **30.** $\frac{1}{8}$. **31.** 2, −4. **32.** $-\frac{1}{4}$.
33. 0, −8. **34.** $\frac{1}{3}$. **37.** $-2; \frac{1}{2}$. **38.** 2; −1; no.
39. 3, 5. **40.** $\frac{9}{4}, \frac{10}{3}$. **41.** No root. **42.** $1, -\frac{3}{2}$.
43. No root. **44.** −7. **45.** $-\frac{14}{3}$. **46.** $\frac{9}{2}$.
47. 14. **48.** $-\frac{8}{3}$. **49.** 19. **50.** $12, \frac{121}{13}$.
51. $\frac{32}{3}$. **52.** $5, -\frac{20}{3}$. **53.** 0.32.
54. 21, 23, or −23, −21. **55.** 32, 34, or −34, −32. **56.** 14, 17, or −17, −14.
57. $10 + 10\sqrt{2}$ or 24.14 in. **58.** 8 in. by 32 in.
59. 66. **60.** 7, 13. **61.** 5 in. **62.** $22, -\frac{122}{9}$.
63. 80 min.; 120 min.

Pages 74–75; Lesson 11

4. $\pm 2, \pm 2i$. **5.** $\pm\sqrt{2}, \pm\sqrt{2}\,i$. **6.** $2, -1 \pm \sqrt{3}\,i$.

7. $-3, \frac{3}{2}(1 \pm \sqrt{3}\,i)$. **8.** $0, \pm\sqrt{5}\,i$. **9.** $0, -\frac{1}{2} \pm \frac{\sqrt{3}}{2}i$.

10. $1, \pm\sqrt{5}$. **11.** $-1, \pm 3$. **12.** $\frac{3}{2}, \pm 2i$.
13. $3, \pm\frac{3}{2}i$. **14.** $1, -\frac{1}{2} \pm \frac{1}{2}\sqrt{3}\,i$. **15.** $1 \pm \sqrt{3}\,i$.
17. $-3, 1 \pm 2i$. **18.** $-\frac{1}{2}, 2 \pm i$. **19.** $\pm\sqrt{2}, 1 \pm i$.

20. $\pm\frac{3\sqrt{2}}{2}, 1 \pm 3i$. **21.** $\pm 2, \pm 5i$. **22.** $\pm 1, \pm 3i$.

23. $\pm 2\sqrt{3}, \pm 3i$. **24.** $\pm 2\sqrt{2}\,i, \pm 2\sqrt{3}\,i$. **25.** $\pm 3, \pm\frac{1}{2}i$.
26. $\pm\sqrt{3}, \pm\frac{5}{2}i$. **27.** $0, 1 \pm \sqrt{3}$. **37.** $3, -\frac{2}{3}, -6$.
38. $-5, -\frac{7}{2}, -7$. **39.** $4, -\frac{3}{2}$. **40.** 6, 3.1, −1.25.

CHAPTER IV

Pages 81–82; Lesson 12

1. $5, 4\sqrt{2}, 13, 5\sqrt{2}$. **2.** $2\sqrt{5}, 3, 5, 6\sqrt{2}$. **3.** 24 sq. units.
4. 13.5 sq. units. **5.** $D(9,4)$; 36 sq. units. **6.** $(-3.5, -2)$.
7. $y = 3$. **8.** $x + 5 = 0$. **9.** $y = x$.

Pages 87–89; Lesson 13

4. $V = \frac{S\sqrt{S}}{6\sqrt{\pi}}$. **32.** $S = x^2 + \frac{1{,}024}{x}$. **33.** $V = x(21 - 2x)^2$.

34. $V = \frac{3}{2}\pi x^2(12 - x)$. **35.** $V = \frac{\pi x^2}{3}(12 - x)$. **36.** $C = 0.08\left(\pi x^2 + \frac{400}{x}\right)$.

CHAPTER V

Pages 94–97; Lesson 14

5. (6,2). **6.** (−1,−4). **7.** $(-\frac{5}{2},\frac{3}{2})$. **8.** $(\frac{4}{3},-2)$.
9. Dependent. **10.** $(-\frac{2}{3},-\frac{20}{3})$. **11.** $(\frac{7}{3},4)$. **12.** $(-\frac{5}{2},\frac{7}{3})$.
13. (5,6). **14.** $(-\frac{5}{2},3)$. **15.** $(\frac{3}{2},-6)$. **16.** (0,−27).
17. (−6,11). **18.** (3.4,2.6). **19.** $(\frac{16}{5},-16)$. **20.** Dependent.
21. $(\frac{16}{3},\frac{12}{35})$. **22.** $(7,\frac{5}{3})$. **23.** 75 lb. of 60 per cent; 25 lb. of 20 per cent.
24. 83. **25.** Numerator 52, denominator 12.
26. 14.5 m.p.h.; 5.5 m.p.h. **27.** 8 hr.; $10\frac{2}{3}$ hr.
28. $y = \frac{1}{3}x + \frac{1}{3}$ or $3y = x + 1$.
29. (*a*) $y = \frac{1}{3}x + \frac{4}{3}$ or $x - 3y + 4 = 0$.
(*b*) $y = -x + 2$ or $x + y = 2$.
(*c*) $y = \frac{1}{5}x - \frac{28}{5}$ or $x - 5y = 28$.
(*d*) $y = \frac{2}{3}x$ or $2x - 3y = 0$.
(*e*) $y = \frac{2}{5}x + 2$ or $2x - 5y + 10 = 0$.
(*f*) $y = -x - 7$ or $x + y + 7 = 0$.

Pages 102–105; Lesson 15

3. $x = y = z =$ any constant. **4.** $x = 11, y = -2, z = -4$.
5. $x = 4, y = -1, z = 5$. **6.** $x = -\frac{5}{2}, y = 4, z = 6$.
7. $x = -\frac{4}{3}, y = \frac{2}{3}, z = -1$. **8.** $x = -7, y = 3, z = -4.5$.
9. $u = -2, v = 0, w = 1$. **10.** Inconsistent.
11. Dependent. **12.** $x = 12, y = -7, z = -2$.
13. $u = 3, v = -\frac{4}{3}, w = 6$. **14.** $x = 4, y = -2, z = 3$.
16. $w = 4, x = -2, y = 5, z = -1$. **17.** $s = 3, t = 0, u = -2, v = 5$.
20. $x = 24.5, y = 63.5, z = 19.5$. **21.** $5\frac{1}{3}$ days.
22. (4,3); (−2,0). **23.** (8,−16); $(-\frac{1}{2},-3\frac{1}{4})$.
24. $(0,2\frac{1}{4})$; (9,0). **25.** (−0.76,0.38); (−5.24,2.62).
26. (1,4).
27. $x = 1 + 3i, y = 3i; x = 1 - 3i, y = -3i$.
28. $x = \dfrac{-1 + \sqrt{7}\,i}{2}, y = 5 + \sqrt{7}\,i; x = \dfrac{-1 - \sqrt{7}\,i}{2}, y = 5 - \sqrt{7}\,i$.
29. (0,−8). **30.** (0,0); (4,2). **31.** (5,0); (−1,6).
32. (−4,0); (−3,1); $x = \frac{1}{2}(-9 + \sqrt{3}\,i)$, $y = \frac{1}{2}(-1 - \sqrt{3}\,i)$; $x = \frac{1}{2}(-9 - \sqrt{3}\,i)$, $y = \frac{1}{2}(-1 + \sqrt{3}\,i)$.
33. (6,8); (−10,0). **34.** $(2\sqrt{2}, 4 + 2\sqrt{2}), (-2\sqrt{2}, 4 - 2\sqrt{2})$.
35. (6,2); (−6,2). **36.** (4,2); (−4,2); (4,−2); (−4,−2).
37. (0,3); (0,−3); $x = -\frac{25}{3}, y = 4i; x = -\frac{25}{3}, y = -4i$.
38. (2,2); (−2,2); $x = \sqrt{3}\,i, y = -12; x = -\sqrt{3}\,i, y = -12$.
39. (3,4); (−3,4); $x = 3i, y = -8; x = -3i, y = -8$.
40. 16 in. by 9 in. **41.** 20 in. by 48 in. **42.** 15 and 3.
43. 12 and 8. **44.** 16 in. by 24 in. **45.** 48 m.p.h.; 240 miles.
46. (3,3); (−3,−3). **47.** (6,−2); $(\frac{22}{27},\frac{2}{27})$.
48. (4,3); (−4,−3); $\left(-\sqrt{2}, -\dfrac{5}{\sqrt{2}}\right)$; $\left(\sqrt{2}, \dfrac{5}{\sqrt{2}}\right)$.

49. $(4,-2)$; $(-4,2)$; $\left(\frac{5}{\sqrt{2}}, \frac{1}{\sqrt{2}}\right)$; $\left(-\frac{5}{\sqrt{2}}, -\frac{1}{\sqrt{2}}\right)$.

50. $x = 2, y = -1, z = 3; x = -\frac{1}{3}, y = \frac{11}{3}, z = \frac{2}{3}$.

51. $x = 1, y = \frac{5}{2}, z = -3; x = \frac{37}{14}, y = -\frac{45}{28}, z = \frac{2}{7}$.

CHAPTER VI

Pages 112–114; Lesson 16

6. $x < 9.5$. **7.** $x < -\frac{65}{6}$. **8.** $x < 11$. **9.** $x < -6$.
10. $x > 66$. **11.** $x > -\frac{36}{13}$. **12.** $2 < x < 8$. **13.** $-7 < x < 3$.
14. $x > \frac{3\sqrt{2}}{2}; x < -\frac{3\sqrt{2}}{2}$. **15.** $-\frac{5}{3} < x < \frac{5}{3}$.
16. $x > 4; x < 0$. **17.** $x > 4.5; x < 0$.
18. $x > 4$. **19.** $0 < x < 3; x > 3$. **20.** $-2.5 < x < 1$.
21. $x > -2.5; x < -4$. **22.** $-2 < x < 0; x > 1$.
23. $-2 < x < 0$. **24.** $-1 < x < 0; x > 2$. **25.** $x < 0; 4 < x < 5$.
26. $x < -1; 1.5 < x < 2; x > 2$. **27.** $x < -2.5; 3 < x < 5$.
28. $-5 < x < \frac{4}{3}$. **29.** $x > 3; x < -4.5$. **30.** $x \neq 2.5$.
31. $-1.5 < x < 7$. **32.** $x < 0; 2 < x < 4$. **33.** $-1 < x < 0; x > 2$.
34. $-2 < x < 8$. **35.** $-1 < x < 6$. **36.** $-\frac{1}{2} < x < \frac{3}{2}$.
37. $3 < x < 9$. **38.** $-4.5 < x < 1.5$. **39.** $x < -7; x > 8$.
40. $x < -3; x > 9$. **41.** $2 < x < 2.5$. **42.** $6 < x < 9.6$.
43. $x < 3; x > 10$. **44.** $\frac{1}{3} < x < \frac{9}{10}$. **45.** $-3 < x < -2; x > 3$.
46. $-1.5 < x < 1.5; x > 3$. **47.** $x < -4; 1.5 < x < 9$.
48. $x < -1.75; 0.5 < x < 2$. **49.** $-\frac{6}{5} < x < \frac{1}{3}; x > 54$.
50. $-\frac{5}{2} < x < \frac{2}{3}; x > 5.1$.

CHAPTER VII

Pages 119–120; Lesson 17

1. $\frac{1}{4x}$. **2.** $\frac{1}{64x}$. **3.** $\frac{x^2}{10}$. **4.** $\frac{y}{3x^2}$.

5. $2ab$. **6.** $5x^3y^4$. **7.** $\frac{x^2}{9y^4}$ **8.** $\frac{4}{x-y}$.

9. $\frac{(x+y)^2}{xy}$. **10.** $4x^2$. **11.** $\frac{x-1}{x}$. **12.** $a(3a^2+1)$.

13. $\frac{1}{x} + \frac{2}{\sqrt{xy}} + \frac{1}{y}$. **14.** $x - y$. **15.** $\frac{2(x+1)}{x-1}$. **16.** $1 - z$.

17. $\frac{82}{65}$. **18.** $\frac{1}{a^3} - \frac{1}{b^3}$. **19.** 6.4×10^6. **20.** 4×10^8.

21. 5.26×10^{10}. **22.** 2.4×10^{-6}. **23.** 7×10^{-8}. **24.** 1.23×10^{-5}.
25. $\frac{25}{4}$. **26.** $\frac{1}{9}$. **27.** $\frac{1}{4}$. **28.** $\frac{8}{27}$.
29. 8. **30.** $-\frac{1}{2}$. **31.** 38. **32.** No root.

34. $\frac{2\sqrt{6}}{3}$. **35.** $\frac{3\sqrt{2}}{8}$. **36.** $\frac{4\sqrt{2}}{5}$. **37.** $\sqrt{6} - \sqrt{2}$.

38. $\frac{8 + 2\sqrt{2}}{7}$. **39.** $\frac{x(\sqrt{x} + \sqrt{y})}{x - y}$. **43.** $\frac{\sqrt{x^2 - 4}}{x}$.

Pages 124–125; Lesson 18

18. $5\sqrt{5}$.
20. Between 1 and 2; between −1 and −2.
22. Between 2 and 3; between −3 and −4.
24. Between 7 and 8; between −3 and −4.

CHAPTER VIII

Pages 130–131; Lesson 19

21. 1.5441; 0.47712 − 2 or −1.52288; 1.9912.
22. 2.0212; 0.1461; −0.5441.
23. 2.5741; 3.7993; 3.6990.
24. 1.2431; 1.5353; −0.4771.
25. 3.4471; 0.2431; 0.6990 − 3 or −2.3010.
26. 0.4650; −0.2798; 0.7993.
27. 3.9030; 1.1761 − 3 or −1.8239; −1.0670
28. 0.0757; 0.2720; 0.2594.
29. 2.6251; 0.3806; 1.2828.
30. 2.0106; −0.3890; 0.6667.

Pages 136–137; Lesson 20

6. 1.3766; 8.8579 − 10; 4.2148.
7. 0.9841; 8.5465 − 10; 3.1761.
8. 2.6749; 9.9455 − 10; 6.5911 − 10.
9. 1.8785; 4.6021; 7.2577 − 10.
10. 0.5877; 6.9731 − 10; 4.6767.
11. 2.9903; 9.6263 − 10; 5.4624.
12. 1.0170; 3.4771; 7.8603 − 10.
13. 1.1653; 9.8595 − 10; 3.5957.
14. 0.9928; 7.6650 − 10; 3.8455.
15. 1.3381; 9.7768 − 10; 5.8348.
16. 0.2132; 8.9563 − 10; 4.6739.
17. 0.4972; 7.3393 − 10; 3.9488.
18. 0.4343; 8.5740 − 10; 4.9094.
19. 0.85986; 7.58218 − 10; 4.85522.
20. 1.16316; 9.91666 − 10; 4.73846.
21. 0.99441; 8.51055 − 10; 3.79134.
22. 1.55084; 6.86243 − 10; 5.39375.
23. 1.21096; 9.58593 − 10; 4.72225.
24. 0.99526; 8.85942 − 10; 5.55090.
25. 0.49715; 7.91682 − 10; 4.66676.
26. $A = 58.10$; $B = 0.01487$.
27. $A = 5761$; $B = 0.0002621$.
28. $A = 2.898$; $B = 0.1882$.
29. $x = 34220$; $y = 0.008436$.
30. $x = 1708$; $y = 0.05824$.
31. $r = 2.331$; $s = 0.04461$.
32. $M = 111.3$; $N = 0.8215$.
33. $M = 1514$; $N = 0.006324$.
34. $M = 22.26$; $N = 0.003513$.
35. $A = 5203.3$; $B = 0.025861$.
36. $A = 29.597$; $B = 0.17207$.
37. $A = 443{,}830$; $B = 0.053508$.
38. $x = 7.5657$; $y = 0.00015235$.
39. $x = 172.64$; $y = 0.0011288$.
40. $x = 27{,}952$; $y = 0.0079625$.
41. Characteristic −3; mantissa 0.5626; $N = 0.003653$.
42. Characteristic −4; mantissa 0.1868; $N = 0.0001537$.

Pages 140–141; Lesson 21

1. 0.0770.
2. 92.3.
3. 711.
4. 2.94.
5. 321.8.
6. 1,639.
7. 27.73.
8. 51.38.
9. 7.579.
10. 8,085.
11. 42,500.
12. 425.
13. 412.
14. 2,891.
15. 8.940.
16. 2.44.
17. 0.602.
18. 28.73.
19. 3.077.
20. 78.9.
21. 28.3.
22. 0.647.
23. 0.493.
24. 26.35.
25. 56.2.
26. 0.521.
27. 52.678.
28. 29.291.
29. 7,780.8.
30. 31.242.
31. $8,005.
32. $1,643.30.
33. 142.9 in.
34. 197,100,000 sq. miles.
35. 15.24 ft.
36. 102.7 sq. ft.
37. 2.31 sec.
38. 6.17 ft.

Pages 146–147; Lesson 22

1. 2.496. **2.** 2.237. **3.** 0.9140. **4.** 2.861.
5. 3.172. **6.** 0.5250. **7.** 0.6932. **8.** 6.908.
9. 0.6098. **10.** 5.951. **11.** 3.908. **12.** 7.111.
16. 2. **17.** 1; $-\frac{3}{2}$. **18.** 2.153. **19.** 0.7472.
20. 28.41. **21.** 2.483. **22.** 17.49. **23.** 3.036.
24. 0.7739. **25.** 7.336. **26.** 1.108. **27.** −4.861.
28. 1.962. **29.** 1.167. **30.** 15. **31.** 8.
32. $\frac{1}{6}$. **33.** 41. **34.** $\frac{1}{15}$. **35.** 5.
38. 1; 0.6309. **39.** 2.095. **40.** $n = \frac{\log A - \log P}{\log (1 + r)}$.
41. $x = \log_e (y \pm \sqrt{y^2 - 1})$. **42.** $x = \frac{1}{2} \log_e \frac{1 + y}{1 - y}$.

CHAPTER IX

Pages 151–152; Lesson 23

1. $23\frac{1}{3}$. **2.** $3\frac{3}{8}$. **3.** 128. **4.** 90.
5. 4687.5. **10.** 32.3 lb. per sq. in.
11. Nine times as much at 60 m.p.h. **12.** 114.
13. 70.4 lb. per sq. in. **14.** 3.14 sec. **15.** 178 lb.
17. Resistance decreased by 4 per cent. **18.** 10,757 days.

CHAPTER X

Pages 156–157; Lesson 24

12. $18 - 5n$. **13.** $\frac{1}{4} - \frac{3}{32}n$. **14.** $\frac{3}{2}(n + 7)$.
15. $1{,}000 + n$. **16.** $2(z + 1 - n)$. **17.** $2z + 2n - 5$.
18. 77. **19.** 100. **20.** 40.5. **21.** −51.
22. $14\frac{2}{3}$. **23.** 4.6. **24.** −3.72. **25.** 84.5.
26. $23a + 13b$. **27.** $-18a - 10b$. **28.** 4,950. **29.** 10,000.
30. 32. **31.** 650. **32.** 301.5. **33.** 142.5.
34. 860. **35.** −72. **36.** 300. **37.** 129.
41. 14. **42.** 10. **44.** 16. **45.** 0; 2.

Pages 162–164; Lesson 25

1. $a_n = 3^n$. **2.** $a_n = 2^{n+1}$. **3.** $a_n = 6(-\frac{1}{3})^{n-1}$. **4.** $a_n = 8(-\frac{1}{2})^{n-1}$.
5. $a_n = 2^{\frac{1}{2}n}$. **6.** $a_n = 100(-0.1)^{n-1}$. **7.** log 16; log 256.
8. $\log \sqrt[3]{2}$; $\log \sqrt[9]{2}$. **10.** 15; 9. **11.** 3; −14.
12. 0; $-\frac{27}{14}$. **13.** 3; $-\frac{1}{3}$. **16.** $13\frac{13}{27}$; 13.5.
17. 62.48. **18.** −3,280. **19.** $12\frac{51}{64}$. **20.** $40\frac{1}{3}$.
21. 255.75. **22.** $\frac{5}{6}$. **23.** $\frac{2}{3}$. **24.** 4.5.
25. 16.2. **26.** 48. **27.** 5. **28.** $\frac{13}{9}$.
29. $\frac{11}{3}$. **30.** $\frac{79}{9}$. **31.** $\frac{131}{9}$. **32.** $\frac{27}{11}$.
33. $\frac{281}{33}$. **34.** $\frac{368}{495}$. **35.** $\frac{47}{330}$. **36.** $\frac{410}{333}$.
37. $\frac{237}{37}$. **38.** 42 ft. **40.** −2, 6, 14, or 46, 6, −34.
41. About 11.4 gal. **42.** 59.049 per cent; 7 strokes.

CHAPTER XII

Pages 175–176; Lesson 27

1. $x^6 + 6x^5y + 15x^4y^2 + 20x^3y^3 + 15x^2y^4 + 6xy^5 + y^6.$
2. $a^5 - 10a^4b + 40a^3b^2 - 80a^2b^3 + 80ab^4 - 32b^5.$
3. $x^4 + 8x^3 + 24x^2 + 32x + 16.$
4. $32x^{10} - 80x^8y + 80x^6y^2 - 40x^4y^3 + 10x^2y^4 - y^5.$
5. $x^3 + 3x^{\frac{5}{2}}y + \frac{15}{4}x^2y^2 + \frac{5}{2}x^{\frac{3}{2}}y^3 + \frac{15}{16}xy^4 + \frac{3}{16}x^{\frac{1}{2}}y^5 + \frac{1}{64}y^6.$
6. $x^4y^8 - 8x^3y^6 + 24x^2y^4 - 32xy^2 + 16.$
7. $8a^6 - 36a^4b^2 + 54a^2b^4 - 27b^6.$
8. $1 - 6z^2 + 15z^4 - 20z^6 + 15z^8 - 6z^{10} + z^{12}.$
9. $a^3 + 6a^{\frac{5}{2}}b^{\frac{1}{2}} + 15a^2b + 20a^{\frac{3}{2}}b^{\frac{3}{2}} + 15ab^2 + 6a^{\frac{1}{2}}b^{\frac{5}{2}} + b^3.$
10. $y^{12} - 12y^9a + 54y^6a^2 - 108y^3a^3 + 81a^4.$
11. $64x^3 - 576x^{\frac{5}{2}}y^{\frac{1}{2}} + 2160x^2y - 4320x^{\frac{3}{2}}y^{\frac{3}{2}} + 4860xy^2 - 2916x^{\frac{1}{2}}y^{\frac{5}{2}} + 729y^3.$
12. $4\sqrt{2}\,a^5 + 10a^4b + 5\sqrt{2}\,a^3b^2 + \frac{5}{2}a^2b^3 + \frac{5}{16}\sqrt{2}\,ab^4 + \frac{1}{32}b^5.$
13. $1 + 16x^{\frac{1}{2}} + 112x + 448x^{\frac{3}{2}} + 1120x^2 + 1792x^{\frac{5}{2}} + 1792x^3 + 1024x^{\frac{7}{2}} + 256x^4.$
14. $a^{12}b^6 - 12a^{10}b^5c^3 + 60a^8b^4c^6 - 160a^6b^3c^9 + 240a^4b^2c^{12} - 192a^2bc^{15} + 64c^{18}.$
15. $x^5 + 10x^{\frac{9}{2}}y^{\frac{1}{2}} + 45x^4y + 120x^{\frac{7}{2}}y^{\frac{3}{2}} + 210x^3y^2 + 252x^{\frac{5}{2}}y^{\frac{5}{2}} + 210x^2y^3 + 120x^{\frac{3}{2}}y^{\frac{7}{2}}$
$+ 45xy^4 + 10x^{\frac{1}{2}}y^{\frac{9}{2}} + y^5.$
16. $a^{12} - 24a^{11}b + 264a^{10}b^2 - 1760a^9b^3 + \cdots.$
17. $x^{14} + 7x^{13} + \frac{91}{4}x^{12} + \frac{91}{2}x^{11} + \cdots.$
18. $x^{10} - 20x^8 + 180x^6 - 960x^4 + \cdots.$
19. $1 + 16\sqrt{x} + 120x + 560x^{\frac{3}{2}} + \cdots.$
20. $1 + \dfrac{16}{x} + \dfrac{120}{x^2} + \dfrac{560}{x^3} + \cdots.$
21. $a^{12} + 36a^{11}\sqrt{b} + 594a^{10}b + 5940a^9b^{\frac{3}{2}} + \cdots.$

22. 1.1046. **23.** 1.2653. **24.** 1.3048. **25.** 0.8508.
26. 0.8330. **27.** 1.9487. **28.** $924x^6y^6$. **29.** $-84a^6b^3$.
30. $3{,}360y^{12}$. **31.** $-792x^3$. **32.** $-560a^8b^6$. **33.** $672a^3b^3$.
34. 1,120. **35.** $6{,}048x^5y^2$. **36.** 210. **37.** 54.
38. 120. **39.** $\frac{33}{14}$. **40.** $\dfrac{1}{n+1}$. **41.** $(k+2)(k+1)$.
42. $n(n+1)$. **43.** $\dfrac{1}{n+1}$. **44.** $\dfrac{n+1}{n}$. **45.** $n^2 - 1$.

Page 180; Lesson 28

1. $1 - \dfrac{1}{2}x - \dfrac{1}{2^2 \cdot 2!}x^2 - \dfrac{1 \cdot 3}{2^3 \cdot 3!}x^3 - \dfrac{1 \cdot 3 \cdot 5}{2^4 \cdot 4!}x^4 - \cdots.$

2. $1 + \dfrac{1}{3}x - \dfrac{2}{3^2 \cdot 2!}x^2 + \dfrac{2 \cdot 5}{3^3 \cdot 3!}x^3 - \dfrac{2 \cdot 5 \cdot 8}{3^4 \cdot 4!}x^4 + \cdots.$

3. $1 - \dfrac{1}{3}x - \dfrac{2}{3^2 \cdot 2!}x^2 - \dfrac{2 \cdot 5}{3^3 \cdot 3!}x^3 - \dfrac{2 \cdot 5 \cdot 8}{3^4 \cdot 4!}x^4 - \cdots.$

4. $1 - 2x + 3x^2 - 4x^3 + 5x^4 - \cdots.$
5. $1 + x + x^2 + x^3 + x^4 + \cdots.$
6. $1 + 2x^2 + 3x^4 + 4x^6 + 5x^8 + \cdots.$

7. $1 - \frac{1}{2}x + \frac{1 \cdot 3}{2^2 \cdot 2!}x^2 - \frac{1 \cdot 3 \cdot 5}{2^3 \cdot 3!}x^3 + \frac{1 \cdot 3 \cdot 5 \cdot 7}{2^4 \cdot 4!}x^4 - \cdots.$

8. $1 + \frac{1}{2}x + \frac{1 \cdot 3}{2^2 \cdot 2!}x^2 + \frac{1 \cdot 3 \cdot 5}{2^3 \cdot 3!}x^3 + \frac{1 \cdot 3 \cdot 5 \cdot 7}{2^4 \cdot 4!}x^4 + \cdots.$

12. 7.071. **13.** 9.899. **14.** 9.165. **15.** 3.072.
16. 4.016. **17.** 2.924. **18.** 1.020. **19.** 1.010.
20. 1.005. **21.** 0.8219. **22.** 0.8375. **23.** 0.8535.
24. 0.8885. **25.** 0.7903. **26.** 0.8131. **27.** 0.6139.

CHAPTER XIII

Pages 184–185; Lesson 29

1. 4.2 per cent. **2.** 5.4 per cent. **3.** 2.3 per cent. **4.** 3.7 per cent.
5. 24.5. **6.** 20.5. **7.** $343.20. **8.** $808.41.
9. $5,718.01. **10.** $745.43. **11.** $1,076.69. **12.** 4.73 per cent.
13. 20 years. **14.** $522.87. **15.** $700.16. **16.** $579.77.
17. 2.9 per cent. **18.** 5.1 per cent.

Pages 189–190; Lesson 30

1. 8.58297. **2.** 6.46841. **3.** 2.77509. **4.** 5.60143.
5. $7,512.90. **6.** $3,753.87. **7.** $1,174.43. **8.** $4,591.70.
9. $170.83. **10.** $1,490.79. **11.** $373.74. **12.** $1,113.27.
13. $6,021.50. **14.** $14,048.88. **15.** $1,365.79. **16.** $10,543.31.

CHAPTER XIV

Pages 197–198; Lesson 31

10. $\frac{2}{3}; -\frac{5}{2}$. **11.** $5; -\frac{3}{2}$. **12.** $-\frac{1}{2}; -\frac{3}{2}$. **13.** $2 \pm 3i$.
14. $x^4 + 3x^3 - 8x^2 - 24x + 3$. **15.** $x^4 + x - 7$.
16. $x^4 - 2x^3 + 4x^2 - 8x + 16$.

Pages 202–204; Lesson 32

1. $A = -2; B = 4$. **2.** $A = 1; B = -6$.
3. $A = 15; B = -10$. **4.** $A = 21; B = 42$.
5. $A = 4; B = 5; C = -1$. **6.** $A = -1; B = 3; C = 2$.
7. $A = -4; B = 1; C = 8$. **8.** $A = 3; B = 1; C = 5$.
9. $A = -\frac{1}{2}; B = 13; C = \frac{3}{2}$. **19.** $x^4 - 3x^3 + 4x = 0$.
20. $12x^4 + 20x^3 + 11x^2 + 2x = 0$. **21.** $9x^3 + 3x^2 - 14x - 8 = 0$.
22. $x^4 - 16 = 0$. **23.** $x^4 - 2x^3 - x^2 + 6x - 6 = 0$.
24. $x^4 + 4x^3 + 2x^2 + 4x + 1 = 0$.
25. $27x^5 + 108x^4 + 171x^3 + 134x^2 + 52x + 8 = 0$.
26. $x^6 + 4x^5 + 14x^4 + 20x^3 + 25x^2 = 0$. **27.** $x^3 + 7x - 6i = 0$.
28. $x^3 - (2 - i)x^2 + (2 - 2i)x + 2i = 0$.
29. $x^4 - 2x^3 + 3x^2 - 2x + 2 = 0$. **30.** $x^3 + 6ix^2 - 5x + 12i = 0$.
35. $\frac{1}{2}(-3 \pm \sqrt{5})$. **36.** $-\frac{1}{2}; -6$. **37.** 1.5.
38. $-2; -\frac{1}{2}$. **39.** $\pm 2i$. **40.** $\pm i$.

Pages 209–210; Lesson 33

1. $-4 + 2i$; $-4 - 2i$. **2.** $-9(13 + i)$; $-9(13 - i)$.
3. $43 - 12i$; $43 + 12i$. **18.** -1; $\frac{1}{2} \pm \frac{1}{2}\sqrt{3}\,i$.
19. ± 1; $\pm i$. **20.** $x^4 + 2x^3 + 4x^2 + 8x + 16 = 0$.
21. Two; one. **22.** None; one. **23.** $2 - i$; $-\frac{2}{3}$.
24. $1 + 4i$; $\frac{5}{2}$. **25.** $-3i$; $\frac{7}{3}$. **26.** $-\sqrt{2}$; $1 \pm \sqrt{3}\,i$.
27. $2 - \sqrt{5}$; 1, 1. **28.** $3 - \sqrt{2}\,i$; $\pm\sqrt{3}$.
29. $1 + \sqrt{2}\,i$; 1; -3. **30.** $2 - 2i$; $-2 \pm 2i$.

Pages 214–215; Lesson 34

1. None. **2.** 4. **3.** None. **4.** -1; -2.
5. -3. **6.** $-\frac{3}{2}$; $\pm\sqrt{7}$. **7.** $-\frac{7}{2}$; $\pm\sqrt{\frac{3}{2}}$.
8. None. **9.** $\frac{1}{2}$; 6; -1. **10.** -1; $\frac{2}{3}$; $2 \pm i$.
11. -4. **12.** $\frac{1}{3}$; -3; $\pm 2\sqrt{3}$. **13.** None.
14. 3. **15.** $-\frac{3}{2}$. **16.** 2; -2; $\frac{5}{3}$; $\pm i$.
17. $\frac{1}{2}$; $-\frac{1}{2}$; $2 \pm \sqrt{7}$. **18.** $-\frac{7}{4}$, 2, $\frac{1}{2}(3 \pm \sqrt{5})$. **21.** $-\frac{3}{2}$.
22. $-i$; $-\frac{7}{2}$; $2 \pm \sqrt{3}$. **23.** $1 - i$; -2; $1 \pm 2\sqrt{2}$.

Pages 220–222; Lesson 35

3. 5.236. **4.** 3.646. **5.** 2.93. **6.** -2.67.
7. 1.48. **8.** 3.732. **9.** 3.684. **10.** 3.43.
11. 2.13. **12.** 2.59. **13.** 1.20 in.; 3.71 in.

Pages 226–229; Lesson 36

1. $x^3 - x = 0$. **2.** $2x^3 + x^2 - 18x - 9 = 0$
3. $18x^3 - 111x^2 + 161x = 0$. **4.** $x^3 - 10x^2 + 29x = 0$.
5. $1 \pm 2i$; $1 \pm \sqrt{5}$. **6.** 1; -5; $-2 \pm i$.
7. $x^4 - 8x^3 + 24x^2 - 32x + 7 = 0$.
9. $3x^3 + 2x^2 - 7x + 2 = 0$; $2x^3 + 7x^2 + 2x - 3 = 0$.
10. $2x^3 - 3x^2 - 5x + 6 = 0$; $6x^3 + 5x^2 - 3x - 2 = 0$.
12. $x^4 - 2x^3 - 8x^2 + 10x + 15 = 0$. **13.** $2x^3 + 11x^2 + 27x + 31 = 0$.
14. $x^3 + 9x^2 + 11x - 6 = 0$. **15.** $2x^4 - 14x^3 + 30x^2 - 19x - 11 = 0$.
16. $x^5 + 10x^4 + 36x^3 + 56x^2 + 40x + 24 = 0$.
17. $x^4 - 8x^3 - 9.5x^2 - 3x - 8.1875 = 0$. **18.** $x^3 + 6.6x^2 + 14.52x - 1.352 = 0$.
19. $x^3 + 9.3x^2 + 28.83x - 1.209 = 0$. **20.** $3x^4 + 13x^3 + 13x^2 - 4x - 10 = 0$.

Pages 232–234; Lesson 37

1. 3.476. **2.** 4.932. **3.** 2.759. **4.** 1.586.
5. 0.268. **6.** 1.646. **7.** 3.081. **8.** -2.464.
9. -3.292. **10.** -0.359. **11.** 7.123. **12.** -4.196.
13. 3.118. **14.** 0.88. **15.** 5.74.

Page 238; Lesson 38

1. -1; $-\frac{1}{2} \pm \frac{1}{2}\sqrt{3}\,i$. **2.** -3; $-3 \pm \sqrt{3}\,i$.
3. 3; $-3 \pm \sqrt{3}\,i$. **4.** 1; -1; 3.
5. 1; $1 \pm \sqrt{3}\,i$. **6.** -1; $-1 \pm 3\sqrt{2}\,i$.
7. 6; $\pm\sqrt{3}$. **8.** 3; $-1 \pm i$.

CHAPTER XV

Pages 245–247; Lesson 39

1. 12. **2.** -5. **3.** $\frac{5}{2}$. **4.** 27.
5. -77. **6.** -36. **7.** -44. **8.** -1.
9. -62. **10.** $(\frac{8}{3},-2)$. **11.** $(-\frac{5}{2},3)$. **12.** $(1.5,-3.5)$.
13. $(-\frac{3}{2},4)$. **14.** $(3,-\frac{3}{4})$. **15.** $(1.5,1.8)$.
16. $x = -1; y = -2; z = 4$. **17.** $x = 3; y = -3; z = 5$.
18. $x = -\frac{7}{4}; y = 3; z = 7$. **19.** $x = \frac{3}{4}; y = -\frac{1}{2}; z = -3$.
20. $x = -\frac{2}{7}; y = 4; z = -1$. **21.** $x = \frac{8}{3}; y = -\frac{2}{3}; z = 6$.
22. $x = -\frac{9}{2}; y = \frac{3}{2}; z = -3$. **23.** $x = 2; y = 0; z = -\frac{4}{3}$.

Pages 249–250; Lesson 40

6. 48. **7.** 14. **8.** -18. **9.** -5. **10.** 20.
11. -2. **12.** 84. **13.** -42. **14.** 0. **15.** 0.
16. 104. **17.** 48. **18.** 16. **19.** 0. **20.** 7.

Pages 255–256; Lesson 41

1. -322. **2.** 978. **3.** 924. **4.** 0. **5.** -138.
6. 350. **7.** -60. **8.** 384. **9.** 624. **10.** 1,120.

Pages 262–263; Lesson 42

1. $x = -4, y = 1, z = 2, w = 5$. **2.** $x = \frac{3}{2}, y = \frac{5}{2}, z = -3, w = 6$.
3. $x = -2, y = \frac{2}{3}, z = -\frac{2}{3}, w = \frac{9}{2}$. **4.** $x = -3, y = 2.5, z = 1.5, w = -1$.
5. $x = 5, y = -2.5, z = -1.5, w = 2$.
6. $x = 2.25, y = -0.75, z = -0.25, w = 2$.
14. $x = 5, y = 2, z = -1$. **15.** $x = 2k + 3, y = 2k - 1, z = k$.
16. $x = k + 6, y = 3 - 2k, z = k$. **17.** $x = k, y = -5k, z = \frac{3}{2}k$.
18. $x = \frac{2}{3}k, y = -\frac{4}{3}k, z = k$.
19. $x = 2k + 1, y = 2k - 1, z = -4k, w = k$.
20. $x = -k, y = 3k, z = -\frac{5}{2}k, w = k$.

CHAPTER XVI

Pages 268–269; Lesson 43

1. $-20 + 8i$. **2.** $2 + i$. **3.** $3 + 10i$. **4.** $2 + 4i$.
5. $2i$. **6.** $2i$. **7.** $3 - 3i$. **8.** $4 - 3i$.
9. $-2 - 4i$. **10.** 2. **39.** $5; 6; 13; 4\sqrt{2}; 7$.
40. $\sqrt{13}; 10; 5; 2\sqrt{2}; 6$. **41.** $\sqrt{2}$. **42.** $5\sqrt{2}$.
43. 2. **44.** $2\sqrt{5}$. **45.** $4 + i$. **46.** $-0.1 - 2.3i$.
47. $-2 + 3i$. **48.** $-3 - 7i$. **49.** $-1 - 0.6i$. **50.** $19 - 8i$.
51. $-2 - 25i$. **52.** $0.8i$. **53.** $0.2i$. **54.** $7 + 2i$.

Pages 273–274; Lesson 44

21. $\sqrt{8}(\cos 45° + i \sin 45°)$; $8(\cos 0° + i \sin 0°)$; $6(\cos 90° + i \sin 90°)$.
22. $\sqrt{2}(\cos 225° + i \sin 225°)$; $4(\cos 180° + i \sin 180°)$; $2(\cos 270° + i \sin 270°)$.
23. $4\sqrt{2}(\cos 135° + i \sin 135°)$; $3(\cos 180° + i \sin 180°)$; $3(\cos 90° + i \sin 90°)$.
24. $5\sqrt{2}(\cos 315° + i \sin 315°)$; $16(\cos 0° + i \sin 0°)$; $8(\cos 270° + i \sin 270°)$.

25. $4(\cos 315° + i \sin 315°)$; $2(\cos 30° + i \sin 30°)$; $(\cos 180° + i \sin 180°)$.
26. $4(\cos 330° + i \sin 330°)$; $\sqrt{2}\,(\cos 45° + i \sin 45°)$; $3(\cos 270° + i \sin 270°)$.
27. $6(\cos 300° + i \sin 300°)$; $4\sqrt{2}\,(\cos 45° + i \sin 45°)$; $32(\cos 0° + i \sin 0°)$.
28. $4(\cos 120° + i \sin 120°)$; $3\sqrt{2}\,(\cos 135° + i \sin 135°)$; $\cos 270° + i \sin 270°$.

NOTE: In writing the answers to Probs. 29–34, we use cis θ to mean $\cos \theta + i \sin \theta$. k denotes any integer.

29. $6 \text{ cis } (90° + k \cdot 360°)$; $\sqrt{2} \text{ cis } (45° + k \cdot 360°)$; $8 \text{ cis } (180° + k \cdot 360°)$.
30. $2\sqrt{2} \text{ cis } (135° + k \cdot 360°)$; $16 \text{ cis } (k \cdot 360°)$; $\text{cis } (270° + k \cdot 360°)$.
31. $2 \text{ cis } (135° + k \cdot 360°)$; $\text{cis } (180° + k \cdot 360°)$; $4 \text{ cis } (270° + k \cdot 360°)$.
32. $2 \text{ cis } (30° + k \cdot 360°)$; $8 \text{ cis } (90° + k \cdot 360°)$; $32 \text{ cis } (180° + k \cdot 360°)$.
33. $6 \text{ cis } (60° + k \cdot 360°)$; $64 \text{ cis } (180° + k \cdot 360°)$; $12 \text{ cis } (90° + k \cdot 360°)$.
34. $6\sqrt{2} \text{ cis } (225° + k \cdot 360°)$; $2 \text{ cis } (150° + k \cdot 360°)$; $4 \text{ cis } (90° + k \cdot 360°)$.
35. $2\sqrt{2} + 2\sqrt{2}\,i$; $3i$.
36. $3\sqrt{3} + 3i$; 2.
37. $-4\sqrt{3} + 4i$; -2.
38. $-6\sqrt{2} + 6\sqrt{2}\,i$; $-4i$.
39. $-\frac{1}{2}\sqrt{2} - \frac{1}{2}\sqrt{2}\,i$; 7.
40. $2 - 2i$; $-4 + 4\sqrt{3}\,i$.
41. $-3\sqrt{2} + 3\sqrt{2}\,i$.
42. -24.
43. -10.
44. $-12i$.
45. 12.
46. $-5.736 + 8.192i$.
47. $17.26 + 16.67i$.
48. $4i$.
49. $4i$.
50. $4.698 + 1.710i$.
51. $-6i$.
52. $-\frac{1}{2} + \frac{1}{2}i$.
53. $-0.1 - 0.1i$.
54. $0.5847 + 1.913i$.

Page 279; Lesson 45

1. 1.
2. $-512i$.
3. $-8i$.
4. 64.
5. $-2{,}500$.
6. $-1{,}024i$.
7. $196.1 + 164.6i$.
8. $-\frac{1}{2} + \frac{1}{2}\sqrt{3}\,i$.
9. $\sqrt{2}\,(1 - i)$; $\sqrt{2}\,(-1 + i)$.
10. $\sqrt{6} + \sqrt{2}\,i$; $-\sqrt{6} - \sqrt{2}\,i$.
11. $\frac{1}{2}(\sqrt{3} + i)$; $\frac{1}{2}(-\sqrt{3} + i)$; $-i$.
12. -3; $\frac{3}{2}(1 \pm \sqrt{3}\,i)$.
13. 1; $-\frac{1}{2} \pm \frac{1}{2}\sqrt{3}\,i$.
14. $\sqrt[3]{6} \text{ cis } 50°$; $\sqrt[3]{6} \text{ cis } 170°$; $\sqrt[3]{6} \text{ cis } 290°$.
15. $2 \text{ cis } 22.5°$; $2 \text{ cis } 112.5°$; $2 \text{ cis } 202.5°$; $2 \text{ cis } 292.5°$.
16. $\pm(\sqrt{3} + i)$; $\pm(1 - \sqrt{3}\,i)$.
17. $\pm\sqrt{3}\,(1 + i)$; $\pm\sqrt{3}\,(1 - i)$.
18. 2; $2 \text{ cis } 72°$; $2 \text{ cis } 144°$; $2 \text{ cis } 216°$; $2 \text{ cis } 288°$.
19. ± 2; $\pm(1 + \sqrt{3}\,i)$; $\pm(1 - \sqrt{3}\,i)$.
20. $\pm i$; $\pm\frac{1}{2}(\sqrt{3} + i)$; $\pm\frac{1}{2}(\sqrt{3} - i)$.
21. $\pm\frac{1}{2}\sqrt{2}\,(1 + i)$; $\pm\frac{1}{2}\sqrt{2}\,(1 - i)$.
22. ± 2; $\pm 2i$.
23. $2i$; $-\sqrt{3} - i$; $\sqrt{3} - i$.
24. -1; $\frac{1}{2}(1 \pm \sqrt{3}\,i)$.
25. $\text{cis } 36°$; $\text{cis } 108°$; -1; $\text{cis } 252°$; $\text{cis } 324°$.
26. $\pm(\sqrt{3} + i)$; $\pm(\sqrt{3} - i)$; $\pm 2i$.
27. $\pm(1 + \sqrt{3}\,i)$.
28. $\pm(\sqrt{3} + i)$; $\pm(1 - \sqrt{3}\,i)$.

CHAPTER XVII

Pages 284–285; Lesson 46

1. 60.
2. 116,280.
3. 16.
4. 12; 128.
5. 576.
6. 8,640.
7. 151,200; 10^6.
8. 180; 648.
9. 120.
10. 61.
11. 20,160.
12. 831,600.
13. 6,720.
14. 181,440.
15. 360.
16. 1,440.
17. 1,956.
18. 207,900.

Pages 291–292; Lesson 47

1. 45. **2.** 220. **3.** 220. **4.** 66; 45.
5. $\dfrac{52!}{13!39!}$. **6.** 90. **7.** 450. **9.** 6,084.
10. $\dfrac{13!\cdot 39!}{6!7!7!32!}$. **11.** 1,296. **12.** 35. **13.** 126,126
14. 126. **15.** 1,260. **16.** 60. **17.** 15.
18. 42. **19.** 7,200. **20.** 57. **21.** 92. **22.** 260; 70.

Pages 299–300; Lesson 48

1. 1/6. **2.** 1/17. **3.** 39/850. **4.** 6/91.
5. 32/133. **6.** \$1.38. **7.** 31 cents.
8. 637 to 375 against. **9.** 518 to 451 in favor. **10.** 2/5.
11. 1/2. **12.** 0.416. **13.** 0.250. **14.** 30 cents. **15.** 75 cents.
16. 5/9. **17.** 1/32. **18.** 2/9. **19.** \$3.20. **20.** \$5.50.

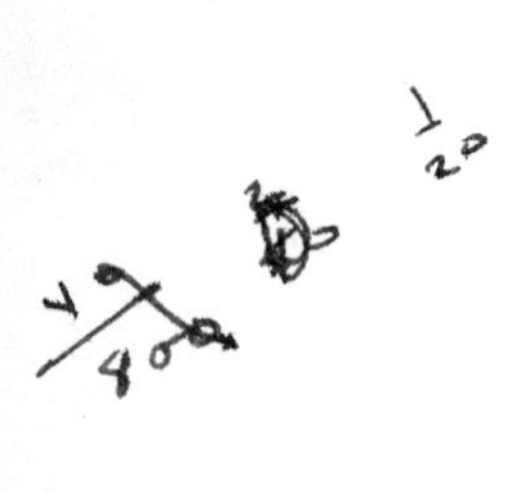

INDEX

V

Z